The Vertical Farm

Amid economic uncertainties, fluctuating oil prices, and a rising environmental consciousness, the need for sustainable and efficient food production has become dire. *The Vertical Farm: Scientific Advances and Technological Developments* systematically navigates the realm of vertical farming (VF), rooted in a robust scientific foundation. Unveiling the intricate convergence of plant biology, environmental science, and agronomy provides a profound understanding of contemporary agriculture. The book spans lighting systems and climate control mechanisms, focusing on sustainability. From small urban initiatives to significant commercial endeavors, real-world case studies showcase VF's adaptability, scalability, and resilience. Addressing multiple challenges, the book explores economic considerations and public perceptions, recognizing their roles in fostering meaningful advancements in agricultural innovation.

A volume in the Nextgen Agriculture series, this book is valuable to scientists, practitioners, and students in urban agriculture and planning, horticulture, engineering, landscape architecture, and plant/technology sciences.

NextGen Agriculture: Novel Concepts and Innovative Strategies
Series Editor:
Chittaranjan Kole

Allele Mining for Genomic Designing of Fruit Crops
Chittaranjan Kole, Kenta Shirasawa, and Anil Kumar Singh

Allele Mining for Genomic Designing of Cereal Crops
Chittaranjan Kole, Sharat Kumar Pradhan, and Vijay K Tiwari

Allele Mining for Genomic Designing of Vegetable Crops
Chittaranjan Kole, Tusar Kanti Behera, and Prashant Kaushik

Allele Mining for Genomic Designing of Oilseed Crops
Chittaranjan Kole, Manish Kumar Pandey, and Naveen Puppala

Sustainable Urban Agriculture: New Frontiers
Kheir Al-Kodmany, Madhav Govind, Sharmin Khan, and Chittaranjan Kole

Allele Mining for Genomic Designing of Grain Legume Crops
Chittaranjan Kole, C. Bharadwaj, and Abhimanyu Sarkar

The Vertical Farm: Scientific Advances and Technological Developments
Kheir Al-Kodmany, Andrew Keong Ng, Abel Tablada, and Chittaranjan Kole

For more information, please visit our website: https://www.routledge.com/Nextgen-Agriculture/book-series/NGA

The Vertical Farm
Scientific Advances and Technological Developments

Edited by
Kheir Al-Kodmany, Andrew Keong Ng,
Abel Tablada, and Chittaranjan Kole

CRC Press
Taylor & Francis Group
Boca Raton London New York

CRC Press is an imprint of the
Taylor & Francis Group, an **informa** business

Designed cover image: Shutterstock

First edition published 2025
by CRC Press
2385 NW Executive Center Drive, Suite 320, Boca Raton FL 33431

and by CRC Press
4 Park Square, Milton Park, Abingdon, Oxon, OX14 4RN

CRC Press is an imprint of Taylor & Francis Group, LLC

Library of Congress Cataloging-in-Publication Data
Names: Al-Kodmany, Kheir, editor. | Ng, Andrew Keong, editor. | Tablada, Abel, editor. | Kole, Chittaranjan, editor.
Title: The vertical farm : scientific advances and technological developments /
edited by Kheir Al-Kodmany, Andrew Keong Ng, Abel Tablada, and Chittaranjan Kole
Other titles: Nextgen agriculture: novel concepts and innovative strategies.
Description: First edition | Boca Raton, FL : CRC Press, 2025 |
Series: Nextgen agriculture: novel concepts and innovative strategies |
Includes bibliographical references and index |
Summary: "Amid economic uncertainties, fluctuating oil prices, and a rising environmental consciousness, the need for sustainable and efficient food production is urgent. The Vertical Farm: Scientific Advances and Technological Developments systematically navigates the realm of vertical farming (VF), rooted in a robust, scientific foundation. Unveiling the intricate convergence of plant biology, environmental science, and agronomy, it provides a profound understanding of contemporary agriculture. The book spans lighting systems and climate control mechanisms focusing on sustainability. From small urban initiatives to significant commercial endeavors, real-world case studies showcase VF's adaptability, scalability, and resilience. Addressing multiple challenges, the book explores economic considerations and public perceptions, recognizing their role in fostering meaningful advancements in agricultural innovation"– Provided by publisher
Identifiers: LCCN 2024013551 | ISBN 9781032437224 (hardback) | ISBN 9781032437217 (paperback) |
ISBN 9781003368588 (ebook)
Subjects: LCSH: Alternative agriculture. | Vertical gardening. | Hydroponics. | Artificial light gardening. |
Crops–Technological innovations. | Sustainable agriculture.
Classification: LCC S494.5.A65 V47 2025 | DDC 338.1/62–dc23/eng/20240711
LC record available at https://lccn.loc.gov/2024013551

ISBN: 978-1-032-43722-4 (hbk)
ISBN: 978-1-032-43721-7 (pbk)
ISBN: 978-1-003-36858-8 (ebk)

DOI: 10.1201/b23309

Typeset in Times
by codeMantra

This series is dedicated to

Dr. Norman Ernest Borlaug

The Father of Green Revolution, Nobel Peace Prize Winner and the Founder of the World Food Prize

Who inspired my research and academic works with his generous appreciation and advice.

Preface to the Series

Agriculture is going through a phase of transformation now. It has both challenges and opportunities. One of the major challenges it is facing is obviously food security. A population of about 9.3 billion projected by 2050 necessitates an increase of crop production by 60%–70%. Another 10%–15% increase would be needed to counter the reduction of crop production due to global warming and climate change. Nutrition security is another challenge, with more than 800 million people suffering from malnutrition. Depletion of fossil fuel and increasing dependence on biofuel from crop residues are another challenge. Loss of biodiversity and land degradation also add to the list of challenges. The cultivable area is also decreasing fast due to housing, transport, and industrialization. Above all, climate change and global warming have a huge impact on agriculture, health, and environment.

Several novel concepts and strategies are emerging to address the above challenges. For example, the practice of precision farming will lead to precise and timely use of inputs, resulting in higher yield and better quality of produce on the one hand and reduction of loss of inputs on the other. Similarly, urban agriculture and vertical agriculture will add a lot of cultivable areas in the future. Protected farming will facilitate growing crops under controlled conditions, leading to a reduction in biotic and abiotic stresses and precise utilization of inputs.

Agriculture requires maintenance of the philosophy farmers have been following since inception over thousands of years which are being sold today in new packages with sophisticatedly coined labels such as traditional agriculture, natural agriculture, sustainable agriculture, conservation agriculture, integrated agriculture, regenerative agriculture, organic agriculture, and more.

In the changing scenarios, next-generation agriculture will need designed crop varieties and breeds produced through designed practices. This book series aims at deliberations on several novel concepts and innovative strategies developed for next-generation agriculture and will hopefully benefit agricultural education, research, and outreach.

Chittaranjan Kole
Kolkata
January 10, 2024

Special Acknowledgement to Phullara Kole

For her outstanding assistance in editing this book series

Contents

About the Editors

Dr. Kheir Al-Kodmany, an internationally reputed professor of urban planning at the University of Illinois Chicago (UIC), USA, boasts a prolific career spanning 30 years. His extensive research and teaching portfolio encompass diverse subjects, including vertical urbanism, sustainable design, geographic information systems (GISs), visualization systems, public participation, crowd management, economic development, and skyscrapers. A prolific author, Dr. Al-Kodmany has published 15 books and over 150 scholarly works, earning widespread acclaim for their comprehensive insights into architecture and urban design. His publications are highly regarded, with over 250,000 reads on ResearchGate. Recognized for his academic excellence, he serves on 20 editorial boards of professional journals. Dr. Al-Kodmany's impactful teaching career spans 30 years at UIC and University of Illinois Urbana-Champaign (UIUC), where he secured grants totaling several hundred thousand dollars and developed innovative visualization software. Beyond academia, Dr. Al-Kodmany's expertise has been sought by governments, mayors, and organizations worldwide. Notably, the Saudi government invited him to contribute to the planning of Hajj, which earned international recognition for enhanced safety. His involvement in Chicago's "Taste of Chicago" event and contributions to Mayor Daley's bid for the 2016 Summer Olympics further demonstrated his impactful civic engagement. Dr. Al-Kodmany delivered over 200 presentations globally at prestigious institutions as a sought-after keynote speaker and trainer. His leadership roles at UIC, including director of Graduate Studies, associate director of the City Design Center, and co-director of the Urban Data Visualization Laboratory, demonstrate his commitment to innovative tools in participatory planning and design. Affiliated with esteemed professional organizations such as the APA, ACSP, CTBUH, and URISA, Dr. Al-Kodmany continues to contribute significantly to the field of urban planning, carrying forward the legacy of his early architectural training from his father, Dr. Abdul Muhsen Al-Kodmany, a Le Corbusier trainee and École des Beaux-Arts graduate.

Dr. Andrew Keong Ng is an associate professor at Singapore Institute of Technology, Singapore. He is the programme leader of the Master of Science in mechanical engineering, the Master of Engineering Technology, the Bachelor of Engineering with honours in sustainable infrastructure engineering (land), and the Bachelor of Engineering with honours in engineering systems. Moreover, Dr. Ng is a chartered engineer with the UK Engineering Council and serves on the committees of various international and local professional engineering institutions, such as the Institute of Electrical and Electronics Engineers (IEEE), Institution of Engineers Singapore (IES), Institution of Engineering and Technology (IET), and Institution of Railway Signal Engineers (IRSE). He is a senior member of IEEE and IES; a consultant and advisor to startups and multinational corporations; and a topic editor for Multidisciplinary Digital Publishing Institute journals. In addition, Dr. Ng is a principal investigator of several grants amounting to more than SGD 1.5 million. He holds one international patent and has over 30 publications as both first and corresponding author. His research and development innovations have also garnered him several prestigious awards, such as Outstanding Researcher Award, Teaching Excellence Award, Amity Researcher Award, Young Investigator Award, and National Instruments Editor's Choice Engineering Impact Award. Furthermore, Dr. Ng has been a keynote and invited speaker at various international conferences. He has also been frequently quoted and interviewed by news media on railway transportation issues and current affairs.

Dr. Abel Tablada is a full professor at the Technological University of Havana (CUJAE), Cuba where he leads the discipline of environmental conditioning and focuses on carbon-neutral architecture and urbanism. He holds a doctorate in engineering and a master's of architecture in human settlements from the Katholieke Universiteit Leuven. He also holds a master's of architecture and conservation from the Cuban Center for Conservation, Restoration and Museology. Dr. Tablada had a significant role in the heritage architecture department of the Office of Havana's Historian, focusing on building restoration and research on natural ventilation and thermal comfort. He was a part of the National University of Singapore in 2011–2019 and cofounded the NUS-CDL Tropical Technologies Lab, where he completed a research project on the integration of solar energy and farming systems in productive facades. Dr. Tablada has published over 20 scientific articles and book chapters on urban and building physics, building integrated agriculture, and solar energy systems. He has been a guest professor/lecturer at over 15 universities and institutions and has been awarded research and design recognitions, including the Japan Prize on urban climate research, two national prizes on design and building restoration, and a first prize at the Caribbean Biennale of Architecture.

Dr. Chittaranjan Kole is an internationally reputed scientist with an illustrious professional career spanning over 40 years and has made original contributions to the fields of plant genomics, biotechnology, and molecular breeding, leading to the publication of more than 160 quality research articles and reviews. He has edited nearly 200 books for leading international publishers including Springer Nature, Wiley-Blackwell, and Taylor & Francis Group. His scientific contributions and editing acumen have been appreciated by seven Nobel Laureates, including Profs. Norman Borlaug, Arthur Kornberg, Werner Arber, Phillip Sharp, Günter Blobel, Lee Hartwell, and Roger Kornberg. He has been honored with a number of fellowships, honorary fellowships, and national and international awards, including the Outstanding Crop Scientist Award conferred by the International Crop Science Society. Recently, he was awarded the Raja Ramanna fellowship by the Department of Energy, Government of India. He has served in many prestigious positions in academia including vice-chancellor of Bidhan Chandra Krishi Viswavidyalaya, project coordinator of Indo-Russian Center of Biotechnology in India, and director of research of the Institute of Nutraceutical Research of Clemson University, USA. He was also a visiting professor at the Pennsylvania State University and Clemson University, USA. Presently, he heads the International Climate-Resilient Crop Genomics Consortium, the International Phytomedomics and Nutriomics Consortium, and the Genome India International as their founding president. He is also the founding chairman of the Prof. Chittaranjan Kole Foundation for Science and Society.

Contributors

Kheir Al-Kodmany
University of Illinois Chicago
Chicago, Illinois

Hui An
Singapore Institute of Technology
Dover, Singapore

Mohd Faizal Mohideen Batcha
Faculty of Mechanical and Manufacturing Engineering
Universiti Tun Hussein Onn Malaysia
Batu Pahat, Johor, Malaysia

Szu-Cheng Chien
Singapore Institute of Technology
Dover, Singapore

Dmitrii Filatov
Nizhny Novgorod State Agrotechnological University
Russia

Kevin M. Folta
Horticultural Sciences Department
University of Florida
Gainesville, Florida

Chittaranjan Kole
Foundation for Science and Society (pckfss.org)
Kolkata, West Bengal, India

Vesna Kosorić
Balkan Energy AG
Starrkirch-Wil, Switzerland

R. Mahkeswaran
Singapore Institute of Technology
Dover, Singapore

Akmal Nizam Mohammed
Faculty of Mechanical and Manufacturing Engineering
Universiti Tun Hussein Onn Malaysia
Batu Pahat, Johor, Malaysia

Andrew Keong Ng
Singapore Institute of Technology
Dover, Singapore

Sulastri Sabudin
Center for Energy and Industrial Environment Studies
Universiti Tun Hussein Onn Malaysia (UTHM)
Batu Pahat, Johor, Malaysia

Zulhazmi Sayuti
Horticulture Research Center
Malaysia Agricultural Research and Development Institute (MARDI)
Serdang, Selangor, Malaysia

Abdul Muin Shaari
Center for Energy and Industrial Environment Studies
Universiti Tun Hussein Onn Malaysia (UTHM)
Batu Pahat, Johor, Malaysia

Chew Beng Soh
Singapore Institute of Technology
Dover, Singapore

Abel Tablada
Faculty of Architecture
Technological University of Havana J.A. Echeverria
Havana, Cuba

Muhd Akhtar Mohd Tahir
Horticulture Research Center
Malaysia Agricultural Research and Development Institute (MARDI)
Serdang, Selangor, Malaysia

Jorge Flores Velazquez
Colegio de Postgraduados, Posgrado en Hidrociencias
Carr. Mex-Tex, Mexico

Shy Chyi Wuang
Temasek Polytechnic
Tampines, Singapore

Preface

The evolution of our exploration into the innovative realm we navigate in this book has stemmed from the pressing need to address economic, social, and environmental challenges. As the intricacies of the global landscape unfolded—marking the impact of economic crises, the surge in oil prices, and an awakening environmental consciousness—the urgency for sustainable and efficient food production became apparent. *The Vertical Farm: Scientific Advances and Technological Developments* is a comprehensive journey into vertical farming (VF) and the forefront of agricultural innovation.

Our story unfolds methodically, revealing a wealth of inventions at the cutting edge of revolutionizing the agricultural paradigm encapsulated within VF. Initiating our journey with a robust scholarly foundation, we explore the scientific principles that form the bedrock of this innovative approach. This intellectual odyssey traverses the realms of plant biology, delving into the intricate processes that govern growth and development; environmental science, unraveling the complex interplay between crops and their surroundings; and agronomy, deciphering the art and science of cultivating the land. Each discipline adds a layer to our understanding, culminating in a profound comprehension of how these diverse scientific realms seamlessly converge within the intricate systems that define and characterize our contemporary agricultural landscape. This holistic approach illuminates the intellectual depth of our exploration and underscores the interdisciplinary nature that defines cutting-edge agricultural innovation within the realm of VF.

Navigating the technological realm, our narrative seamlessly spans a spectrum encompassing leading lighting systems and delving into the intricacies of sophisticated climate control mechanisms. Throughout this exploration, a consistent emphasis is placed on sustainability, shedding light on the inherent eco-friendly aspects that characterize this transformative agricultural approach. Going beyond the confines of the present, our journey extends into the future, probing not only the emerging technologies that are reshaping the landscape but also potential avenues for scientific exploration yet to be charted.

The fabric of our narrative is woven with real-world case studies, offering a tapestry of invaluable insights that illuminate both the triumphs and the challenges encountered in the diverse implementations of VF. From the intimate scale of small urban initiatives to the expansive realms of commercial endeavors, these meticulously curated case studies serve as windows into the multifaceted landscape of agricultural innovation. Each case study serves as a vibrant vignette, inviting readers to immerse themselves in a firsthand exploration of the myriad models that contribute to the ongoing transformation of the agricultural landscape. These experiences provide nuanced perspectives on VF's adaptability, scalability, and resilience, showcasing its dynamic applications across varied settings. These case studies, whether unfolding in the warehouses or vertical expanses of commercial agriculture, offer a panoramic view of VF's diverse and evolving role in reshaping how we cultivate and sustainably produce food.

Yet, in our literary endeavor, we steadfastly refuse to shy away from the inherent challenges posed by this profound agricultural transformation. Spanning the spectrum from intricate economic considerations to the nuanced terrain of public perception, we courageously confront the hurdles that must be navigated and overcome to realize a sustainable and efficient agricultural future. This careful exploration forms a critical aspect of our examinations, acknowledging that understanding and addressing challenges are integral to fostering meaningful advancements in agricultural innovation.

This book is a knowledge repository and an invitation to explore the ever-changing field of VF. It is aimed at scholars, practitioners, and enthusiasts, providing deep insights into the scientific and technical advancements driving the future of agriculture. Readers are invited to embark on an educational journey via the pages of *The Vertical Farm*, where they may discover the groundbreaking possibilities of VF in terms of its role in promoting sustainable and efficient food production.

List of Abbreviations

AdaBoost	Adaptive boosting
AI	Artificial intelligence
AWAV	Area-weighted average velocity
BIA	Building integrated agriculture
BIM	Building information modeling
CEA	Controlled environment agriculture
CFD	Computational fluid dynamics
CIGS	Copper indium gallium selenide
CVF	Closed vertical farm
DFT	Deep Flow Technique
DLI	Daily light integral
DO	Dissolved oxygen
DUA	Digital urban agriculture
EC	Electrical conductivity
ECM	Energy conservation measures
FCU	Fan coil unit
FL	Fluorescent lamps
GHG	Greenhouse gases
GUI	Graphical user interface
HDB	Housing Development Board
HVAC	Heating, ventilation, and air conditioning
IAQ	Indoor air quality
INC	Incandescent lamps
IoAT	Internet of agricultural things
IoT	Internet of things
IVFs	Indoor vertical farms
LCA	Life cycle assessment
LED	Light-emitting diode
MARDI	Malaysian Agricultural Research and Development Institute
MCDM	Multi-criteria decision-making
MIP	Mixed-integer linear programming
ML	Machine learning
ML	Metal halide lamps
MV	Mercury vapor lamps
NLP	Natural language processing
NUS	National University of Singapore
O&M	Operation and maintenance
ORS	Optimal required sunlight
PAR	Photosynthetically active radiation
PF	Productive façade(s)
PFALs	Plant factories with artificial lighting
PM	Particulate matter
PPFD	Photosynthetic photon flux density
PR	Plot ratio
PV	Photovoltaic
RE	Renewable energy
ROI	Return on investment

SAH	Superabsorbent hydrogel
SEC	Specific energy consumption
SEU	Significant energy user
SIT	Singapore Institute of Technology
SMEs	Small and medium enterprises
T^2 Lab	Tropical Technologies Laboratory
TUA	Traditional urban agriculture
UA	Urban agriculture
UDI	Useful daylight intensity
UFAD	Underfloor air distribution
UGS	Urban green spaces
UmFm	Urban-metabolic farming-module
UV	Ultraviolet
VF	Vertical farming
VFEEMP	Vertical farming elevator energy minimization problem
VFS	Vertical farming system(s)
VGS	Vertical greenery systems

1 Introduction
Pioneering the Future of Agriculture with Vertical Farming

Kheir Al-Kodmany, Andrew Keong Ng, Abel Tablada, and Chittaranjan Kole

A convergence of economic, social, and environmental factors has driven the adoption, conversation, and deployment of vertical farming (VF) systems worldwide. According to Munoz-Liesa et al. (2020) and Tablada and Kosorić (2022), some of these factors are the tangible effects of the global economic crisis on food availability, the rise in oil prices that raises the cost of commercial transportation, the public's increased awareness of safe and environmentally friendly food products, and the availability of unused office, commercial, and industrial spaces. Furthermore, VF has evolved into a calculated response to the decline in cultivable land per capita (Olsson et al., 2023; World Bank, 2023). Experts like Despommier (2010, 2012, 2013) suggest a significant shift in food production, distribution, and consumption due to the harm that current agricultural practices cause to the environment and their inability to meet future food needs (Thomaier et al., 2015). It is imperative to increase agricultural yields and introduce novel crops and dietary habits to guarantee food security for a world population that is expanding and has differing economic capabilities. A paradigm shift in society and technology can accomplish this (Olsson et al., 2023). Amid growing global challenges, it is imperative to guarantee food production that is both sustainable and efficient. This puts VF in a position to significantly impact how agriculture develops in the future (Al-Kodmany, 2018; Benke & Tomkins, 2017).

This book serves as an exhaustive expedition into the forefront of innovation within VF, unraveling the intricate tapestry of advancements reshaping this agricultural paradigm. The book explores the intersection of science and technology at its core, casting a luminous spotlight on the latest breakthroughs propelling VF into the vanguard of sustainable agriculture. Commencing with a scholarly foundation, the narrative delves into the scientific principles intricately woven into the fabric of VF. A profound exploration of plant biology, environmental science, and agronomy unfolds, providing readers with a deep understanding of the convergence of these disciplines within the intricate systems of VF. From cutting-edge lighting systems to sophisticated climate control mechanisms, the book navigates the technological landscape shaping modern vertical farms, emphasizing sustainability and elucidating the eco-friendly aspects of VF. The discussion offers a glimpse into the future, exploring emerging technologies and potential avenues for further scientific exploration, making it an indispensable resource for researchers, practitioners, and enthusiasts in the dynamic field of VF.

Real-world case studies punctuate the narrative, engaging readers in a firsthand look into the triumphs of successful VF implementations. These case studies, ranging from small-scale urban initiatives to expansive commercial endeavors, provide invaluable insights into the diverse models shaping the landscape of VF. Navigating the industry's complexities, the book courageously confronts the challenges faced by VF, spanning economic considerations to public perception. *The Vertical Farm: Scientific Advances and Technological Developments* stands as an indispensable reservoir of knowledge, beckoning researchers, practitioners, and enthusiasts into the dynamic and evolving realm of VF. Whether one is a scientist, farmer, or policymaker, this literary opus unfurls a comprehensive overview, offering profound insights into the scientific and technological forces shaping the future of agriculture.

DOI: 10.1201/b23309-1

BOOK OUTLINE

Chapter 2 provides a comprehensive overview of VF, examining its core ideas, benefits, and drawbacks, emphasizing its potential to maximize space, preserve resources, and promote environmental sustainability. The integration of cutting-edge technologies, such as precision agriculture and the Internet of things (IoT), is central to contemporary agricultural innovation, increasing efficiency and effectiveness to previously unheard-of levels. The chapter examines obstacles such as significant upfront costs and prevalent consumer perceptions by studying the intricacies hindering VF's broad implementation. It emphasizes how collaboration and education may have a transformative effect on creating an atmosphere supporting VF growth. Moreover, the analysis broadens its scope to include the worldwide impact of VF on food security, emphasizing the latter's resilience due to its year-round production, climate flexibility, and regional distribution networks. The chapter culminates in a futuristic vision of VF as a keystone in the profound upheaval of the world's food systems.

Embarking on a journey through the frontiers of horticultural illumination, **Chapter 3** unravels the latest strides in lighting systems, offering a nuanced exploration fueled by an exhaustive literature review. Artificial luminaires, including incandescent, metal halide, fluorescent, and high-pressure sodium lamps, have long been enlisted to enhance lighting efficiency in various settings. However, light-emitting diodes (LEDs) have ushered in a transformative era, specifically engineered to elevate plant growth and energy efficiency. Amid the spectrum of LEDs, the spotlight shines brightly on blue and red LEDs, chosen for their prowess in absorbing chlorophyll a and b wavelengths. The chapter delves into the intricate decision-making process behind LED selection, intricately balancing yield and energy cost considerations. A fascinating exploration ensues, uncovering the impact of red-blue light augmentation on the growth of lettuce, microgreens, and fruit crops. At the same time, even green LEDs make a noteworthy contribution to lettuce and microgreen cultivation. The Emerson effect, a venerable phenomenon spanning over half a century, takes center stage, elucidating the heightened photosynthetic efficiency induced by light in the 670 and 700 nm wavelengths. As the discourse extends to far-red light's potential to amplify photosynthesis, the chapter meticulously navigates the intricate interplay of different spectra, with red light emerging as the unequivocal champion in influencing energy-efficient lighting systems for diverse plant species. Anchoring these illuminating insights is a strategic imperative—maintaining the daily light integral (DLI)—as the chapter unveils how judicious light pulse frequency and duration adjustments can catalyze heightened energy utilization in this luminous journey through cutting-edge plant illumination.

Chapter 4 delves into the transformative potential of building envelopes as hosts for vertical farming systems (VFS). Unraveling as a response to the heightened energy consumption associated with indoor VF factories, this exploration unveils VFS as a strategic alternative, remarkably poised to address challenges in low-income regions and urban settings. A comprehensive taxonomy of design and implementation options takes center stage, each option meticulously categorized based on its performance and interface with interior functions. The chapter serves as a critical discourse, navigating the advantages and potential threats VFS poses on building envelopes, supported by a discerning analysis of case studies and research investigations centered on rooftops and facades. Amid this analytical backdrop, the chapter discerns the acute challenges impeding the widespread adoption of VFS, providing a roadmap for overcoming obstacles. In its essence, the design of VFS on building envelopes emerges as a delicate balancing act, striving not only to meet but enhance traditional functions—serving as a barrier against external conditions while ensuring protection, safety, identity, and privacy. Simultaneously, integrating new functions, from rainwater collection to electricity generation and food production, charts a course toward sustainable urban agriculture that harmonizes with the built environment.

While environmental and economic factors continue to receive more attention, the social dimension and social acceptance are crucial for the development of urban agriculture. **Chapter 5** seeks to illuminate the often-overlooked yet dynamically crucial role of the social dimension and social acceptance. Specifically, honing in on integrating urban farming within building skins and structures, the discourse revolves around the myriad commercial, ecological, and social benefits these initiatives offer as indispensable contributors to future food security and sustainable urban development. The chapter delves into the intricate fabric of stakeholders' needs, expectations, and preferences, dissecting various aspects such as utilization, upkeep, and aesthetics across diverse farming systems—from collectively and privately organized rooftops to façades and indoor VF. A meticulous exploration of user-perceived benefits and risks forms the basis for identifying critical drivers and obstacles. By scrutinizing the advantages and disadvantages of social sustainability and acceptability across different farming systems, the chapter draws upon specific yet invaluable studies and practices from various corners of the globe. The insights gleaned and proposed future directions aim to pave the way for the seamless adoption of farming in and on buildings. This strategic approach aligns with the broader objective of decarbonization in buildings and cities and endeavors to render them sustainable, livable, and lovable.

Chapter 6 explores the burgeoning intersection of urban agriculture and digital tools, presenting a strategic alliance to navigate the complexities of burgeoning cities. The rising demand for innovative solutions prompts the integration of tools like building information modeling, performance simulations, and micro-climate controls during the design phase of digital urban agriculture. This symbiotic relationship aims to elevate planning, design, and implementation, optimizing the configuration of farm enclosures, amplifying yields, and mitigating environmental impacts. A profound emphasis is placed on leveraging these tools to overcome urban constraints related to space, resources, and environmental considerations. By fine-tuning the design of farm enclosures, these tools adapt farming practices to the unique contexts of urban landscapes, fostering resilience against climate change. Illustrated through a compelling case study in Singapore, this chapter underscores the indispensability of incorporating digital tools in the design phase of urban farming systems. The presented tools optimize resource utilization and crop yields and contribute to the overarching goal of minimizing environmental impacts, thus championing the potential and promise of digital urban agriculture.

Chapter 7 delves into the transformative realm of the Internet of Agricultural Things (IoAT), a convergence of intelligent technologies and the IoT specifically tailored for agricultural applications. At its core, IoAT employs an intricate network of sensors, devices, and software to intricately collect data about crops, soil, weather, and other variables that intricately influence plant growth and yield. The efficacy of IoAT becomes particularly pronounced in indoor farming, where crops flourish within carefully controlled environments, encompassing greenhouses, vertical farms, and other indoor facilities. Integrating IoT technologies into indoor farming offers farmers unprecedented control over the growing environment. The ability to monitor and regulate crucial factors such as temperature, humidity, lighting, and irrigation optimizes plant growth conditions and is a potent tool for reducing waste. As this chapter unfolds, it unveils the intricate workings and far-reaching implications of IoAT, offering insights into how this technological convergence is reshaping the landscape of modern agriculture, with a specific focus on its application in indoor farming.

Chapter 8 embarks on a comprehensive exploration of the technological landscape shaping aquaponics systems, a pioneering approach that fosters a symbiotic relationship between fish and plants within a closed-loop ecosystem. Delving into the integration of emergent technologies such as the IoT, automation, artificial intelligence, green technology, extended reality, and advanced materials, the chapter unravels the intricate tapestry of innovations propelling aquaponics toward enhanced commercial viability. Within this discourse, aquaponics' environmental and health benefits take center stage, juxtaposed against the challenges and implications posed by these disruptive technologies. The chapter offers a nuanced evaluation of the advantages and disadvantages inherent in these technological strides, culminating in insightful recommendations for future research and development endeavors. As the boundaries of agricultural innovation expand, this chapter serves as a beacon, illuminating the transformative potential of aquaponics and its integration with cutting-edge technologies in the pursuit of sustainable and secure food production.

Embarking on the frontier of sustainable aquaculture practices, **Chapter 9** delves into the multifaceted applications of microalgae, traditionally recognized as pivotal live feeds in aquaculture. A paradigm-shifting, microalgae-based recirculating aquaculture system takes center stage, showcasing its efficacy in removing key pollutants such as ammonia, nitrite, and nitrate from fish water. As an innovative twist, the harvested microalgae biomass finds a secondary purpose in the formulation of fish feeds, elevating the nutritional profile while promising significant reductions in reliance on traditional fishmeal. The technical prowess of the microalgae-based system is meticulously established, with a simplified cost analysis offering insights into its potential integration within vertical aquaculture farms. Beyond its tangible benefits, such low-cost systems emerge as catalysts for enhanced profitability and sustainability, ushering in a new energy and cost-savings era. The chapter concludes by shedding light on the inherent advantages of deploying high-productivity microalgae systems within vertical farms, amplifying their production capacity, and further substantiating their role in revolutionizing the aquaculture landscape.

As vertical farms continue to harness technological advancements in their infrastructures, the plant has somewhat overshadowed one critical facet of the system. In controlled environments, the conventional approach of relying on existing crop varieties primarily selected for field conditions falls short of unlocking the full potential of vertical farm production. **Chapter 10** navigates the unexplored terrain of plant breeding tailored for closed, controlled agricultural spaces. Unlike traditional breeding methods aimed at overcoming field challenges, breeding for controlled environments presents a unique canvas where established traits lose relevance and novel opportunities for innovation arise. The chapter ventures into tailoring crops to the intricacies of indoor cultivation. This shift not only holds the promise of increased profitability for vertical farmers but also yields more desirable products for consumers. The narrative unfolds around the concept of breeding to align with specific environments, delving into the intricate ways plants interpret light signals and the convergence of genetics and environment to produce traits that benefit both growers and consumers. By scrutinizing this symbiotic interplay, the chapter sheds light on the transformative potential of adopting a targeted approach to plant breeding in vertical farms, unlocking new dimensions of control over crop growth and product quality.

Despite its prodigious potential, urban agriculture remains conspicuously underutilized in Mexico, a nation with abundant natural conditions yet to fully embrace this agricultural revolution. **Chapter 11** unfolds against the backdrop of this untapped potential, delving into the intricacies of developing a cost-effective vertical farm project as a catalyst to propel urban agriculture into mainstream horticulture practices in Mexico. Acknowledging the unique challenges posed by closed-building cultivation, the chapter navigates the crucial variables essential for indoor crops—water, artificial light, and carbon dioxide—highlighting their pivotal role in supplementing crop requirements. The discourse extends beyond mere food production, emphasizing the cultivation of plants that contribute to a healthier environment. In dissecting the core components of a low-cost vertical farm project, the chapter sheds light on critical parameters such as agronomic management, climate control, and spatial considerations. Central to this exploration is the indispensable role of artificial light as a determining factor in production. By unraveling the intricate web of variables—from water consumption to oxygen levels in nutrient solutions—the chapter not only outlines the requirements for successful implementation but also advocates for the integration of urban agriculture, including plant factories and vertical farms, as an imperative response to the existing needs and favorable natural conditions in Mexico.

The financial viability of indoor farming is intricately linked to the formidable costs associated with providing artificial lighting and maintaining precise micro-climatic conditions. In this context, energy efficiency becomes a linchpin, offering a pathway to reduce production costs and ensure the sustainability of indoor farming practices. **Chapter 12** unfolds against the backdrop of this imperative, presenting a compelling case study focused on energy efficiency in Malaysia's distinct indoor farming facilities—a 40-foot container-type plant factory and a large-scale plant factory, both owned by the Malaysian Agricultural Research and Development Institute in Serdang. The case study goes beyond establishing the unique energy profile of indoor farming, delving into the granular specifics of energy consumption for harvested crops. Through a comprehensive approach that includes walk-throughs and detailed energy audits, the chapter identifies and dissects energy efficiency opportunities. Noteworthy strategies that have proven successful include the strategic scheduling of lighting and air-conditioning systems based on crop growth stages, optimization of internal airflow, and the integration of solar energy—a triumphant step toward achieving sustainable and cost-effective indoor farming practices.

REFERENCES

Al-Kodmany, K. (2018). The vertical farm: A review of developments and implications for the vertical city. *Buildings*, 8(2), 24.

Benke, K., Tomkins, B. (2017). Future food-production systems: Vertical farming and controlled-environment agriculture. *Sustain. Sci. Pract. Policy*, 13, 13–26.

Despommier, D. (2010). *The Vertical Farm: Feeding the World in the 21st Century*. Thomas Dunne Books: New York.

Despommier, D. (2012). Advantages of the vertical farm. In: S. Th. Rassia and P. M. Pardalos (eds.), *Sustainable Environmental Design 259 in Architecture: Impacts on Health, Springer Optimization and Its Applications 56*. Springer: Springer Science+Business Media, LLC, pages 259–275. https://doi.org/10.1007/978-1-4419-0745-5_16.

Despommier, D. (2013). Farming up the city: The rise of urban vertical farms. *Trends Biotechnol.*, 31, 388–389.

Munoz-Liesa J, Royapoor M, Lopez-Capel E, Cuerva E, Rufí-Salís M, Gasso-Domingo S, Josa A. (2020). Quantifying energy symbiosis of building-integrated agriculture in a mediterranean rooftop greenhouse. *Renew. Energy*, 156, 696–709. https://doi.org/10.1016/j.renene.2020.04.098.

Olsson, L., Cotrufo, F., Crews, T., Franklin, J., King, A., Mirzabaev, A., Scown, M., Tengberg, A., Villarino, S., Wang, T. (2023). Annual review of environment and resources the state of the World's Arable Land. *Annu. Rev. Environ. Resour.*, 48: 451–75. https://www.annualreviews.org/doi/pdf/10.1146/annurev-environ-112320-113741. Accessed on 9-12-2023.

Tablada, A., Kosorić, V. (2022). Vertical farming on facades: Transforming building skins for urban food security. In: E. Gasparri, A. Brambilla, G. Lobaccaro, F. Goia, A. Andaloro, A. Sangiorgio. *Woodhead Publishing Series in Civil and Structural Engineering, Rethinking Building Skins*. Woodhead Publishing, pp. 285–311. ISBN 9780128224779. https://doi.org/10.1016/B978-0-12-822477-9.00015-2.

Thomaier, S., Specht, K., Henckel, D., Dierich, A., Siebert, R., Freisinger, U.B., Sawicka, M. (2015). Farming in and on Urban buildings: Present practice and specific novelties of zero-acreage farming (ZFarming). *Renew. Agric. Food Syst.*, 30, 43–54.

World Bank. (2023). https://data.worldbank.org/indicator/AG.LND.ARBL.HA.PC. Accessed on 13-12-2023.

2 Cultivating Tomorrow
An Overview of Vertical Farming

Kheir Al-Kodmany

INTRODUCTION

Vertical farming emerges as a transformative and forward-thinking approach to agriculture, ushering in a new era that redefines the very essence of cultivating crops. This innovative method orchestrates the systematic growth of plants in vertically stacked layers or on inclined surfaces, primarily within meticulously controlled environments such as purpose-built structures or repurposed shipping containers. Diverging from conventional outdoor farming practices, vertical farming unfolds predominantly indoors, harnessing state-of-the-art technologies like hydroponics or aeroponics to nurture plants without reliance on traditional soil. Instead, the cultivation medium pivots to nutrient-rich water solutions, serving as the primary conduit for delivering essential elements to the plants.

Sustainability lies at the heart of vertical farming's evolution. By minimizing land use, reducing water consumption, and eliminating the need for synthetic pesticides and fertilizers, vertical farming aims to reduce its environmental footprint. The controlled environments mitigate soil degradation and water pollution, contributing to the development of more ecologically responsible farming practices. Additionally, the proximity of vertical farms to urban centers shortens the supply chain, reducing transportation-related emissions and fostering the growth of resilient local food systems.

The primary premise underlying vertical farming is the optimization of space. This method effectively utilizes vertical stacking of crops to maximize the use of available space, offering a viable solution for agricultural practices in metropolitan areas or other regions where traditional farming faces limitations owing to restricted arable land. The utilization of spatial ingenuity proves to be particularly advantageous in metropolitan areas with high population density, as it facilitates the cultivation of food locally and reduces the environmental consequences associated with long-distance food transportation.

Vertical farms incorporate advanced technologies such as sensors, automation, and artificial intelligence (AI) to monitor and optimize growing conditions. This integration enhances efficiency, precision, and the overall sustainability of the farming process. Hydroponics and aeroponics, integral components of the vertical farming paradigm, epitomize soilless cultivation techniques. Hydroponics involves the direct delivery of nutrient-rich water to plant roots, while aeroponics delicately mists the roots with a nutrient solution. These sophisticated methods enhance resource efficiency, affording precise control over nutrient levels and environmental conditions and facilitating optimal plant growth. The controlled indoor environment protects crops from unpredictable weather conditions, pests, and diseases, ensuring a more dependable and consistent harvest.

A primary objective of vertical farming is to maximize production yields while minimizing the environmental impact traditionally associated with agriculture. This is accomplished by circumventing the necessity for vast expanses of arable land, reducing water consumption, and eliminating reliance on synthetic pesticides and fertilizers. The strategic placement of vertical farms near urban centers aligns with the concept of "local food," curtailing the carbon footprint linked to the transportation of food over long distances and addressing concerns tied to the environmental impact of "food miles."

DOI: 10.1201/b23309-2

While vertical farming undeniably offers promising solutions to several challenges in modern agriculture, it is essential to acknowledge and address certain considerations that may impact its widespread adoption. Notably, the initial setup costs associated with establishing vertical farms can be substantial, encompassing investments in specialized infrastructure, technology, and operational systems. Additionally, the energy consumption of indoor vertical farming facilities, particularly for maintaining optimal growing conditions such as lighting and climate control, is a factor that requires careful management to ensure the overall sustainability of the practice.

Overall, vertical farming represents a forward-thinking and sustainable approach to agriculture, addressing the growing need for efficient, resilient, and environmentally friendly food production systems in the face of increasing urbanization and climate change. Vertical farming signifies a revolutionary paradigm shift in agriculture, capitalizing on technology to surmount the spatial limitations inherent in traditional farming. This approach champions a sustainable, efficient, and localized methodology for food production, showcasing the potential to reshape the future of agriculture (Levi & Robin, 2020; McCarthy et al., 2018).

A Brief History

Vertical farming has evolved, blending innovation in agriculture, technology, and sustainability. Here's a brief history of vertical farming (Despommier, 2013; Despommier, 2019; Van Gerrewey et al., 2021):

- Early Concepts (20th century): Growing crops in vertically stacked layers or structures is not entirely new. Early concepts date back to the 20th century, with visionary thinkers exploring the possibilities of multi-story agricultural systems.
- Dickson Despommier's Proposal (1999): "Vertical farming" gained prominence with Dr. Dickson Despommier, a professor of environmental health sciences at Columbia University. In 1999, Despommier proposed the concept as a potential solution to the challenges of traditional agriculture, including land scarcity and environmental concerns.
- Skyfarms and Vertical Harvest Farms (Early 2000s): Various architects and visionaries explored vertical farming concepts during the early 2000s. Skyfarms, designed by Gordon Graff, and Vertical Harvest Farms, conceptualized by Chris Jacobs, were among the early architectural designs showcasing vertical farming's potential.
- Hydroponic and Aeroponic Systems (2000s): The 2000s saw the increased adoption of hydroponic and aeroponic systems integral to vertical farming. These soilless cultivation methods became essential in creating efficient and controlled environments for plant growth.
- Dickson Despommier's Book (2010): In 2010, Dickson Despommier published *The Vertical Farm: Feeding the World in the 21st Century*. The book delves into the details of vertical farming, emphasizing its potential to address global food security challenges, reduce environmental impact, and optimize resource use.
- Sky Greens in Singapore (2012): Sky Greens, a vertical farm in Singapore, became one of the early commercial vertical farms to use a patented rotating vertical farming system. This system maximizes space utilization and promotes efficient crop production.
- Vertical Farms Worldwide (2010s–Present): The 2010s marked the proliferation of vertical farms worldwide. Various projects and commercial enterprises emerged in cities such as Japan, the Netherlands, the United States, and the United Arab Emirates. These vertical farms incorporated advanced technologies, including controlled-environment agriculture (CEA), automation, and artificial lighting.
- AeroFarms in Newark, New Jersey (2015): AeroFarms, a vertical farming company, opened a high-tech indoor farm in Newark, New Jersey, in 2015. Using aeroponic systems and light-emitting diode (LED) lighting, AeroFarms showcased the potential for vertical farming to produce crops efficiently in urban environments.

- Spread Co. in Kyoto, Japan (2019): Spread Co. in Kyoto, Japan, opened the world's first fully automated vertical farm for commercial production in 2019. The farm, known as the Kameoka Plant, focuses on lettuce cultivation using robotic systems and advanced technologies.
- Continued Technological Advancements (2020s): In the 2020s, vertical farming continues to witness technological advancements. Innovations include integrating Internet of things (IoT) devices, machine learning, and smart farming technologies to enhance efficiency, optimize resource use, and further address sustainability goals.

Vertical farming, once a visionary concept, has become a tangible and growing industry, with ongoing research and development aimed at refining the model for sustainable and efficient food production. It represents a dynamic intersection of agriculture, architecture, and technology to address the challenges of feeding a growing global population.

Crops

Vertical farming has emerged as a versatile and efficient method for cultivating various crops, ranging from nutrient-packed microgreens to staple root vegetables. Here's an in-depth exploration of some of the most-grown crops in vertical farms (Buchner, 2022; Jaeger et al., 2022):

- Microgreens: Microgreens are the young, tender shoots of vegetables, falling between baby leaf vegetables and sprouts in their growth stage. These miniature greens span a spectrum of textures and hues and boast potent fragrances and rich vitamin content. Among the most popular microgreens are arugula, radish, beets, kale, sprouts, shoots, watercress, and many tiny edible plants.
- Leafy Greens: Leafy greens encompass a variety of vegetables known for their edible leaves, offering low-calorie, nutrient-dense profiles. High in essential nutrients such as vitamins, minerals, antioxidants, and fiber, leafy greens contribute to preventing or managing chronic diseases. Spinach, kale, lettuce, cabbage, arugula, and collard greens are prominent leafy greens widely cultivated in vertical farms.
- Herbs: Herbs are plants known for their aromatic or medicinal properties, serving diverse purposes such as flavoring food, making tea, or even treating ailments. Herbs can be employed in various forms—fresh or dried, whole or chopped—and may involve different plant parts like leaves, flowers, seeds, roots, or stems. Basil, mint, rosemary, thyme, lavender, chamomile, ginger, and parsley are among the herbs flourishing in the controlled environments of vertical farms.
- Root Vegetables: Root vegetables are staples in global cuisines, characterized by their underground edible parts such as tubers, bulbs, rhizomes, or corms. Root vegetables are nutritious, offering carbohydrates, fiber, vitamins, minerals, and antioxidants. Potatoes, carrots, onions, garlic, beets, radishes, and turnips are prevalent root vegetables cultivated in vertical farms.
- Fruit-Bearing Crops: Leafy fruit-bearing crops yield edible fruits and feature edible leaves. They often thrive in tropical or subtropical regions, providing a year-round food source. Papaya, banana, pineapple, moringa, tomatoes, chaya, strawberries, peppers, cucumbers, and cassava exemplify the diverse spectrum of fruit-bearing crops thriving in vertical farming environments.

In the dynamic realm of vertical farming, the cultivation of these diverse crops showcases the adaptability of this agricultural method. It emphasizes its potential to contribute to sustainable, local, and year-round food production (Badami & Ramankutty, 2015).

THE LOGIC AND BENEFITS OF VERTICAL FARMING

The logic behind vertical farming revolves around addressing critical challenges in traditional agriculture and creating a more sustainable, efficient, and resilient system for food production. The core principles and logic of vertical farming are discussed in the following subsections.

Space Efficiency

By definition, vertical farming is a method of agriculture that utilizes the vertical dimension of space to grow crops. By stacking layers of plants or using structures that support vertical growth, vertical farming can produce more food per unit area than conventional horizontal farming. This technique is particularly beneficial in urban settings, where land availability and soil quality are often scarce. In other words, the main advantage of vertical farming is its space efficiency. Vertical farming can produce up to 100 times more food per square meter of land than traditional farming methods. This means less land is needed to feed the same number of people, reducing the pressure on natural resources and biodiversity. Additionally, vertical farming can use abandoned or underutilized buildings, such as warehouses, skyscrapers, or parking lots, to create productive and sustainable urban farms. Vertical farming can also reduce the transportation costs and carbon emissions associated with food distribution, as the crops can be grown closer to the consumers (Despommier, 2013; Song et al., 2015).

Resource Efficiency

Vertical farms aim to optimize the use of resources such as water, nutrients, and energy. By using closed-loop systems that recycle water and nutrients, precise control over growing conditions such as temperature, humidity, and light, and soilless cultivation methods such as hydroponics or aeroponics that deliver nutrients directly to the plant roots, vertical farms can reduce water consumption, minimize nutrient waste, and use energy more efficiently. As such, one of the main advantages of vertical farming is that it can produce more food per unit area than traditional farming methods. By stacking multiple layers of crops in a vertical structure, vertical farms can increase the yield per square meter of land. This can help meet the growing demand for food in urban areas, where space is limited, and land is expensive. Vertical farms can also reduce the environmental impact of agriculture by lowering greenhouse gas emissions, decreasing transportation costs, and enhancing food security (Al-Kodmany, 2020).

Climate Independence

Vertical farming is a method of growing crops indoors, using artificial lighting and climate control systems. Unlike traditional agriculture, vertical farming does not depend on the natural weather conditions outside. This means vertical farmers can produce food all year round, regardless of the seasons. This is especially beneficial for areas with harsh or unstable climates, such as deserts, polar regions, or disaster-prone zones. These regions can achieve food security by using vertical farming and reducing their reliance on imports from other countries. Vertical farming also has the potential to reduce greenhouse gas emissions and water consumption, as it uses less land and resources than conventional farming methods. As such, controlled environments in vertical farms enable consistent and predictable growing conditions. Year-round crop production ensures a continuous and reliable fresh produce supply, reducing dependency on seasonal harvests and external suppliers (Bhattacharya, 2019; Despommier, 2013; Kozai et al., 2019; Praveen & Sharma, 2019).

Reduced Environmental Impact

Vertical farming is driven by the goal of significantly decreasing its environmental footprint through a multifaceted approach. This involves not only minimizing land use but also actively seeking to curtail water consumption and eradicate the reliance on synthetic pesticides and fertilizers. By adopting these practices, vertical farming aims to substantially contribute to sustainable agriculture. This comprehensive strategy addresses pressing concerns associated with soil degradation, water pollution, and biodiversity loss. Further, in contrast to the conventional model, where the journey from farm to consumer spans considerable distances, the strategic placement of vertical farms substantially shortens this logistical gap. This reduction in transportation distance not only ensures that consumers have access to fresher and more nutritionally potent produce but also results in a noteworthy decline in transportation-related carbon emissions, commonly quantified as "food miles." The concept of localized production, often synonymous with the burgeoning movement of "local food," extends beyond mere environmental considerations; it catalyzes fostering the evolution of resilient local food systems. As a result, the endeavor toward reduced environmental impact in vertical farming extends beyond mere efficiency to encompass a broader commitment to fostering ecologically responsible and sustainable agricultural practices, making significant strides in mitigating the detrimental effects of traditional farming on our planet (De Bernardi & Azucar, 2020; Eigenbrod and Gruda, 2015).

Technological Innovation

Vertical farming embraces technological innovations, including automation, artificial lighting, and data analytics, to optimize the entire cultivation process. Vertical farms leverage advanced technologies, including IoT, sensors, and automation, for precise monitoring and control. Precision agriculture optimizes resource use, minimizes waste, and enables data-driven decision-making for crop management, contributing to higher efficiency and productivity. Ongoing technological advancements contribute to increased efficiency, reduced operational costs, and the continuous improvement of vertical farming practices (Al-Kodmany, 2020; Kalantari et al., 2017; Siregar et al., 2022).

Therefore, the logic underpinning vertical farming revolves around a strategic response to the limitations inherent in traditional agriculture. This forward-thinking approach embraces technological innovations to overcome challenges, prioritizing efficiency in space and resource utilization. The fundamental aim is to reduce the environmental impact of food production while concurrently fostering resilience and sustainability. By cultivating crops in stacked layers or controlled indoor environments, vertical farming transcends the constraints of traditional methods, ensuring year-round crop production, minimizing land use, and promoting locally sourced, fresh produce. In summary, vertical farming represents a paradigm shift that harmonizes agricultural practices with cutting-edge technology, aiming to create a more sustainable, resilient, and environmentally conscious future for food production.

The benefits and underlying logic of vertical farming coalesce into a comprehensive vision aimed at revolutionizing food production. This transformative force addresses the inherent limitations of traditional agriculture, championing innovation, sustainability, and resilience as its guiding principles. Vertical farming, through its ingenious use of space, controlled environments, and advanced technologies, charts a course toward a future where food systems are not only more efficient but also more sustainable and secure on a global scale. This holistic approach represents a paradigm shift, signaling the potential for a food production model that aligns harmoniously with the needs of the present and the imperatives of a resilient and sustainable future.

Vertical farming stands as a beacon of resilience within the agricultural sector, introducing a transformative model that adeptly navigates challenges and positions itself as a dynamic solution in the face of evolving global pressures. Its capacity to overcome spatial and seasonal constraints is a testament to its adaptability, liberating food production from traditional limitations. The spatial efficiency of vertical farming, achieved through the stacking of crops in controlled environments, addresses the challenge of land scarcity, especially in urban areas. Utilizing vertical space maximizes productivity per square foot, offering a practical solution to the diminishing availability of arable land.

Furthermore, vertical farming's ability to operate year-round, independent of external seasons, is a strategic response to the unpredictable impacts of climate change. This continuous production mitigates the vulnerabilities associated with weather fluctuations and ensures a consistent supply of fresh produce. In terms of environmental impact, vertical farming takes a proactive stance by significantly reducing the need for synthetic pesticides and fertilizers. The controlled indoor environments not only minimize the risk of soil degradation but also contribute to cleaner and more sustainable agricultural practices, aligning with the imperative of environmental stewardship.

Technological advancements play a pivotal role in enhancing the adaptability of vertical farming. The integration of automation, AI, and data analytics allows for real-time monitoring and adjustments, optimizing resource use and crop management. This technological synergy fosters a responsive farming model that can swiftly adapt to emerging challenges, from changes in market demand to unforeseen disruptions in the global supply chain.

VERTICAL FARMING TECHNOLOGIES

Vertical farming uses advanced technologies to create controlled environments that optimize crop growth. The integration of these technologies in vertical farming reflects a concerted effort to overcome traditional agricultural challenges, offering a more efficient, sustainable, and controllable approach to crop cultivation. Some key technologies commonly employed in vertical farming include the following (Krishnan, & Swarna, 2020; Saad et al., 2021).

Hydroponics

Hydroponics is a soilless cultivation method that revolutionizes traditional agriculture by growing plants in nutrient-rich water solutions without using soil. In hydroponic systems, plants receive nutrients directly from water, enriched with a balanced mix of minerals and fertilizers. This innovative approach allows for precise control over growing conditions, optimizing nutrient absorption and fostering plant growth in controlled environments. In vertical farming, hydroponics is crucial in optimizing resource efficiency and fostering plant growth through precise control over nutrient levels, pH, and water delivery to plant roots. Applying hydroponics in vertical farming maximizes resource efficiency, provides precise control over growing conditions, and contributes to sustainable and productive agriculture in the context of stacked cultivation (Pascual et al., 2018; Touliatos et al., 2016).

Aeroponics

Aeroponics is an advanced method of plant cultivation in which plants are suspended in air, and their roots are exposed to a nutrient-rich mist. Unlike traditional soil-based or hydroponic systems, aeroponics relies on delivering nutrients in the form of a fine mist sprayed directly onto the plant roots. This innovative approach provides plants with an oxygen-rich environment while optimizing nutrient absorption, promoting faster growth, and allowing for precise control over the plant's nutrient intake. Aeroponics technology is applied in vertical farms to promote optimal nutrient absorption and oxygenation, enhancing plant growth and facilitating resource efficiency. Therefore, applying aeroponics technology in vertical farms enhances nutrient absorption, promotes oxygenation, and facilitates resource efficiency. This method is well-suited for the unique challenges and objectives of vertical farming, contributing to increased productivity, reduced environmental impact, and sustainable agricultural practices (Eldridge et al., 2020).

Aquaponics

Aquaponics is a sustainable agricultural system that integrates fish farming, known as aquaculture, with hydroponics. This innovative approach creates a symbiotic relationship between fish and plants, where the waste produced by the fish serves as a nutrient source for the plants, and the plants help filter and purify the water for the fish. It is a closed-loop system that maximizes resource efficiency and promotes the co-cultivation of fish and crops in a mutually beneficial environment. In aquaponics, the waste produced by fish serves as a nutrient source for plants, and plants filter and purify the water, creating a sustainable and integrated ecosystem. Therefore, aquaponics involves the synergistic relationship between fish and plants, where fish waste becomes a nutrient source for plants, and plants, in turn, contribute to water purification. This integrated and closed-loop system creates a sustainable ecosystem, promoting efficient nutrient cycling and supporting the dual cultivation of fish and crops (Khandaker & Kotzen, 2018).

LED Lighting

LEDs are semiconductors that emit light when an electric current passes through them. In agriculture, particularly vertical farming, and CEA, LEDs serve as artificial light sources designed to meet plants' specific light requirements throughout their various growth cycles. In vertical farming, LEDs are strategically employed to provide the correct spectrum and intensity of light, playing a crucial role in promoting photosynthesis and ensuring year-round cultivation independent of natural sunlight. As such, the strategic application of LEDs in vertical farming ensures that crops receive the right light conditions for optimal growth throughout various stages, facilitating year-round cultivation and reducing dependency on natural sunlight (Nájera et al., 2022).

Vertical Farming Towers and Racks

Specially designed vertical structures, such as towers and racks, refer to purpose-built frameworks that efficiently stack multiple crop layers vertically tiered. These structures are designed to optimize available vertical space, a defining characteristic of vertical farming. The specially designed vertical structures, including towers and racks, have significant applications in optimizing space utilization within vertical farming. Therefore, applying specially designed vertical structures in vertical farming is integral to creating a space-efficient environment that maximizes cultivation potential. These structures contribute to the core principle of vertical farming—cultivating crops in vertical layers—as they optimize space utilization, enhance resource efficiency, and support sustainable and productive farming practices (Al-Kodmany, 2018).

Automated Monitoring and Control Systems

Advanced sensors and control systems refer to sophisticated technologies integrated into agricultural environments, particularly in CEA systems like vertical farming or greenhouse setups. These systems utilize sensors to monitor key environmental parameters, such as temperature, humidity, and nutrient levels. Additionally, they employ advanced control mechanisms to adjust these parameters in real time, ensuring optimal conditions for crop growth. Integrating advanced sensors and control systems in vertical farming brings about several applications focusing on automation. In summary, applying automation through advanced sensors and control systems in vertical farming enhances precision by enabling real-time adjustments to environmental parameters. This ensures optimal conditions for plant growth and reduces the dependency on constant manual oversight, allowing growers to focus on strategic aspects of cultivation (Chin & Audah, 2017; Chuah et al., 2019).

Smart Farming Software

Intelligent farming software is a comprehensive system integrating advanced technologies such as data analytics, machine learning, and AI into agricultural practices. This software is designed to collect, process, and analyze vast data from various sources within the farming ecosystem, aiming to optimize crop management and improve overall agricultural efficiency. By leveraging sophisticated data analytics, machine learning, and AI, intelligent farming software brings transformative capabilities to the agricultural landscape. Applying intelligent farming software in agriculture signifies a shift toward data-driven, precision-based practices. By harnessing the power of analytics and AI, farmers can make more informed decisions, optimize resource usage, and ultimately enhance productivity and sustainability in modern agriculture (Mohamed et al., 2021).

Biotechnology and Genetic Engineering

Biotechnology and genetic engineering encompass scientific processes that involve manipulating and altering an organism's genetic material, particularly in the context of plants, to enhance or introduce desirable traits. In agriculture, this often involves the modification of plant genetics to achieve characteristics such as increased resistance to pests, diseases, or environmental stress, as well as improvements in nutritional content or other agronomic traits. Biotechnological tools, such as genetic modification and gene editing, are employed to precisely engineer the genetic makeup of plants, allowing for targeted enhancements that align with specific agricultural goals. Biotechnology and genetic engineering find valuable applications in vertical farming, offering tailored solutions to the unique challenges and opportunities presented by controlled indoor environments. Applying biotechnology and genetic engineering in vertical farming empowers growers to create crops adapted to controlled indoor environments. This targeted approach addresses the challenges of vertical farming while aligning with market demands for sustainable, nutritious, and customizable food options (Despommier, 2019).

Internet of Things

The IoT is a concept that refers to the interconnection of everyday devices and sensors to the Internet, enabling them to collect, transmit, and exchange data. In an IoT ecosystem, physical objects—from household appliances and industrial machinery to wearable devices and environmental sensors—are embedded with Internet connectivity and computing capabilities. This connectivity facilitates seamless communication between devices, allowing them to share information and respond to real-time data, ultimately creating a network of interconnected and intelligent systems. In vertical farming, IoT technologies are pivotal in transforming traditional agricultural practices into intelligent, data-driven systems. In summary, IoT technologies in vertical farming revolutionize conventional agriculture by providing farmers with actionable insights, increased control, and the ability to create optimized and responsive growing environments. This contributes to improved crop yields, resource efficiency, and overall sustainability in vertical farming (Krishnan & Swarna, 2020).

Vertical farming harnesses cutting-edge technologies to establish controlled environments that are meticulously designed to optimize every aspect of crop growth. The integration of these advanced technologies represents a deliberate and strategic response to the challenges ingrained in traditional agriculture. This forward-thinking approach is geared toward providing a more efficient, sustainable, and controllable method of cultivating crops. The controlled environments in vertical farms are a testament to the precision that technology brings to the agricultural landscape. Variables such as temperature, humidity, light intensity, and nutrient levels are carefully monitored and regulated to create ideal conditions for plant growth. Automation, AI, and data analytics play pivotal roles in this process, ensuring real-time adjustments and continuous optimization.

One of the key motivations behind incorporating advanced technologies into vertical farming is the need to overcome limitations associated with traditional agriculture. These limitations include seasonal constraints, land scarcity, and susceptibility to pests and diseases. By leveraging technology, vertical farming extends beyond these constraints, enabling year-round crop production, minimizing land use, and providing a controlled environment that significantly reduces the reliance on pesticides and herbicides. The efficiency of resource use in vertical farming is a hallmark of its technological integration. Precise control over inputs such as water, nutrients, and energy results in higher resource-use efficiency compared to traditional farming methods. This not only contributes to the economic viability of vertical farming but also aligns with broader sustainability goals by reducing environmental impact.

In essence, the integration of advanced technologies in vertical farming symbolizes a departure from conventional agricultural practices. It represents a bold stride toward a future where the synergy between technology and agriculture not only addresses longstanding challenges but also propels the industry into a new era of efficiency, sustainability, and control over the cultivation process. Vertical farming stands as a testament to the transformative power of technology in redefining the possibilities and potential of modern agriculture.

VERTICAL FARMING CHALLENGES

Vertical farming presents novel answers to various issues encountered in conventional agriculture but encounters obstacles. Several prominent hurdles can be identified, which are discussed in the following sections.

High Initial Investment

One of the main challenges of vertical farming is the high capital expenditure required to establish and operate a facility. Unlike traditional farming, which relies on natural resources and manual labor, vertical farming uses sophisticated technologies such as artificial lighting, climate control, hydroponics, and robotics to optimize crop production in a limited space. These technologies are expensive to acquire, install, and maintain, and consume a lot of energy. Therefore, vertical farmers need to secure a lot of funding before starting their business. This can deter many potential entrants who do not have access to sufficient financial resources or are unwilling to risk investing in an emerging and uncertain market. As a result, the adoption of vertical farming remains limited to a few regions and crops, and its potential to transform the global food system is not fully realized (Al-Kodmany, 2020).

Operational Costs

Another challenge of vertical farming is the high operational costs of running such facilities. Vertical farms require a lot of energy to power artificial lighting, heating, cooling, ventilation, and irrigation systems. They also need regular maintenance to ensure optimal growing conditions and prevent pests and diseases. Additionally, vertical farms may require skilled labor to manage the complex technology and monitor the crops. These factors can make vertical farming more expensive than conventional farming, especially in areas where energy or labor costs are high. Therefore, vertical farming may not be economically viable or competitive in some markets unless it reduces operational costs or increases productivity and profitability (O'sullivan et al., 2019).

Energy Consumption

Vertical farms rely on artificial lighting, heating, ventilation, and air conditioning (HVAC) systems, and other energy-intensive technologies. The energy consumption of vertical farms can contribute to environmental concerns and operational expenses, requiring the development of energy-efficient technologies. However, vertical farms also offer potential benefits, such as reducing land use, water use, and transportation emissions. Therefore, it is essential to evaluate the environmental impacts of vertical farms compared to conventional agriculture, considering the trade-offs between energy use and other factors. Additionally, vertical farms should adopt best practices and innovative solutions to minimize their energy demand and optimize their productivity (Touliatos et al., 2016).

Technical Complexity

One of the barriers to adopting advanced technologies in agriculture is the need for technical expertise. Automation, IoT, and precision agriculture are examples of technologies that can improve productivity, efficiency, and sustainability. Still, they also require high skills and knowledge to implement and manage. Small-scale farmers or those who lack access to specialized knowledge may find it challenging to use these technologies effectively or troubleshoot problems that arise. Therefore, there is a need for more training, education, and support for farmers who want to leverage these technologies in their operations (Al-Kodmany, 2020).

Limited Crop Variety

Some crops may be more challenging to grow in vertical systems, limiting the variety of produce that can be cultivated. Focusing on specific crops may limit vertical farms' market appeal and hinder agricultural product diversification. For example, crops that require a lot of space, such as corn or wheat, may not be suitable for vertical farming. Additionally, crops with different light, temperature, or humidity requirements may be challenging to accommodate in a single vertical structure. Therefore, vertical farms may have to specialize in certain types of produce, such as leafy greens or herbs, which may reduce their competitiveness and profitability in the global market. Furthermore, the lack of crop diversity may negatively impact the produce's soil health, pest management, and nutritional value (Despommier, 2019).

Environmental Impact of Construction Materials

One of the challenges of vertical farming is to select appropriate materials for constructing the buildings that house the crops. The materials used in constructing vertical farm structures may have environmental implications. Considering the sustainability of construction materials is crucial to ensuring that the negative impact of materials does not outweigh the environmental benefits of vertical farming. For example, some materials may require a lot of energy or water to produce, transport, or maintain, which could offset the savings from reduced land use and transportation of food. Other materials may emit greenhouse gases or toxic substances during their life cycle, which could harm the environment and human health. Therefore, it is crucial to evaluate the environmental performance of different materials and choose those with low embodied energy, carbon footprint, water footprint, and toxicity levels (Al-Kodmany, 2018).

Scaling Challenges

Scaling vertical farming operations to meet increased demand can be challenging. As demand grows, vertical farms may face difficulties in expanding their operations to produce larger quantities of food cost-effectively. Some of these difficulties include finding suitable locations, securing adequate funding, hiring skilled workers, and complying with environmental and safety regulations. Vertical farms must also invest in advanced technologies, such as artificial lighting, climate control, hydroponics, and automation, to optimize production and reduce waste. These technologies can be expensive and complex to maintain, requiring constant monitoring and adjustment. Therefore, vertical farming requires careful planning, management, and innovation to overcome these challenges and achieve scalability (Despommier, 2013).

Regulatory Hurdles

One of the barriers that vertical farming faces is the lack of appropriate regulations that can accommodate its specific features and needs. Vertical farming differs from conventional farming in many aspects, such as artificial lighting, hydroponics, aeroponics, and controlled environments. These aspects may challenge existing land use regulations, food safety, and agricultural practices designed for traditional forms of agriculture. Regulatory challenges can impede the development and expansion of vertical farming initiatives, requiring adjustments and collaboration with regulatory bodies. For example, vertical farms may need permits for zoning, building, water, and electricity use, which may not be readily available or suitable for their operations. Additionally, vertical farms may face difficulties in complying with food safety standards and certifications, which may not account for the different methods and risks involved in vertical farming. Therefore, vertical farming must engage with regulators and policymakers to create a supportive and enabling regulatory environment to foster its growth and innovation (Al-Kodmany, 2018).

Effectively addressing the challenges inherent in vertical farming necessitates a sustained commitment to research, innovation, collaboration among stakeholders, and the unwavering adherence to sustainable and responsible practices within the industry. Ongoing research is fundamental to deepen our understanding of the complexities involved and to develop solutions that enhance the efficiency, productivity, and environmental friendliness of vertical farming. Innovation plays a pivotal role in overcoming obstacles. This encompasses advancements in technology, from more efficient lighting and climate control systems to the development of novel growing mediums and sustainable farming practices. Continuous innovation ensures that vertical farming remains at the forefront of agricultural sustainability, adapting to emerging challenges and opportunities.

Collaboration among stakeholders is essential for creating a comprehensive and inclusive approach to vertical farming. This involves partnerships between farmers, technology developers, policymakers, and the wider community. By fostering a collaborative ecosystem, we can leverage diverse perspectives and expertise to address challenges holistically, from regulatory frameworks to the implementation of best practices on the farm. Moreover, a commitment to sustainable and responsible practices is paramount. This includes minimizing resource consumption, optimizing energy efficiency, reducing waste, and implementing eco-friendly growing methods. Sustainable practices not only mitigate the environmental impact of vertical farming but also contribute to the industry's long-term viability and acceptance.

In summary, tackling the challenges in the vertical farming industry requires a multifaceted approach. It involves a dedication to ongoing research to expand our knowledge, continuous innovation to drive technological advancements, collaboration to integrate diverse perspectives, and a steadfast commitment to sustainable and responsible practices. Through this collective effort, the vertical farming industry can navigate challenges successfully, contributing to a more sustainable and resilient future for global food production.

CONCLUSIONS: EMBRACING THE FUTURE OF AGRICULTURE THROUGH VERTICAL FARMING

In summary, the examination of vertical farming in the framework of this chapter demonstrates a paradigm shift in agriculture that has the potential to tackle the significant issues encountered by conventional farming techniques effectively. Exploring the ideas, applications, and issues associated with vertical farming highlights its promise as a viable and effective response to the continuously increasing worldwide need for food while promoting sustainability. As we approach the conclusion, numerous significant insights become apparent.

1. Sustainable Agriculture for the Future: Vertical farming emerges as a beacon of sustainable agriculture, responding to the urgent need for resource-efficient and environmentally conscious food production. Its ability to maximize space, optimize resource use, and reduce environmental impact positions it as a crucial player in the quest for a more resilient and sustainable global food system.

2. Technological Integration and Precision Agriculture: The integration of advanced technologies, such as IoT, automation, and intelligent farming software, showcases the role of precision agriculture in the success of vertical farming. The marriage of cutting-edge technologies with agriculture enhances efficiency and marks a significant leap toward a data-driven and optimized approach to crop management.
3. Overcoming Challenges for Growth: While the potential of vertical farming is vast, it is essential to acknowledge the challenges accompanying its adoption. The issues of high initial investment, operational costs, and technical complexity underscore the need for ongoing innovation, collaboration, and supportive regulatory frameworks to ensure the sustainable growth of vertical farming ventures.
4. Addressing Global Food Security: Vertical farming's year-round cultivation, climate independence, and proximity to urban centers align with the vision of localized food production. These attributes contribute to the reduction of "food miles" and hold the promise of bolstering local food security by providing consistent access to fresh produce.
5. Diversification of Crops and Consumer Perception: The ability of vertical farming to diversify crops while maintaining precision in cultivation techniques addresses the demand for a variety of fresh, locally grown produce. Overcoming consumer skepticism and fostering trust in vertically grown produce will ensure widespread acceptance and market success.
6. The Role of Education and Collaboration: Education plays a crucial role in fostering an understanding of the benefits and methodologies of vertical farming. Collaboration between stakeholders, including researchers, policymakers, farmers, and consumers, is essential for creating an ecosystem that supports the growth and integration of vertical farming into mainstream agriculture.

In essence, the journey through vertical farming beckons us to embrace innovation, sustainability, and a forward-thinking approach to food production. Positioned at the crossroads of agriculture and technology, vertical farming emerges as a dynamic and promising trajectory. It holds the potential to reshape the entire panorama of global food systems, offering a pathway to a more resilient and sustainable future for generations yet to unfold. As we navigate this frontier, the principles of vertical farming beckon us to reimagine the way we grow and consume food, envisioning a world where efficiency, environmental consciousness, and forward-thinking practices converge to create a nourished and sustainable planet.

REFERENCES

Al-Kodmany, K. (2018). The vertical farm: A review of developments and implications for the vertical city. *Buildings*, *8*(2), 24.

Al-Kodmany, K. (2020). The vertical farm: Exploring applications for peri-urban areas. In Srikanta Patnaik, Siddhartha Sen, Magdi S. Mahmoud (eds), Smart Village Technology: Concepts and Developments, Springer, NYC, 203–232.

Badami, M. G., & Ramankutty, N. (2015). Urban agriculture and food security: A critique based n an assessment of urban land constraints. Global *F*ood *S*ecurity, *4*, 8–15.

Bhattacharya, A. (2019). Global climate change and its impact on agriculture. In Amitav Bhattacharya (ed.), Changing Climate and Resource *u*se Efficiency in Plants. Academic Press: Cambridge, MA, 1–50.

Buchner, L. (2022). Investigating the potential use of existing buildings for vertical farming. Master's thesis, Faculty of Engineering and the Built Environment.

Chin, Y. S., & Audah, L. (2017). Vertical farming monitoring system using the internet of things (IoT). In *AIP Conference Proceedings* (Vol. 1883, No. 1, p. 020021). AIP Publishing LLC, Johor, Malaysia.

Chuah, Y. D., Lee, J. V., Tan, S. S., & Ng, C. K. (2019). Implementation of smart monitoring system in vertical farming. In IOP Conference Series: Earth and Environmental Science (Vol. 268, No. 1, p. 012083). IOP Publishing, West Java, Indonesia.

De Bernardi, P., Azucar, D., (2020). The food system grand challenge: a climate smart and sustainable food system for a healthy Europe. In Paola De Bernardi and Danny Azucar (eds), Innovation in Food Ecosystems: Entrepreneurship for a Sustainable Future, Springer, NYC, 1–25.

Despommier, D. (2013). Farming up the city: The rise of urban vertical farms. *Trends in Biotechnology*, *31*(7), 388–389.

Despommier, D. (2019). Vertical farms, building a viable indoor farming model for cities. Field actions science reports. The Journal of Field Actions, *12*(Special Issue 20), 68–73.

Eigenbrod, C., & Gruda, N. (2015). Urban vegetable for food security in cities. A review. Agronomy for Sustainable Development, *35*, 483–498.

Eldridge, B. M., Manzoni, L. R., Graham, C. A., Rodgers, B., Farmer, J. R., & Dodd, A. N. (2020). Getting to the roots of aeroponic indoor farming. *New Phytologist*, *228*(4), 1183–1192.

Jaeger, S. R., Chheang, S. L., & Ares, G. (2022). Text highlighting as a new way of measuring consumers' attitudes: A case study on vertical farming. *Food Quality and Preference*, *95*, 104356.

Kalantari, F., Mohd Tahir, O., Mahmoudi Lahijani, A., & Kalantari, S. (2017). A review of vertical farming technology: A guide for implementation of building integrated agriculture in cities. In Advanced *E*ngineering *F*orum (Vol. 24, pp. 76–91). Trans Tech Publications Ltd, Baech, Switzerland.

Khandaker, M., & Kotzen, B. (2018). The potential for combining living wall and vertical farming systems with aquaponics with special emphasis on substrates. *Aquaculture Research*, *49*(4), 1454–1468.

Kozai, T., Niu, G., & Takagaki, M. (Eds.). (2019). Plant *F*actory: *A*n *I*ndoor *V*ertical *F*arming *S*ystem for *E*fficient *Q*uality *F*ood *P*roduction. Academic Press, NYC.

Krishnan, A., & Swarna, S. (2020, October). Robotics, IoT, and AI in the automation of agricultural industry: A review. In *2020 IEEE Bangalore Humanitarian Technology Conference (B-HTC)* (pp. 1–6). IEEE, Karnataka, India.

Levi, E., & Robin, T. (2020). COVID-19 did not cause food insecurity in Indigenous communities but it will make it worse. Yellowhead Institute *W*ebsite, 29.

McCarthy, U., Uysal, I., Badia-Melis, R., Mercier, S., O'Donnell, C., & Ktenioudaki, A. (2018). Global food security-issues, challenges and technological solutions. Trends in Food Science & Technology, *77*, 11–20.

Mohamed, E. S., Belal, A. A., Abd-Elmabod, S. K., El-Shirbeny, M. A., Gad, A., & Zahran, M. B. (2021). Smart farming for improving agricultural management. The Egyptian Journal of Remote Sensing and Space Science, *24*(3), 971–981.

Nájera, C., Gallegos-Cedillo, V. M., Ros, M., & Pascual, J. A. (2022, April). The 1st International Electronic Conference on Horticulturae. LED lighting in vertical farming systems enhances bioactive compounds and productivity of vegetables crops. In Biology and Life Sciences Forum (Vol. 16, No. 1, p. 24). MDPI.

O'sullivan, C. A., Bonnett, G. D., McIntyre, C. L., Hochman, Z., & Wasson, A. P. (2019). Strategies to improve the productivity, product diversity and profitability of urban agriculture. *Agricultural Systems*, *174*, 133–144.

Pascual, M. P., Lorenzo, G. A., & Gabriel, A. G. (2018). Vertical farming using hydroponic system: Toward a sustainable onion production in Nueva Ecija, Philippines. Open Journal of Ecology, 8(1), 25.

Praveen, B., & Sharma, P. (2019). A review of literature on climate change and its impacts on agriculture productivity. Journal of Public Affairs, 19(4), e1960.

Saad, M. H. M., Hamdan, N. M., & Sarker, M. R. (2021). State of the art of urban smart vertical farming automation system: advanced topologies, issues and recommendations. *Electronics*, *10*(12), 1422.

Siregar, R. R. A., Seminar, K. B., Wahjuni, S., & Santosa, E. (2022). Vertical farming perspectives in support of precision agriculture using artificial intelligence: A review. *Computers*, *11*(9), 135.

Song, W., Pijanowski, B. C., & Tayyebi, A. (2015). Urban expansion and its consumption of high-quality farmland in Beijing, China. *Ecological Indicators*, *54*, 60–70.

Touliatos, D., Dodd, I. C., & McAinsh, M. (2016). Vertical farming increases lettuce yield per unit area compared to conventional horizontal hydroponics. Food and Energy Security, *5*(3), 184–191.

Van Gerrewey, T., Boon, N., & Geelen, D. (2021). Vertical farming: The only way is up? *Agronomy*, *12*(1), 2.

3 Vertical Farm Lighting Systems
Ways and Means to Increase Their Efficiency

Dmitrii Filatov

INTRODUCTION

Today, the world has passed the 8 billion mark. The well-being of any person depends, first and foremost, on his or her health. One of the critical factors influencing health is nutrition. Food should be not only of the necessary quantity but also of the necessary quality. At the same time, the food must be affordable. The affordability of food depends on costs, including the production and transportation of products.

Modern glass greenhouses are usually located outside the urban environment. This is due to a lack of the necessary land area for city greenhouses. The situation is exacerbated by the high cost of land in cities. At the same time, continued urbanization increases the density of the urban population. As a result, the places where people produce and live are removed. To solve the problem, the transportation of products to places of consumption with high transport costs is used. High transportation costs are due to the high price of motor fuel.

An essential part of the horticulture production cost is the purchase of electricity and heat. These energy resources are spent to create optimal conditions for high plant productivity.

Electric energy, heat, and motor fuel are derived from hydrocarbons, which are currently expensive. At the same time, electricity from renewable energy sources (solar, wind, water) is also expensive due to the instability of resources, imperfect technology, and lack of capacity.

The triangle of problems (urbanization, high energy prices, food security) can be solved using high-tech vertical farms. A vertical farm is essentially a multi-story production facility. Strictly speaking, vertical farms can be divided into closed and open.

In terms of artificial lighting, open vertical trusses do not differ much from the famous glass greenhouses. At the same time, even for open vertical trusses, it is difficult to talk about sufficient natural light in the conditions of dense urban development.

The situation is complicated by global warming: frequent freezing rains, floods, droughts, and other unfavorable phenomena of nature. Reducing the impact and dependence of vegetable and berry production on nature is possible by using artificial systems in a fully controlled environment. Therefore, this chapter will focus on closed vertical farms (CVFs), where artificial lighting plays a more critical role in food production and the financial costs of energy.

ENERGY EFFICIENCY OF CVFS AND GREENHOUSES

Unlike a greenhouse, a CVF is a system of hermetically sealed materials with high insulation. Sometimes, enclosed vertical farms are also called plant factories (PFs). Such a plant is characterized by maximum production density and high productivity by creating the most favorable microclimate with minimal interaction with the external environment. In this direction, the data on the energy consumption of greenhouses and PF are of interest.

The initial data and technologies available today allow scientists to model plant production in greenhouses and CVF to understand the prospects for urban agriculture and food security issues. Graamans

DOI: 10.1201/b23309-3

and colleagues simulated lettuce production at three sites located in different latitudes and climates (Graamans et al., 2018): Kiruna in Sweden (SWE: 67.8° n.l., 20.2° e.l.), Amsterdam in the Netherlands (NLD: 52.0° n.l., 5.7° e.l.), and Abu Dhabi in the United Arab Emirates (UAE: 24.3° n.l., 54.2° e.l.).

A simulation was done of greenhouses and PFs with a footprint of 100 × 100 m, a size deemed sufficient to negate border effects. The model for the plant factory assumes five production layers, and that for the greenhouse, just one. The greenhouse is modeled as a Venlo type, consisting of 25 spans of 4 m, with a north-south gutter, a gutter height of 6 m, and a roof slope of 23°. The plant factory is modeled as a highly insulated opaque box illuminated by LEDs with an overall height of 6 m. In PFs, it is assumed that no air is exchanged with the exterior climate. The results of the study led to several important conclusions.

First, the study showed a clear advantage of PF over greenhouses regarding plant productivity, regardless of the region. It can be said that the optimization and homogeneity of the microclimate in PF facilities lead to a higher and more stable yield compared to greenhouses.

Second, PF requires more purchased electrical energy than greenhouses for plant production. This is due to artificial technology compensating for the lack of natural energy sources (light, heat, or cool).

The electric load of artificial lighting in PF exceeds all other loads regardless of the region. It amounts to 80–90% of all electricity consumed, about four times the electricity for greenhouse lighting in Sweden (GH+). Lastly, the economic efficiency of PF depends on the sources and costs of electricity and heat. For the UAE, the cost of purchased electricity for air conditioning seems to be more reasonable than the cost of artificial lighting.

Thus, using open (like greenhouses) or closed vertical trusses is ambiguous. The decision depends on the task: food security, low cost of production, or other goals. In their works, scientists dealing with vertical farms (Kozai, 2013; Weidner et al., 2021; Avgoustaki and Xydis, 2020) agree that PF can be practical and economically feasible nowadays in countries with cold climates or low solar insolation (cloudy weather, technological smog, etc.). An obvious disadvantage of enclosed vertical farms is the high demand for electrical energy for artificial lighting required for photosynthesis. Improving the efficiency of lighting systems is a significant challenge to expanding the geographic scope of CVFs.

LIGHTING SYSTEMS

The active use of artificial light sources in crop production began in the 1980s of the 20th century. Over 40 years, light technology has been continuously improved: incandescent lamps, metal halide lamps, fluorescent lamps (FLs), and high-pressure sodium (HPS) lamps.

HPS lamps are currently most common in crop production because of their high luminous efficacy. However, these types of lamps have disadvantages. First, there is an ineffective spectrum of radiation for plants. Plant development is based on the process of photosynthesis. The spectrum of optical radiation, which coincides with the absorption spectrum of chlorophyll, is in the blue and red region of photosynthetically active radiation. However, an HPS lamp is characterized by a small fraction of energy in the blue region of the spectrum, a substantial fraction in the green region, and an insufficient fraction in the red region. The technology of HP's manufacturing does not allow them to change their emission spectrum. Second, the HPS lamp has a high heating temperature (250–270°C). When the lamps are close to the plants, they cause tissue damage from photo stress. Therefore, the use of HPS lamps for vertical farms is limited.

FLs are widely used in Asian PFs because of their low heat but low light efficiency. The new light-emitting diode (LED) sources do not have the disadvantages of discharge lamps. Also, luminaires based on LEDs are more efficient regarding electromagnetic compatibility than those based on FLs and HPS lamps. When the supply voltage deviates, LED luminaires have stable values of electric power consumption and photosynthetic photon flux (Figure 3.1). This is due to the change in the current consumption (Figure 3.2). Since the voltage level in the power grid is constantly

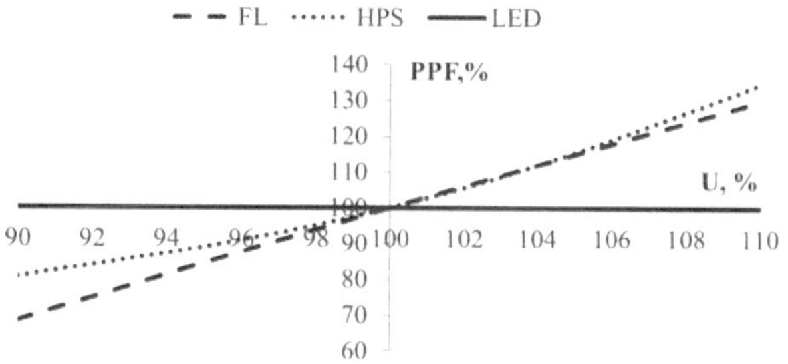

FIGURE 3.1 Relative changes in photosynthetic photon flux (PPF) with relative changes in supply voltage (U) for different light sources.

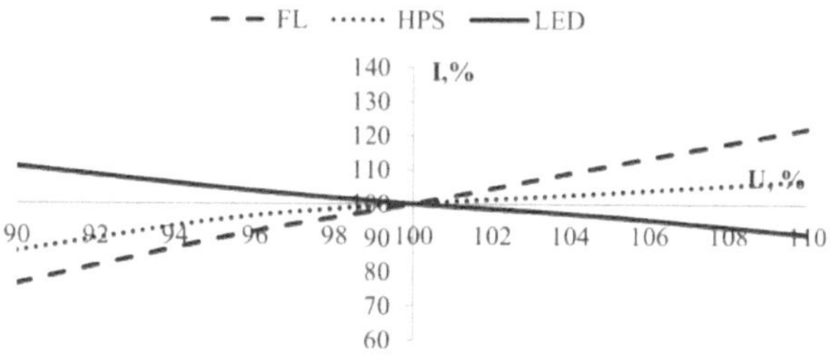

FIGURE 3.2 Relative changes in intensity electric current (I) with relative changes in supply voltage (U) for different light sources.

changing, it leads to instability of FL and HPS lamps' output parameters, which complicates the predictability of the yield and increases the energy intensity of the final product.

At the end of the 20th century, LEDs were first introduced at the University of Wisconsin, Purdue University, and the Kennedy Space Center NASA to test their effect on plant growth for feeding during spaceflight.

The effect of LED light on plants should be considered from two main points of view: energy efficiency and quality food (food rich in vitamins and valuable elements). This is due to the population's need for accessibility to food. Thus, the task is to use the minimum cost of purchased electricity with the highest possible production of quality products

Electricity consumed by lighting systems to produce a unit of product (kWh/kg):

$W_{kg} = P \cdot T \cdot m^{-1}$

P – consumed electric power of lighting systems, kW; T – operating time of lighting systems, h; m – crude/dry weight of produced goods, kg.

Thus, reducing the unit production cost by reducing electricity consumption or increasing yields is possible.

LIGHT SPECTRUM

One of the main advantages of LEDs compared to other light sources is the ability to generate light energy at a wavelength corresponding to plant photoreceptors to influence plant morphology and metabolism. Specialized photoreceptors perform plant responses to light. Five photosensory systems have been found: phytochromes, cryptochromes, phototropins, members of the Zeitlup family, and a locus of resistance to ultraviolet light. Phytochromes maximally absorb light in the red (600–700 nm) and far-red (700–750 nm) regions of the spectrum (Chen and Chory, 2011). Sensitivity peaks are in the red region at 660 nm and the far-red region at 730 nm (Smith, 2000; Demotes-Mainard et al., 2016). Cryptochromes, phototropins, and members of the Zeitlup family maximally absorb light in the spectrum's blue (390–500 nm) region (Ahmad and Cashmore, 1993; Christie, 2007; Suetsugu and Wada, 2013). Blue and red LEDs are the most promising because chlorophyll a and b (chl-a and chl-b) effectively absorb blue and red wavelengths. The absorption maxima for chl-a are 430 and 663 nm and for chl-b are 453 and 642 nm (Chory, 2010). Determining the optimal emission spectrum to maximize the yield of different plant species is an important task.

White vs Red-Blue Light

Today's main discussion is whether white or red-blue (RB) LEDs should be used. The answer to this question is complicated. Proponents of white light often make the argument that such light is more beneficial to the eyes of farm staff. However, modern technology solves the problem for people. For example, the use of special glasses that correct the spectrum. In addition, automated controls are evolving and require less human involvement. Therefore, we will be guided by yield and energy costs.

The results of comparing lettuce yields under white and RB lights are inconsistent (Lu et al., 2019; Kobayashi et al., 2013; Pennisi et al., 2019b; Zhou et al., 2022). The ratio of red (R) to blue (B) LEDs is important when irradiating lettuce. LEDs with an R:B = 3 ratio are the most efficient option in terms of total mass and resource utilization. Application of R:B = 3 resulted in maximum yield and increased chlorophyll and flavonoid concentrations in the leaves, increased nitrogen, phosphorus, potassium, and magnesium uptake.

Similar trends in the blue-red spectrum can be observed in basil cultivation. Experience of growing basil at an intensity of 215 μmol/(m^2 s) and a photoperiod of 16 hours/day under RB LEDs (Pennisi, Blasioli et al., 2019a), characterized by different R:B ratios ranging from 0.5 to 4 and FLs as control showed the best results with R:B = 3. Application of R:B = 3 resulted in higher yields and chlorophyll content, improving water and energy efficiency. The combination of R and B light (especially 70R:30B) in purple and green basil caused basil plants' favorable growth and pigmentation parameters. B light is necessary with R light to increase the vegetable plant's pigment content and antioxidant capacity in a controlled environment (Naznin et al., 2019).

Spectrum has a great influence on the yield and quality of microgreens. In (Brazaityte et al., 2021) mustard and cabbage microgreens were grown hydroponically at different B:R light ratios (447 and 660 nm): 0:100, 10:90, 25:75, 50:50, 75:25, and 100:0 (intensity 220 μmol/(m^2s), photoperiod 18 hours/day). Plants treated with B25R75 and B100R0 light had the highest crude and dry weight. In mustard, pea, and green basil, the yield was higher at R:B = 9 compared to R:B = 2 and R:B = 5 (Bantis, 2021). Increasing R light also positively affects broccoli microgreens (Liang et al., 2022). Five light treatments including white, single red, R:B = 5:1 (R5B1), R:B = 5:3 (R5B3), and R:B = 5:5 (R5B5) were developed in an experiment. The results showed that hypocotyl length and fresh weight increased with R and R5B1. As the proportion of B light increased, the soluble protein content increased and the soluble sugar content decreased. R's total phenolics and flavonoids were significantly lower than the other treatments. The advantage of combining RB LEDs over white, B, and R light was also obtained when growing microgreens of two varieties of amaranth (Meas et al., 2020).

CVFs can be used to grow quality seedlings of fruit crops. The right light also plays an important role. The advantage of RB light (R:B = 9:1) over white light was obtained when growing cucumbers (Jie et al., 2020). The raw and dry weight of cucumber plants was 25% higher. This is presumably due to the high proportion of R light over B light. A study (Hamedalla et al., 2022) supports this assumption by comparing LEDs with different ratios of red to blue light (30:70, 50:50, and 70:30) at an intensity of 150 μmol/(m^2s) and a photoperiod of 12 hours/day. White light was used as a control. Cucumber plants with an R:B ratio of 70:30 had the highest mass.

RB and red-blue-green (RBG) LEDs generally favor growth and photosynthesis in cherry tomato seedlings (Xiaoying, 2012). Plants under RB and RBG are shorter and more vigorous than plants under white (C), orange (O), green (G), and red (R) light. Tomato plants under 70R:30B and 50R:50B had 39% and 36% more dry weight than under white light, respectively. In addition, 70R:30B and 50R:50B LEDs had 172% higher energy efficiency (g/kWh) than under white light of FLs (Hernández et al., 2016). When exposed to white and red light, Tomato yields are comparable (LU et al., 2012).

Comparing white and RB light, we can say that the second option is more efficient in using energy resources to produce quantitative and qualitative yields. The proportion of R light can range from 70% to 90% of the total photosynthetic photon flux. The value of R light is refined depending on the plant variety and the desired ratio of biomass to minerals. In this case, the most universal for different plants is the ratio of red to blue LEDs R:B = 75:25.

Adding Green to the RB LEDs

Green (G) light is active in the plant's regulatory processes. G light has an increased penetrating ability to deliver energy to light-depleted areas of plants. This means that where B and R rays stop working, it is G light that makes a significant contribution to photosynthesis. The pigments involved in the assimilation of G light are carotenoids. One of the roles of these pigments, along with chlorophyll, is to trap light, converting it into energy within the plant. Carotenoids work in two polar areas. On the one hand, they protect chlorophyll from excessive light; on the other hand, they help collect light energy more efficiently when it is scarce.

In many cases, physiological changes in plants induced by G light are opposite to the effect of blue light. For example, the accumulation of anthocyanins (pigments that color fruits and vegetables in bright colors) induced by blue light is suppressed by G light. Blue light promotes the opening of estuaries, while G light promotes their closure. Blue light inhibits early stem elongation during germination, whereas G light promotes it. Consequently, G light interacts very closely with blue light, so it is crucial to the gross amount of these two spectra separately and the ratio (B:G) between them in the projected spectrum.

Additional green LED irradiation based on a combination of red and blue LEDs can also improve lettuce growth (Son and Oh, 2015). Combinations of six LED sources (R:B = 9:1, 8:2, 7:3; R:G:B = 9:1:0, 8:1:1, 7:1:2) were made based on red (655 nm), blue (456 nm) or green (518 nm) light at photosynthetic photon flux densities of 173 μmol/(m^2 s). Replacing blue with green LEDs in the presence of a fixed fraction of red LEDs enhanced lettuce growth. In particular, fresh masses of red leaf lettuce shoots under R8G1B1 were approximately 61% higher than under R8B2. In addition, analysis of leaf morphology, transmission coefficient, cell division rate, and leaf anatomy when treated with green LEDs confirmed enhanced growth of the two lettuce cultivars tested.

At the same time, the addition of G light to the RB LEDs reversed the spectrum efficiency in growing microgreens. In (Kamal et al., 2020) microgreens were grown at four different LED ratios (%); R:B 80:20 and 20:80 (R80:B20 and R20:B80), or R:G:B 70:10:20 and 20:10:70 (R70:G10:B20 and R20:G10:B70) at 150 μmol/(m^2 s) intensity. The results showed that partial replacement of red light with green LEDs (R70:G10:B20) increased vegetative growth and morphology, while blue LEDs (R20:B80) increased mineral and vitamin content. At the same time, the proportion of G light is supposed to be significant (10–20%) because 5% G light does not increase vitamins (Samuolienė et al., 2019).

Partial replacement of red and blue light with G light increases the biomass and yield of tomatoes (Kaiser et al., 2019). As the percentage of G light increased, there was a tendency for a linear increase in leaf biomass, specific leaf area, stem biomass, stem length, and the number of internodes. Increasing the proportion of G light (+32%) in the spectrum increased plant biomass and yield (+6.5%).

Partial replacement of red light with G light improves photosynthesis and increases the dry biomass of cucumber (Claypool and Lieth, 2021). It should be noted that the increase in biomass is due to an increase in leaf area. This negatively affects the cucumber yield because the plant shading increases with the increase in leaf area. When the blue light is replaced by G light, the opposite of red light reactions occur for the cucumber.

Adding G light to red and blue LEDs can increase plant productivity. However, green LEDs are energetically less efficient than red and blue LEDs. Using green LEDs instead of red or blue LEDs will result in higher energy costs. Therefore, the value of green LEDs should be determined by the yield effect obtained. Adding green LEDs should help to reduce the energy intensity of production.

Adding Far-Red Light to the RB LEDs

Science has known for over half a century such a phenomenon as the Emerson effect – an increase in photosynthetic efficiency when chloroplasts are simultaneously exposed to light with a wavelength of 670 nm (shortwave) and 700 nm (longwave red light). In this case, they will have a more significant effect than the sum of their effects separately. Thus, far-red (FR) light can enhance photosynthesis. However, this influence is limited.

A study (Lee et al., 2016) investigated the value of adding FR spectrum to B-R LEDs when irradiating lettuce. The peaks of B, R, and FR LED lights were 440, 660, and 735 nm, respectively. For all treatments, the photosynthetic photon flux was 130 $\mu mol/(m^2\ s)$. It was found that adding FR increases the yield regardless of the value. The highest value was recorded at a B+R/FR ratio of 1.2 (B:R = 2:8). The content of antioxidant phenolic compounds per lettuce plant was higher at a low B+R/FR ratio (0.7 and 1.2) than in control and B+R. In another study (Meng and Runkle, 2017), the best results were obtained when irradiating lettuce at the ratio R:FR = 3:1 (B:R = 1:1).

Adding FR light to RB LEDs increases basil yield, including a doubling of basil yield compared to white LEDs (Rahman et al., 2021). Adding FR light to RB LEDs when irradiating microgreens increases crude biomass (Gerovac et al., 2016), increases antioxidants, and decreases nitrate content (Giménez et al., 2021).

Increased photosynthesis and early flowering with the addition of the FR spectrum form a large mass of tomato fruit. This is due to the increased absorption power of the fruit because the FR light enhances sugar transport and metabolism. Adding 20% and 40% FR to the red spectrum increases tomato fruit biomass by 6.8% and 11.7%, respectively (Kim et al., 2018). Adding 36%, 80%, and 100% FR to the blue-red spectrum increases tomato fruit biomass by 19.3%, 30.9%, and 30%, respectively (Kalaitzoglou et al., 2019). Thus, the highest value of tomato yield was recorded with a B+R/FR ratio of 1.24 (B:R = 5:95).

Energy Efficiency and Degradation of LEDs

After considering the effects of different spectra on plant productivity, we can say that red light dominates in the most effective radiation spectrum for different plant species. This positively affects the energy efficiency of lighting systems since red LEDs are more energy efficient than blue, green, and white LEDs.

From the example of TL PROM FITO plant illuminators (Light Technologies, Russia) using OSRAM OSLON SSL LEDs, we can see (Figure 3.3) that to generate equal photosynthetic photon flux (PPF), the illuminator with predominant red light requires less electrical power (P) than the illuminator with predominant blue light. In this example, the red LED is 25% more efficient than the blue LED. Depending on the manufacturers, the energy efficiency of red LEDs can be 30–40% higher than other colored LEDs, the energy efficiency of blue and white LEDs is comparable, and green LEDs have the lowest energy efficiency (Prikupets, 2019).

At the same time, red LEDs have a low degradation rate of luminous flux compared to other colored LEDs. For example, a study (Narendran et al., 2001) shows that after 6,000 hours, green LEDs need more electric energy consumption by 20%, blue by 30%, and white by 50% to generate the initial PPF. At the same time, red LEDs will need 10% more electrical energy.

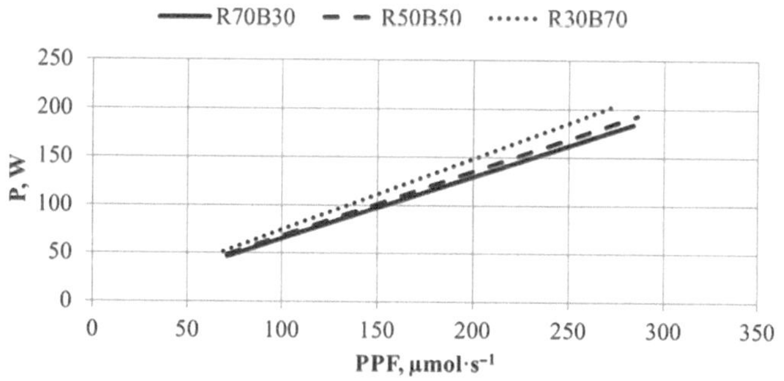

FIGURE 3.3 Dependence of electrical power consumption (P) on the required generation of equal photosynthetic photon flux (PPF) of TL PROM FITO plant illuminators with different ratios of red and blue LEDs.

Depending on the manufacturers, the energy efficiency of red LEDs compared to blue, green, and white LEDs may vary, but the trend will remain.

PPF DENSITY (INTENSITY) OF LIGHT

Light intensity or photosynthetic photon flux density (PPFD, µmol/(m^2 s)) also affects the energy intensity of production. More electrical energy must be expended to create a higher intensity (with the same spectrum). The PPF emitted by LEDs is proportional to their electrical power consumption.

Studies show (Zhou et al., 2022; Yan et al., 2019; Chen et al., 2020; Yi et al., 2020) that a sharp increase in the biomass of lettuce and basil is observed when the light intensity is increased to 200 µmol/(m^2 s). With a further increase in light intensity, biomass growth slows down. After 300 µmol/(m^2 s), biomass is slightly increased or decreased. Meanwhile, the percentage of soluble sugar, total anthocyanins, phenolic compounds, and flavonoids per plant positively correlates with light intensity. At a light intensity of 220 µmol/(m^2 s), lettuce had the best biomass, vitamin C, and nitrate combination in lettuce leaves (Fu et al., 2017). Combining the results for sweet basil growth, yield, and nutritional quality identified 224 µmol/(m^2 s) as the optimum light intensity (Dou et al., 2018). Thus, the best quantity and product quality ratio in cultivating lettuce and basil can be obtained at a light intensity of 220 µmol/(m^2 s). This value of light intensity for lettuce and basil is resource-efficient. In (Pennisi et al., 2020), experiments were performed to irradiate lettuce and basil with red (669 nm) and blue (465 nm) LEDs with R:B = 3:1 ratio. Plants were grown under 5 light intensities of 100, 150, 200, 250, and 300 µmol/(m^2 s) at 16 hours/day photoperiod. The best results for water, light, and power efficiency were obtained at light intensity 250 µmol/(m^2 s). There was little difference between 200 and 250 µmol/(m^2 s) for lettuce and a pronounced difference at 250 µmol/(m^2 s) for basil.

For microgreens, the optimal light intensity depends on the crop being grown. The preferred intensity for microgreens arugula, mustard, and kohlrabi is 200 µmol/(m^2 s) (Johnson et al., 2019; Jones-Baumgardt et al., 2019; Samuolienė et al., 2013; Brazaitytė et al., 2015), The effects of LED illumination spectra and intensity on carotenoid content in Brassicaceae microgreens. The preferred intensity for microgreens, radishes, cabbage, and broccoli is 100 µmol/(m^2 s) (Vetchinnikov et al., 2021; Gao et al., 2021). For all species of microgreens, an increase in light intensity led to a decrease in shoot height and leaf area but an increase in dry biomass. Increasing light intensity in the 100–200 µmol/(m^2 s) range did not significantly affect the crude biomass. The amount of vitamins and minerals depends on the intensity and the variety of microgreens. Thus, the content of soluble protein, sugar, free amino acids, flavonoids, vitamin C, and glucosinolates in broccoli microgreens was higher when exposed to 70 µmol/(m^2 s). Thus, the value of light intensity to obtain the maximum biomass of microgreens and beneficial substances can vary significantly.

Considering the growth and quality of cucumber seedlings, the optimal light intensity is 150 μmol/(m^2 s). At this light intensity, the highest number of flowers is observed (Kwack et al., 2014). As the light intensity increases, light use efficiency decreases with increasing dry weight (An et al., 2021). Thus, the decrease in light use is greater than the increase in dry biomass.

In Fan et al. (2013), tomato seedlings were irradiated with blue-red LEDs with R:B = 1:1 ratio at light intensities of 50, 150, 200, 300, 450, and 550 μmol/(m^2 s). The results showed that fresh weight, dry weight, stem diameter, and health index were higher in plants grown at 300, 450, and 550 μmol/(m^2 s). Energy efficiency was highest at 300 μmol/(m^2 s). The same level of irradiation was found to be optimal (He et al., 2019). In Huber et al. (2021), tomato seedlings were irradiated with blue-red LEDs with R:B = 58:42 ratio under light intensities of 100, 150, and 200 μmol/(m^2 s). An optimal intensity of 200 μmol/(m^2 s) is recommended based on the combination of energy costs and seedling quality. In Hwang et al. (2020), tomato seedlings were irradiated with white LEDs at light intensities of 100, 150, 200, 250, and 300 μmol/(m^2 s). No apparent difference in efficiency coefficient between the treatments with different intensities was found.

When discussing optimum light intensity, giving an exact value for a particular plant is impossible. The value of effective light intensity for different varieties of the same plant species can differ by a factor of 1.5–2. Considering biomass, nutrients, and energy expenditure, one can speak of an effective 100–250 μmol/(m^2 s) range. However, every plant species inevitably has a point of light saturation. Since the practical light intensity values were obtained at different operating times (12–18 hours/day), it is evident that the light saturation of plants is most likely related to the daily light integral (DLI).

DAILY LIGHT INTEGRAL

The DLI refers to the amount of light during the day required to achieve maximum productivity of quality plants with minimal resources. The value of the DLI is defined as the effect of a given intensity of light in a period of a given time:

$$DLI = PPFD \cdot H \cdot 3{,}600 \cdot 10^{-6}$$

PPFD – photosynthetic photon flux density (light intensity), μmol/(m^2 s)); H – photoperiod, hours/day.

From the expression, we see that the same DLI can be obtained either by high light intensity at a short or low light intensity at a long photoperiod.

Previously, we showed that increasing DLI by increasing light intensity positively affects the quantity and quality of grown products. The photoperiod also plays an important role. Increased light time correlates positively with lettuce biomass. In addition, longer illumination time increases the concentration of crude fiber and soluble sugar, reduces the concentration of nitrates, and increases the vitamin C content (Shen et al., 2014). In the study (Vaštakaitė-Kairienė et al., 2022) mustard microgreens were grown. The days were 8, 12, 16, 20, and 24 hours at a constant photon flux density of 300 μmol/(m^2 s). The shortest 8-hour photoperiod resulted in increased leaf area and fresh mass of microgreens. Better potassium, magnesium, phosphorus, and zinc uptake was observed in mustard grown under 8-hour photoperiod conditions. Increasing the photoperiod from 8 to 20 hours/day under LED lighting with an R:B = 70:30 ratio and an intensity of 130 μmol/(m^2 s) significantly increased the biomass of two varieties of amaranth microgreens (Meas et al., 2020). Increasing the photoperiod from 12 to 20 hours/day and an intensity of 70 μmol/(m^2 s) of two varieties of cabbage microgreens showed that a photoperiod of 14–16 hours/day was more effective in biomass production (Liu et al., 2022).

Thus, increasing DLI increases fresh and dry shoot mass, leaf width and number, and chlorophyll concentration, regardless of the combination of PPFD and photoperiod. The relationship between photoperiod and intensity within the same DLI is of interest from the perspective of efficient use of energy resources.

In Kelly et al. (2020), two varieties of lettuce, "Rex" and "Ruxay," were grown at photoperiods of 16, 20, and 24 hours/day. Under the intensity/photoperiod regime of 180/24, the crude biomass was 14% and 18% higher than the light regimes of 216/20 and 270/16. In another study (Palmer and van Iersel, 2020), aboveground biomass increased by 16.0% in lettuce and 18.7% in mizuna in response to an increase in photoperiod from 10 to 20 hours. The best dry weight results were obtained with an intensity/photoperiod ratio of 278/16. Both studies obtained the best results with a close DLI (15.6 and 16 mol/(m^2 d)). At these DLIs, plants grown at a lower PPFD, and a more extended photoperiod had more excellent fresh and dry weight than those grown at a higher PPFD and a shorter photoperiod, while at a lower DLI, the plant response was less pronounced.

In the study (Hwang et al., 2020), tomato seedlings were grown at 100, 150, 200, 250, 300, μmol/(m^2 s) and 12, 16, and 20 hours/day photoperiod. When comparing treatments providing the same DLI (12/200 and 16/150 treatment; 12/250 and 20/150 treatment; 16/250 and 20/200 treatment), tomato seedlings had significantly higher dry weight when grown under more extended photoperiod and lower PPFD conditions. The highest light efficiency based on the dry weight of tomato seedlings was obtained at a photoperiod of 20 hours/day and a light intensity of 200 μmol/(m^2 s).

In Yan et al. (2021), cucumber seedlings were grown for 21 days at 457, 320, 246, 200, 168, and 145 μmol/(m^2 s) and photoperiods of 7, 10, 13, 16, 19, and 22 hours/day at a constant DLI of 11.5 mol/(m^2 d). The more extended photoperiod combined with lower PPFD at the same DLI resulted in higher photosynthetic pigment content, thicker leaves, and compact cucumber seedlings with a smaller height and shorter hypocotyl. With the same DLI, a more extended photoperiod improved the quality of cucumber seedlings, compensating for the decrease in PPFD. Increasing the photoperiod to 16–19 hours/day was favorable for sucrose, starch, and cellulose accumulation, which may improve the mechanical strength of cucumber seedlings for transplanting. Light efficiency based on fresh and dry weight of cucumber seedlings increased with increasing photoperiod from 7 to 16 hours/day, and no significant differences were observed when the photoperiod was increased from 16 to 22 hours/day with the same DLI. Increasing the photoperiod while decreasing the light intensity within the same DLI does not decrease the yield. In Lanoue et al. (2021), mini cucumber yields were equal under lighting regimes with intensities/photoperiod ratios of 175/16 and 113/24.

As we can see, plants grown at a lower PPFD and a more extended photoperiod under the same DLI had a higher fresh and dry weight than plants grown at a higher PPFD and a shorter photoperiod. This is because longer photoperiods with lower PPFD increased light interception, chlorophyll content index, and photosystem quantum yield.

Even more important is the energy effect of the ratio of light intensity to photoperiod. First, a reduction in the energy intensity of the obtained products is. This is due to an increase in the volume of production obtained with the same amount of electricity consumed. Electricity consumed by lighting systems under DLI (kWh):

$$W = P \cdot t$$

P – electrical power of lighting systems required to create a given light intensity, W; t – lighting system operating time coinciding with the photoperiod, h.

The expression shows that the electricity consumption during the day is like DLI. Regardless of the power-to-time ratio of the lighting system within the same DLI, the power consumption will be the same. Since a more extended photoperiod with a lower PPFD leads to an increase in yield, the energy intensity of production will decrease.

Second, the initial capital cost of lighting systems is reduced because lower wattages or fewer illuminators are required to provide a lower PPFD. An example of the relationship between the power required and the PPFD can be seen in Figure 8.8.

Third, using a lower PPFD reduces the instantaneous heat generated by light sources, thereby reducing air conditioning requirements. Less intensive operation of air conditioning and ventilation units helps reduce energy consumption, which helps reduce the energy intensity of products.

It should be noted that the marginal DLI value is vital for maximum resource efficiency, as research shows. A further increase or decrease in DLI leads to lower yields and higher costs per unit produced. Therefore, when growing plants, it is helpful to maintain DLI throughout the period from planting to harvesting

MAINTAINING THE DLI

The plant height increases during cultivation and the distance to the lighting systems decreases. This can hurt the energy content because the PPF increases as the distance to the plants decreases.

When the power of the controlled LED light source decreases, along with a decrease in luminous flux, the heating temperature of the light source decreases as well (Kondratieva et al., 2020). For LEDs, the heating temperature affects the lifetime. Reducing the heating temperature of the LED increases the service life. Thus, the forced reduction of luminous flux to maintain DLI helps to reduce power consumption and increase the lifetime of light sources. With a reduction of electric power consumption by 5%, the lifetime is increased by 20%. The lifetime is doubled at a reduction of electric power consumption by 20% (Figure 3.4). Generally, the change in service life from the reduction of electric power consumption has an exponential character.

Changing the PPFD to maintain DLI can be used in two ways: continuous current reduction (CCR) and pulse-width modulation (PWM). Each method has its advantages and disadvantages. When controlling the light output, the energy efficiency of LEDs with CCR regulation is 10–60% higher than that of LEDs with PWM regulation (Gu et al., 2006). When changing the supply voltage, the energy characteristics of LED lamps with CCR regulation are generally less susceptible to adverse changes than LED lamps with PWM regulation (Filatov et al., 2022). Undervoltage is more common in electrical networks. Undervoltage hurts PWM-regulated LED lamps and positively affects CCR-regulated LED lamps (Figures 3.5 and 3.6).

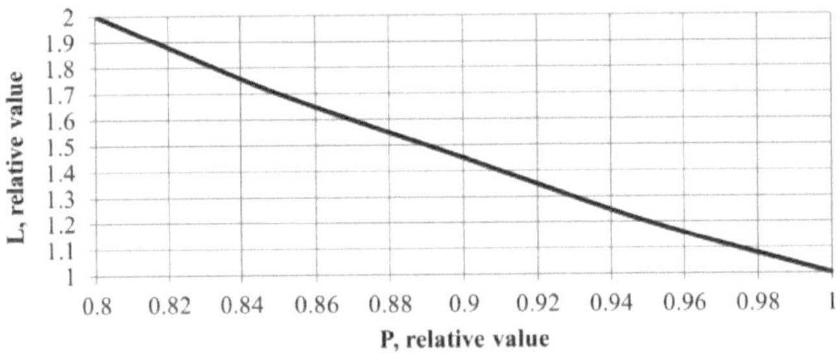

FIGURE 3.4 Dependence of the lifetime (L) of a controlled LED light source on the power consumption (P).

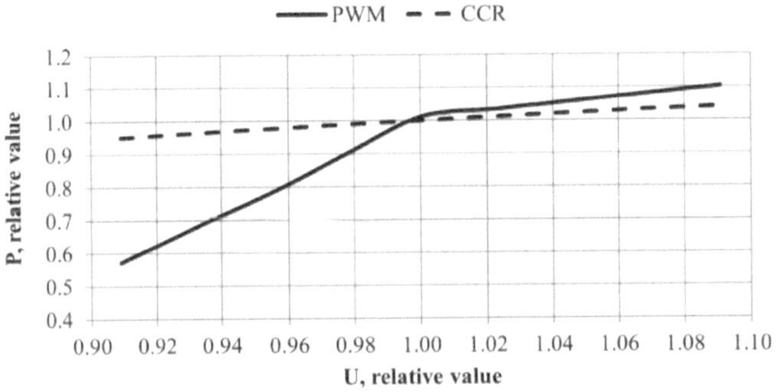

FIGURE 3.5 Variation of power consumption when changing the supply voltage for LEDs with different types of dimming.

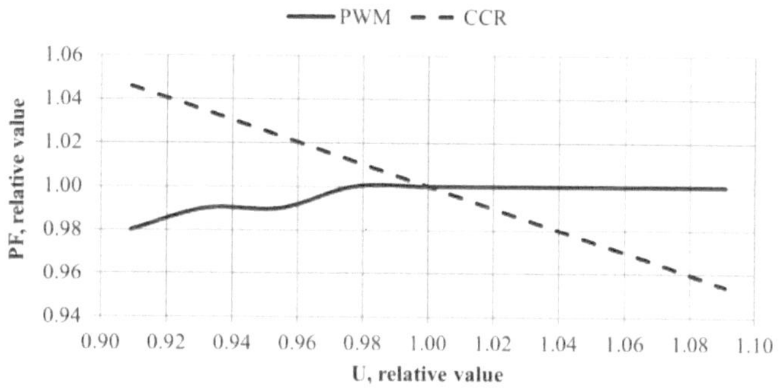

FIGURE 3.6 Change in power factor when changing the supply voltage for LEDs with different types of dimming.

PWM regulation of light intensity (PPF) is energetically less efficient than CCR. Despite these disadvantages, PWM can be used to regulate the light pulses of LEDs, affecting plants' morphological characteristics.

OPTIMIZATION OF LED LIGHT PULSES

Plant photosynthesis consists of alternating light reactions: a first phase, when light energy is collected and converted into chemical energy products, and a second phase of dark reactions, when the products of the first phase are used to assimilate CO_2. Light reactions occur in the nanosecond to millisecond range, while dark reactions occur from seconds to minutes. A possible strategy to increase energy use is to replace continuous lighting with pulsed lighting by modulating the LEDs' frequency and pulse duration ratio (when the light is on).

Due to the advantages provided by LED technology, light pulses can also be extremely short and frequent since LEDs can be turned completely off and entirely on very quickly, even at millisecond intervals. Today, the main challenge for using pulsed light technology with LED lighting systems that can provide swift flashes is determining light frequencies that increase energy efficiency in plant cultivation. By increasing the duration of the dark periods between pulses, energy can be saved, thereby reducing the cost of overall crop production. In Avercheva et al. (2016), the period of light pulses was varied from 30 to 501 μs in the frequency range from 0 to 30 kHz when growing Chinese cabbage (*Brassica chenensis* L.). When PPFD values did not exceed 400 μmol/(m^2 s), pulsed light suppressed plant growth compared to continuous light. At cycle durations longer than 350 μs, the inhibition was most pronounced. However, the dry mass was higher when PPFD was increased to 500 μmol/(m^2 s) under pulsed light conditions.

In Carotti et al. (2021), a response to two different blue diode switching frequencies in an LED lighting system was determined when growing lettuce (PPFD 215 μmol/(m^2 s)). Lettuce was irradiated with light at a high switching frequency (850 kHz) and a low switching frequency (293 kHz) for the blue diode. In all treatments, the switching frequency for the red diode was close (437 and 443 kHz, respectively). As a result of the experiment, lettuce plants grown at a low switching frequency had 42% higher biomass per unit of electricity consumed.

In Kanechi et al. (2016), pulse-cycling irradiances of 0.5, 1, 1.3, 2.5, 5, 50, 500 Hz, and 1, 1.3, 2, 4, 10, 20 kHz at 50% light duration/cycle at 80 and 200 μmol/(m^2 s) were studied. Fresh shoot weight and leaf area of lettuce increased to 20% at high frequencies compared with low frequencies and continuous illumination at a PPFD of 200 μmol/(m^2 s). The rate of photosynthesis was kept relatively constant throughout the entire range of measurements under pulsed light at a PPFD of 80 μmol/(m^2 s).

Thus, pulsed light can have both stimulating and inhibiting effects on the accumulation of plant dry mass depending on PPFD. Determination of the optimal light irradiation frequency and PPFD for different plants remains to be specified in further studies.

VARIABLE (INTERMITTENT) LIGHTING

In the natural environment, light has a changeable nature. During the day, day changes to night, and sunny weather can be interrupted by clouds and rain. The spectrum of natural light also changes during the day as the sun's position changes. Therefore, scientists have speculated about the possible stimulation of plant growth during irradiation by variable light, both in time and spectrum.

A study (Shimokawa et al., 2014) examined the effect of light conditions created by alternating red and blue light (R/B) on the growth of leaf lettuce of three different varieties (Summer Surge, Black Rose, Green Span). Two experiments were conducted. In the first experiment, lettuce plants were grown with simultaneous RB (12 hours of light and 12 hours of darkness) and alternating R/B (12 hours of red and 12 hours of blue) irradiation. The light intensity was 160 μmol/(m^2 s) for control, 100 μmol/(m^2 s) for red, and 60 μmol/(m^2 s) for blue. As a result, the crude weight under R/B irradiation was higher than under RB conditions by 66% and 40% for Summer Surge and Black Rose, respectively. Dry weight under R/B irradiation was higher than under RB conditions by 90% and 120% for the variety Summer Surge and Black Rose, respectively. For the Green Span variety, raw and dry weights were comparable.

The second experiment investigated the interval between each light for the Summer Surge variety. The alternating red and blue light period was 1/1 (1 hour of red and 1 hour of blue light throughout the growth period), 3/3, 6/6, 12/12, 24/24, and 48/48. The maximum raw and dry weight was recorded at the irradiation mode R/B = 12/12. However, this experiment determined the best spectrum alternation interval when the plants were illuminated 24 hours a day, while the optimal photoperiod is considered 16 hours a day.

In Chen et al. (2017), four alternating light (R/B) variants were studied for lettuce of the Green Oak leaf variety, which were 8, 4, 2, and 1 hours during the 16-hour photoperiod (200 μmol/(m^2 s) for red, and 100 μmol/(m^2 s) for blue), respectively. Overall, with the same energy inputs, R/B (8h) and R/B (1h) resulted in higher yields, while R/B (4 hours) and R/B (2 hours) provided higher nutritional value compared to simultaneous illumination.

In Chen and Yang (2018), the effect of continuous and intermittent light on lettuce was investigated. Combined RB light illuminated 16 hours daily in all variants. A continuous RB with a single light/dark cycle (L/D) of 16/8 hours in the L/D(1) variant was considered the control. The five intermittent RB treatments were, respectively, L/D(2) with two L/D cycles of 8/4 hours over a 24-hour period, L/D(3) with three L/D cycles of 6/3 hours or 4/2 hours, L/D(4) with four L/D cycles of 4/2 hours, L/D(6) with six L/D cycles of 3/1.5 hours or 1/0.5 hours, and L/D(8) with eight L/D cycles with eight 2/1 hours. The results showed that lettuce biomass increased with intermittent RB treatments with even cycles but decreased significantly with L/D(3) compared to controls. Higher soluble sugars and lower fiber and starch content were observed in plants treated with L/D(2).

Intermittent irradiation mode 8/4 can be effective in growing microgreens (Filatov et al., 2022). The photoperiod study was performed in two variants. The control was a light system operating on a light/dark schedule of 16/8 hours per day. The experiment was the lighting system operating on a schedule of light/darkness 8/4 hours a day for two periods. The 8/4(2) irradiation increased the crude weight of radish microgreens by 6% compared to the control, while the crude weight of cabbage microgreens was comparable in both experiments (Figure 3.7).

In Avgoustaki et al. (2020), the effects of two lighting systems were tested on the basil cultivar Genovese: under 16 hours of continuous light and a 14 hours of photoperiod with intermittent lighting. Intermittent lighting was conducted using three cycles of 14 hours per day. One cycle included the first 4 hours of continuous illumination, followed by four light/dark repetitions of 10/50 min. This resulted in a volume of dry basil biomass with no statistical difference. At the same time, the electrical energy requirement was 20% lower for intermittent lighting than for continuous lighting.

FIGURE 3.7 Appearance microgreens obtained as a result of the experiment under different irradiation modes: (a) radish and (b) cabbage.

Variable spectrum and time lighting systems effectively increase plant biomass and micronutrients, lowering energy costs per unit. At the same time, intermittent lighting systems allow for the lowest cost hours during the day. Potentially, it allows for a reduction in the financial costs for equal power consumption.

USE OF DIFFERENTIATED PRICES FOR ELECTRICITY

One of the challenges for PFs is the price of electricity due to the high proportion of artificial light in the total balance of electricity consumption. In some countries, electricity prices vary over a day or even months. This can be a valuable factor for PFs. The lighting cycle for plants can be planned out of periods when electricity costs are high to avoid the enormous operating costs associated with the financial cost of electricity. The premise of optimizing plant electricity costs while maximizing biomass depends on three factors: electricity prices, knowledge of how plants respond to lighting, and knowledge of the energy consumption of the lighting system.

In Filatov et al. (2022), a study was conducted on two lighting options for growing microgreens. The control was a lighting system operating on a light/dark schedule of 16/8 hours per day. The operating time of the lighting system was 21.00–13.00. The control was a lighting system running on a light/darkness schedule of 8/4 hours a day for 2 periods. The operating time of the lighting system was 23.00–7.00 (first period) and 11.00–19.00 (second period). Changes in the price of electricity during the day are shown in Figure 3.8 (white bars). Together with the yield results obtained, the cost of electricity for growing 1 kg of microgreens decreased by 10%.

In Lork et al. (2020) the possibility of using LEDs of different spectra at different times of the day was investigated, as red LEDs are more energy efficient than blue. Optimization of energy consumption in growing lettuce for 15 days was carried out. PPF from red LEDs was two times higher than from blue LEDs. The blue LEDs worked intermittently for a full day. Red light-emitting diodes

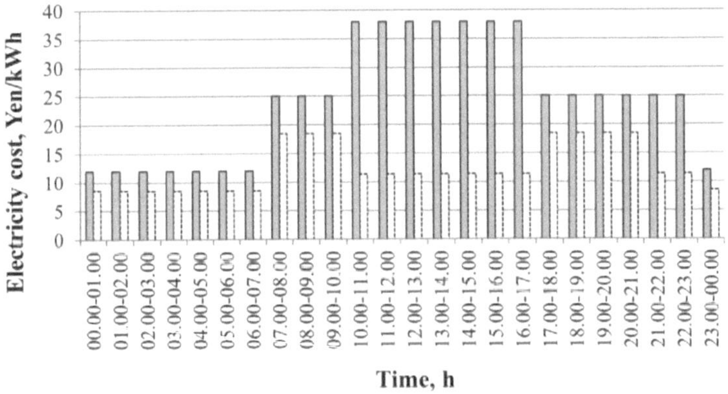

FIGURE 3.8 Electricity price during the day: gray – Tokyo Electric Power Company (Japan) and white – TNS Power NN (Russia). Currencies of different countries are reduced to one: 1 yen = 0.51 rub.

worked intensively during the period of minimum electricity price. The price of electricity during the day is in Figure 3.8 (gray bars). Optimized lighting system achieves 40–52% savings in energy costs while increasing yields by 6%.

A differentiated electricity tariff is the most significant way to reduce energy costs (10–50%) when growing plants. This way of growing plants can perform different tasks: producing cheap products with reduced vitamin content or inexpensive quality products with certain varieties. This method is ideal for vertical farms and is not feasible for glass greenhouses because the minimum electricity prices coincide with the absence of natural light.

CONCLUSIONS

CVFs are one solution to food security in an increasingly populated world and the continuing urbanization of cities. Already today, they can produce more products per square meter than greenhouses. This is due to the fully controlled environment of the growing process. However, more electricity must be spent on lighting systems than in greenhouses, which is the main disadvantage of vertical farms. Therefore, it is vital to use and develop means and methods to improve the efficiency of lighting systems of CVFs to reduce production costs.

An essential way of increasing the efficiency of lighting systems is using LEDs, which allow the formation of a spectrum of radiation for a particular plant. RB spectrum is more efficient in using energy resources to produce quantitative and qualitative yields. The proportion of red light can range from 70% to 90% of the total PPF. The value of red light is refined depending on the plant variety and the desired ratio of biomass and minerals. Depending on the manufacturers, the energy efficiency of red LEDs can be 30–40% higher than that of other colored LEDs. The energy efficiency of blue and white LEDs is comparable, and green LEDs have the lowest energy efficiency. At the same time, red LEDs have a low degradation rate of luminous flux compared to other colored LEDs. Partial substitution of blue for green or red for FR light can increase production efficiency. The spectrum is a qualitative assessment of light.

The amount of light during the day – the DLI – also affects the energy intensity of production. The same DLI can be obtained by either high light intensity with a short photoperiod or low light intensity (PPFD) with a long photoperiod. The second option has more advantages. First, a more extended photoperiod with a lower PPFD results in higher yields for the same energy input. Second, the initial capital cost of lighting systems is reduced because lower power or fewer lights are required to provide a lower PPFD. Third, using a lower PPFD reduces light sources' instantaneous heat output, reducing air conditioning requirements. Less air conditioning and ventilation units help reduce energy consumption, contributing to lower product energy intensity.

Non-standard modes of operation of lighting systems can increase yield or reduce energy consumption. For example, pulsed light can have a stimulating effect on plants. Plant photosynthesis consists of alternating light reactions. Due to the advantages provided by LED technology, light pulses can also be extremely short and frequent since LEDs can be turned completely off and entirely on very quickly, even at millisecond intervals. In the natural environment, light has a changeable nature. During the day, day alternates with night, and sunny weather can be interrupted by clouds and rain. The spectrum of natural light also changes during the day as the sun's position changes. Therefore, it is possible to stimulate plant productivity by irradiating with variable light, both in time and spectrum.

One of the challenges for PFs is the price of electricity due to the high proportion of artificial light in the total balance of electricity consumption. In some countries, electricity prices vary over a day or even months. This can be a valuable factor for PFs. The lighting cycle for plants can be planned out of periods when electricity costs are high to avoid the enormous operating costs associated with the financial cost of electricity.

Currently, CVFs use white light 16 hours a day with varying intensities for a particular plant as the most common technology. The means and methods described in this chapter to improve the efficiency of lighting systems can increase yields by up to 5–120% or reduce energy costs by up to 10–50% compared with the most common lighting system.

Today, lighting systems for CVFs are based on algorithms that work the same way throughout the growing period. The result is evaluated only by the final yield. A promising area of research is dynamic lighting systems in which changes in the quantity and quality of light depend on the condition of the plants at a particular point in time. This requires more in-depth plant and lighting system data analysis throughout the cultivation period. This problem can be solved by applying machine learning to complex robotic systems based on artificial intelligence that controls production. The need for large capital expenditures currently constrains the speed of diffusion of such research.

REFERENCES

Ahmad, M., & Cashmore, A.R., (1993), HY4 gene of A. thalianaencodes a protein with characteristics of a blue-light photoreceptor. *Nature* 366, 162–166.

An, S., Hwang, H., Chun, C., Jang, Y., Lee, H.J., Wi, S.H., Yeo, K.-H., Yu, I.-h., & Kwack, Y., (2021), Evaluation of air temperature, photoperiod and light intensity conditions to produce cucumber scions and rootstocks in a plant factory with artificial lighting. *Horticulturae*, 7, 102. https://doi.org/10.3390/horticulturae7050102.

Avercheva, O. V., Berkovich, Y. A., Konovalova, I. O., Ralhenko, S. G., Lapach, S. N., Bassarskaya, E. M., & Tarakanov, I. G., (2016), Optimizing LED lighting for space plant growth unit: Joint effects of photon flux density, red to white ratios and intermittent light pulses. *Life sciences in Space Research*, 11, 29–42.

Avgoustaki, D. D., & Xydis, G., (2020), Indoor vertical farming in the urban nexus context: business growth and resource savings. *Sustainability*, 12(5), 1965. https://doi.org/10.3390/su12051965.

Avgoustaki, D. D., Li, J., & Xydis, G., (2020), Basil plants grown under intermittent light stress in a small-scale indoor environment: Introducing energy demand reduction intelligent technologies. *Food Control*, 118, 107389.

Bantis, F., (2021), Light spectrum differentially affects the yield and phytochemical content of microgreen vegetables in a plant factory. *Plants*, 10, 2182. https://doi.org/10.3390/plants10102182.

Brazaityte, A., Miliauskiene, J., Vaštakaite-Kairiene, V., Sutuliene, R., Laužike, K., Duchovskis, P., & Małek, S., (2021), Effect of different ratios of blue and red LED light on brassicaceae microgreens under a controlled environment. *Plants*, 10, 801. https://doi.org/10.3390/plants10040801.

Brazaitytė, A., Sakalauskienė, S., Samuolienė, G., Jankauskienė, J., Viršilė, A., Novičkovas, A., & Duchovskis, P., (2015), The effects of LED illumination spectra and intensity on carotenoid content in Brassicaceae microgreens. *Food Chemistry*, 173, 600–606.

Carotti, L., Potente, G., Pennisi, G., Ruiz, K.B., Biondi, S., Crepaldi, A., Orsini, F., Gianquinto, G., & Antognoni, F., (2021) Pulsed LED light: Exploring the balance between energy use and nutraceutical properties in indoor-grown lettuce. *Agronomy*, 11, 1106. https://doi.org/10.3390/agronomy11061106.

Chen, M., & Chory, J., (2011), Phytochrome signaling mechanisms and the control of plant development. *Trends in Cell Biology*, 21, 664–671. https://doi.org/10.1016/j.tcb.2011.07.002.

Chen, X. L., & Yang, Q. C., (2018), Effects of intermittent light exposure with red and blue light emitting diodes on growth and carbohydrate accumulation of lettuce. *Scientia Horticulturae*, 234, 220–226.

Chen, X. L., Yang, Q. C., Song, W. P., Wang, L. C., Guo, W. Z., & Xue, X. Z., (2017), Growth and nutritional properties of lettuce affected by different alternating intervals of red and blue LED irradiation. *Scientia Horticulturae*, 223, 44–52.

Chen, Z., Shah Jahan, M., Mao, P., Wang, M., Liu, X., & Guo, S., (2020), Functional growth, photosynthesis and nutritional property analyses of lettuce grown under different temperature and light intensity. *The Journal of Horticultural Science and Biotechnology*, 1–9. https://doi.org/10.1080/14620316.2020.1807416.

Chory, J., (2010), Light signal transduction: An infinite spectrum of possibilities. *The Plant Journal*, 61, 982–991. https://doi.org/10.1111/j.1365-313X.2009.04105.x.

Christie, J.M., (2007), Phototropin blue-light receptors. *Annual Review of Plant Biology*, 58, 21–45. https://doi.org/10.1146/annurev.arplant.58.032806.103951.

Claypool, N.B., & Lieth, J.H., (2021), Green light improves photosystem stoichiometry in cucumber seedlings (cucumis sativus) compared to monochromatic red light. *Plants*, 10, 824. https://doi.org/10.3390/plants10050824.

Demotes-Mainard, S., Péron, T., Corot, A., Bertheloot, J., Le Gourrierec, J., PelleschiTravier, S., Crespel, L., Morel, P., Huché-Thélier, L., Boumaza, R., Vian, A., Guérin, V., Leduc, N., Sakr, S., (2016), Plant responses to red and far-red lights, applications in horticulture. *Environmental and Experimental Botany*, 121, 4–21. https://doi.org/10.1016/j.envexpbot.2015.05.010.

Dou, H., Niu, G., Gu, M., & Masabni, J. G., (2018), Responses of sweet basil to different daily light integrals in photosynthesis, morphology, yield, and nutritional quality. *HortScience*, 53(4), 496–503. https://doi.org/10.21273/hortsci12785-17.

Fan, Xiao-Xue, Xu, Zhi-Gang, Liu, Xiao-Ying, Tang, Can-Ming, Wang, Li-Wen, & Hanc, Xue-lin, (2013), Effects of light intensity on the growth and leaf development of young tomato plants grown under a combination of red and blue light. *Scientia Horticulturae*, 153, 50–55.

Filatov, D. A., Terentyev, P. V., Sokolova, A. A., & Devyatkin, I. S., (2022a), Comparative analysis of the energy efficiency of dimmable LED lamps with different types of luminous flux control. *IOP Conference Series: Earth and Environmental Science*, 1045(1), 012146.

Filatov, D. A., Vetchinnikov, A. A., Olonina, S. I., & Olonin, I. Y., (2022b), Intermittent LED lighting helps reduce energy costs when growing microgreens on vertical controlled environment farms. *IOP Conference Series: Earth and Environmental Science*, 979(1), 012096.

Fu, Y., Li, H., Yu, J., Liu, H., Cao, Z., Manukovsky, N.S., & Liu, H., (2017), Interaction effects of light intensity and nitrogen concentration on growth, photosynthetic characteristics and quality of lettuce (Lactuca sativa L. Var. youmaicai). *Scientia Horticulturae*, 214, 51–57. https://doi.org/10.1016/j.scienta.2016.11.020

Gao, M., He, R., Shi, R., Zhang, Y., Song, S., Su, W., & Liu, H., (2021), Differential effects of low light intensity on broccoli microgreens growth and phytochemicals. *Agronomy*, 11, 537. https://doi.org/10.3390/agronomy11030537.

Gerovac Joshua, R., Craver Joshua, K., Boldt Jennifer, K., & Lopez Roberto, G., (2016), Light intensity and quality from sole-source light-emitting diodes impact growth, morphology, and nutrient content of brassica microgreens. *Hortscience*, 51(5), 497–503.

Giménez, A., Martínez-Ballesta, M.d.C., Egea-Gilabert, C., Gómez, P.A., Artés-Hernández, F., Pennisi, G., Orsini, F., Crepaldi, A., & Fernández, J.A., (2021), Combined effect of salinity and LED lights on the yield and quality of purslane (Portulaca oleracea L.) microgreens. *Horticulturae*, 7, 180. https://doi.org/10.3390/horticulturae7070180.

Graamans, L., Baeza, E., van den Dobbelsteen, A., Tsafaras, I., & Stanghellini, C., (2018), Plant factories versus greenhouses: Comparison of resource use efficiency. *Agricultural Systems*, 160, 31–43. https://doi.org/10.1016/j.agsy.2017.11.003.

Gu, Y., Narendran, N., Dong, T., & Wu, H., (2006), Spectral and luminous efficacy change of high-power LEDs under different dimming methods. *Sixth International Conference on Solid State Lighting*, 6337, 78–84, SPIE.

Hamedalla, A.M., Ali, M.M., Ali, W.M. et al., (2022), Increasing the performance of cucumber (Cucumis sativus L.) seedlings by LED illumination. *Scientific Reports*, 12, 852. https://doi.org/10.1038/s41598-022-04859-y.

He, W., Huang, Z.-W., Li, J.-P., Su, W.-X., Gan, L., & Xu, Z.-G., (2019), Effect of different light intensities on the photosynthate distribution in cherry tomato seedlings. *The Journal of Horticultural Science and Biotechnology*, 1–9. https://doi.org/10.1080/14620316.2019.1575775.

Hernández, R., Eguchi, T., Deveci, M., & Kubota, C., (2016), Tomato seedling physiological responses under different percentages of blue and red photon flux ratios using LEDs and cool white fluorescent lamps. *Scientia Horticulturae*, 213, 270–280. https://doi.org/10.1016/j.scienta.2016.11.005.

Huber, B. M., Louws, F. J., & Hernández, R., (2021), Impact of different daily light integrals and carbon dioxide concentrations on the growth, morphology, and production efficiency of tomato seedlings. *Frontiers in Plant Science*, 12, 615853. https://doi.org/10.3389/fpls.2021.615853.

Hwang, H., An, S., Pham, M. D., Cui, M., & Chun, C., (2020), The combined conditions of photoperiod, light intensity, and air temperature control the growth and development of tomato and red pepper seedlings in a closed transplant production system. *Sustainability*, 12(23), 9939. https://doi.org/10.3390/su12239939.

Jie, Z., Cheng-bo, Z., Hong, X., Rui-feng, C., Qi-chang, Y., & Tao, L., (2020), The effect of artificial solar spectrum on growth of cucumber and lettuce under controlled environment. *Journal of Integrative Agriculture*, 19(8), 2027–2034.

Johnson, R. E., Kong, Y., & Zheng, Y., (2019), Elongation growth mediated by blue light varies with light intensities and plant species: A comparison with red light in arugula and mustard seedlings. *Environmental and Experimental Botany*, 103898.

Jones-Baumgardt, C., Llewellyn, D., Ying, Q., &Zheng, Y., (2019), Intensity of sole-source light-emitting diodes affects growth, yield, and quality of brassicaceae microgreens. *Hortscience*, 54(7), 1168–1174. https://doi.org/10.21273/HORTSCI13788-18.

Kaiser, E., Weerheim, K., Schipper, R., & Dieleman, J. A., (2019), Partial replacement of red and blue by green light increases biomass and yield in tomato. *Scientia Horticulturae*, 249, 271–279.

Kalaitzoglou, P., van Ieperen, W., Harbinson, J., van der Meer, M., Martinakos, S., Weerheim, K., Marcelis, L. F. M., (2019), Effects of continuous or end-of-day far-red light on tomato plant growth, morphology, light absorption, and fruit production. *Frontiers in Plant Science*, 10. https://doi.org/10.3389/fpls.2019.00322.

Kamal, K. Y., Khodaeiaminjan, M., El-Tantawy, A. A., Abdel Moneim, D., Abdel Salam, A., Ash-shormillesy, S. M. A. I., Fawzy Ramadan, M., (2020), Evaluation of growth and nutritional value of Brassica microgreens grown under red, blue and green LEDs combinations. *Physiologia Plantarum* https://doi.org/10.1111/ppl.13083.

Kanechi, M., Maekawa, A., Nishida, Y., & Miyashita, E., (2016), Effects of pulsed lighting based light-emitting diodes on the growth and photosynthesis of lettuce leaves. *Acta Horticulturae* 1134, 207–214.

Kelly, N., Choe, D., Meng, Q., & Runkle, E. S., (2020), Promotion of lettuce growth under an increasing daily light integral depends on the combination of the photosynthetic photon flux density and photoperiod. *Scientia Horticulturae*, 272, 109565. https://doi.org/10.1016/j.scienta.2020.109565.

Kim, H.-J., Lin, M.-Y., & Mitchell, C. A., (2018), Light spectral and thermal properties govern biomass allocation in tomato through morphological and physiological changes. *Environmental and Experimental Botany*. https://doi.org/10.1016/j.envexpbot.2018.10.019.

Kobayashi, K., Amore, T., & Lazaro, M., (2013), Light-emitting diodes (LEDs) for miniature hydroponic lettuce. *Optics and Photonics Journal*, 1(3), 74–77. https://doi.org/10.4236/opj.2013.31012.

Kondratieva, N., Filatov, D., & Terentiev, P., (2020), Study of operating modes of a controllable lighting system consisting of a triak dimmer and a LED light source with a controllable driver, *Light & Engineering*, 4(28), 84–90.

Kozai, T., (2013), Sustainable plant factory: Closed plant production systems with artificial light for high resource use efficiencies and quality produce. *Acta Horticulturae*, 1004, 27–40. https://doi.org/10.17660/actahortic.2013.1004.2.

Kwack, Y., Park, S. W., & Chun, C., (2014), Growth and development of grafted cucumber transplants as affected by seedling ages of scions and rootstocks and light intensity during their cultivation in a closed production system. *Horticultural Science & Technology*, 32(5), 600–606.

Lanoue, J., Zheng, J., Little, C., Grodzinski, B., & Hao, X., (2021), Continuous Light Does Not Compromise Growth and Yield in Mini-Cucumber Greenhouse Production with Supplemental LED Light. *Plants*, 10, 378. https://doi.org/10.3390/plants10020378.

Lee, M. J., Son, K. H. & Oh, M. M., (2016), Increase in biomass and bioactive compounds in lettuce under various ratios of red to far-red LED light supplemented with blue LED light. *Horticulture, Environment, and Biotechnology*, 57, 139–147. https://doi.org/10.1007/s13580-016-0133-6.

Liang, W., Xue, A., Hao, Y., & Luo, L., (2022), Effects of led light quality on the growth and phenolic compounds of broccoli microgreens. Available at SSRN: https://ssrn.com/abstract=4057879 or https://doi.org/10.2139/ssrn.4057879.

Liu, K., Gao, M., Jiang, H., Ou, S., Li, X., He, R., Li, Y., & Liu, H., (2022), Light intensity and photoperiod affect growth and nutritional quality of brassica microgreens. *Molecules*, 27, 883. https://doi.org/10.3390/molecules27030883.

Lork, C., Cubillas, M., Kiat Ng, B. K., Yuen, C., & Tan, M., (2020), Minimizing electricity cost through smart lighting control for indoor plant factories. IECON 2020 The 46th Annual Conference of the IEEE Industrial Electronics Society.

Lu, N., Maruo, T., Johkan M., Hohjo, M., Tsukagohi, S., Ito, Y., & Shinohara, Y., (2012), Effects of supplemental lighting with light-emitting diodes (LEDs) on tomato yield and quality of single-truss tomato plants grown at high planting density. *Environment Control in Biology*, 50(1), 63–74.

Lu, N., Saengtharatip, S., Takagaki, M., Maruyama, A., & Kikuchi, M., (2019), How do white LEDs' spectra affect the fresh weight of lettuce grown under artificial lighting in a plant factory? – A statistical approach. *Agricultural Sciences*, 10, 957–974. https://doi.org/10.4236/as.2019.107073.

Meas, S., Luengwilai, K., & Thongket, T., (2020), Enhancing growth and phytochemicals of two amaranth microgreens by LEDs light irradiation. *Scientia Horticulturae*, 265, 109204. https://doi.org/10.1016/j.scienta.2020.109204.

Meng, Q., & Runkle, E. S., (2017), Far red is the new red. *Inside Grower*, 26–30. https://www.canr.msu.edu/floriculture/uploads/files/far-red-on-lettuce.pdf

Narendran, N., Maliyagoda, N., Deng, L., & Pysar, R. M., (2001, December), Characterizing LEDs for general illumination applications: mixed-color and phosphor-based white sources. *In Solid State Lighting and Displays*, 4445,137–147. SPIE.

Naznin, M. T., Lefsrud, M., Gravel, V., & Azad, M. O. K., (2019), Blue light added with red LEDs enhance growth characteristics, pigments content, and antioxidant capacity in lettuce, spinach, kale, basil, and sweet pepper in a controlled environment. *Plants*, 8, 93. https://doi.org/10.3390/plants8040093.

Palmer, S., & van Iersel, M. W., (2020), Increasing growth of lettuce and mizuna under sole-source LED lighting using longer photoperiods with the same daily light integral. *Agronomy*, 10(11), 1659. https://doi.org/10.3390/agronomy10111659.

Pennisi, G., Blasioli, S., Cellini, A., Maia, L., Crepaldi, A., Braschi, I.,Gianquinto, G., (2019a), Unraveling the role of red: Blue LED lights on resource use efficiency and nutritional properties of indoor grown sweet basil. *Frontiers in Plant Science*, 10. https://doi.org/10.3389/fpls.2019.00305.

Pennisi, G., Orsini, F., & Blasioli, S. et al., (2019b), Resource use efficiency of indoor lettuce (Lactuca sativa L.) cultivation as affected by red: Blue ratio provided by LED lighting. *Scientific Reports*, 9, 14127. https://doi.org/10.1038/s41598-019-50783-z.

Pennisi, G., Pistillo, A., Orsini, F., Cellini, A., Spinelli, F., Nicola, S., & Marcelis, L. F. M, (2020), Optimal light intensity for sustainable water and energy use in indoor cultivation of lettuce and basil under red and blue LEDs. *Scientia Horticulturae*, 272, 109508. https://doi.org/10.1016/j.scienta.2020.109508.

Prikupets, L. B., (2019), On international scientific and technical conference on the use of LED phyto irradiators. *Light & Engineering*, 27(S1), 9–14.

Rahman, M. M., Vasiliev, M., & Alameh, K., (2021), LED illumination spectrum manipulation for increasing the yield of sweet basil (Ocimum basilicum L.). *Plants*, 10, 344. https://doi.org/10.3390/plants10020344.

Samuolienė, G., Brazaitytė, A., Jankauskienė, J., Viršilė, A., Sirtautas, R., Novičkovas, A., Duchovskis, P., (2013), LED irradiance level affects growth and nutritional quality of Brassica microgreens. *Open Life Sciences*, 8(12). https://doi.org/10.2478/s11535-013-0246-1.

Samuolienė, G., Brazaitytė, A., Viršilė, A., Miliauskienė, J., Vaštakaitė-Kairienė, V., & Duchovskis, P., (2019), Nutrient levels in brassicaceae microgreens increase under tailored light-emitting diode spectra. Frontiers in Plant Science, 10. https://doi.org/10.3389/fpls.2019.01475.

Shen, Y. Z., Guo, S. S., Ai, W. D., & Tang, Y. K., (2014), Effects of illuminants and illumination time on lettuce growth, yield and nutritional quality in a controlled environment. *Life Sciences in Space Research*, 2, 38–42. https://doi.org/10.1016/j.lssr.2014.06.001.

Shimokawa, A., Tonooka, Y., Matsumoto, M., Ara, H., Suzuki, H., Yamauchi, N., & Shigyo, M., (2014), Effect of alternating red and blue light irradiation generated by light emitting diodes on the growth of leaf lettuce, bioRxiv, 003103. https://www.biorxiv.org/content/10.1101/003103v1

Smith, H., (2000), Phytochromes and light signal perception by plants – An emerging synthesis. *Nature* 407, 585-591. https://doi.org/10.1038/35036500.

Son, K. H., & Oh, M. M., (2015), Growth, photosynthetic and antioxidant parameters of two lettuce cultivars as affected by red, green, and blue light-emitting diodes. *Horticulture, Environment, and Biotechnology*, 56, 639–653. https://doi.org/10.1007/s13580-015-1064-3.

Suetsugu, N., & Wada, M., (2013), Evolution of three LOV blue light receptor families in green plants and photosynthetic stramenopiles: Phototropin, ZTL/FKF1/LKP2 and aureochrome. *Plant Cell Physiology*, 54, 8–23. https://doi.org/10.1093/pcp/pcs165.

Vaštakaitė-Kairienė, V., Brazaitytė, A., Samuolienė, G., Viršilė, A., Miliauskienė, J., Jankauskienė, J., Novičkovas, A., & Duchovskis, P., (2022), The influence of LED light photoperiod on growth and mineral composition of Brassica microgreens indoors. *Acta Horticulturae*, 1337, 143–150. https://doi.org/10.17660/ActaHortic.2022.1337.19.

Vetchinnikov, A. A., Filatov, D. A., Olonina, S. I., Kazakov, A. V., & Olonin, I. Y., (2021), Influence of the radiation intensity of LED light sources of the red-blue spectrum on the yield and energy consumption of microgreens. *IOP Conference Series: Earth and Environmental Science*, 723(3), 032046. https://doi.org/10.1088/1755-1315/723/3/032046.

Weidner, T., Yang, A., & Hamm, M. W., (2021), Energy optimisation of plant factories and greenhouses for different climatic conditions. *Energy Conversion and Management*, 243, 114336. https://doi.org/10.1016/j.enconman.2021.114336.

Xiaoying, L., (2012). Regulation of the growth and photosynthesis of cherry tomato seedlings by different light irradiations of light emitting diodes (LED). *African Journal of Biotechnology*, 11(22). https://doi.org/10.5897/ajb11.1191.

Yan, Z., He, D., Niu, G., & Zhai, H., (2019). Evaluation of growth and quality of hydroponic lettuce at harvest as affected by the light intensity, photoperiod and light quality at seedling stage. *Science Horticulturae*, 248, 138–144. https://doi.org/10.1016/j.scienta.2019.01.002.

Yan, Zhengnan, Wang, Long, Dai, Jiaxii, Liu, Yufeng, Lin, Duo, & Yang, Yanjie, (2021), Morphological and physiological responses of cucumber seedlings to different combinations of light intensity and photoperiod with the same daily light integral. *Hortscience*, 56(11), 1430–1438. https://doi.org/10.21273/HORTSCI16153-21.

Yi, C.J., Fu, Y., Yu, J., & Liu, H., (2020), Optimization of light intensity and nitrogen concentration in solutions regulating yield, vitamin C, and nitrate content of lettuce. *The Journal of Horticultural Science and Biotechnology*, 96(1), 62–72.

Zhou, C., Wang, Q., Liu, W., Li, B., Shao, M., & Zhang, Y., (2022a), Efects of red/blue versus white LED light of diferent intensities on the growth and organic carbon and autotoxin secretion of hydroponic lettuce. *Horticulture, Environment, and Biotechnology*, 63, 195–205 https://doi.org/10.1007/s13580-021-00394-3.

Zhou, J., Li, P., & Wang, J., (2022b), Effects of light intensity and temperature on the photosynthesis characteristics and yield of lettuce. *Horticulturae*, 8, 178. https://doi.org/10.3390/ horticulturae8020178.

4 Vertical Farming Systems on Building Envelopes

Abel Tablada and Vesna Kosorić

INTRODUCTION

The urban metabolism of contemporary cities reflects the dependence of urban conglomerations on energy, water and food imports. This dependence on basic resources explains most of the greenhouse gas (GHG) emissions from cities (Chen et al., 2020). Despite the advances in reducing urban carbon footprint, the mitigation measures are still insufficient (Samaniego et al., 2023), poorly distributed (Gazzotti et al., 2021) and failing to respond to diverse global conditions (van den Bergh, 2023). Therefore, the current paradigm in which large urban areas are mostly consumers of resources and goods fabricated far away should change rapidly by accounting for the huge city potential for producing food, conserving water and generating electricity.

Vertical farming systems (VFSs), building integrated agriculture (BIA), and urban agriculture (UA) offer solutions for reducing urban dependence on food imports and maximizing production inside cities. The use of advanced cultivation systems in symbiosis with energy and water systems in buildings contributes to a higher surface efficiency in urban areas compared to more conventional farming systems, which require more surface area to produce the same amount of food (Thomaier et al., 2014). However, the high prices per square meter in central urban areas and the additional energy required to operate VFSs inside buildings may reduce the advantages and incentives for implementing these technologies, except in decommissioned industrial and warehouse buildings.

A compromise solution for urban central areas and most of the building typologies could be the implementation of vertical farming (VF) technologies on building envelopes or in the interior area closest to façades. In this way, higher-profit indoor activities are unaffected and could benefit from the VFS. This approach takes advantage of one of the most underutilized resources cities have, the free solar energy incident on building envelopes – façades and rooftops. According to the latitude and typical climatic conditions of each city and depending on the building's level of exposure, orientation and context, the envelope surfaces receive a quantity of rainwater and sunlight, which can contribute to buildings' energy and food self-sufficiency. In low-density or industrial urban areas, rooftops receive the highest amount of solar radiation. However, in some high-density urban areas, the ratio of façade versus roof surfaces is much higher. This makes building façades a key investigation topic in terms of thermal and visual comfort as well as the possible application of VFSs.

VFSs on building envelopes could maximize the symbiosis not only with building technological systems but also with building envelope's roles as thermal, visual, sound and air pollution regulators. The design of the VFS on building envelopes should fulfill and contribute to both the conventional envelope functions as a climatic buffer from exterior conditions ensuring security, safety, identity and privacy and the new productive functions, which may also include rainwater collection and electricity generation apart from food production (Tablada and Kosorić, 2022). In addition, at the urban level, some variants of VFSs on the exterior skin of buildings could contribute to biodiversity and other ecosystem services.

This chapter explores the potential of the building envelope for integrating VFS as an alternative to the use of indoor VF. Various design and implementation options are classified considering their performance and relationship as an interface with the interior functions. The advantages and

 DOI: 10.1201/b23309-4

potential threats of VFSs on building envelopes are discussed by critically presenting several case studies and investigations about rooftops and façades. Some of the challenges to their widespread implementation are also identified.

CLASSIFICATION OF VFSs ON BUILDING ENVELOPES

VFSs on building envelopes can be defined as any VFS located on the façade and rooftop of buildings as well as in interior areas close to the façades and rooftops where plants receive most of the sunlight needed to grow.

VFSs on building envelopes can be classified into several categories depending on their location, typology, technology, arrangement and management.

Figure 4.1 shows a schematic of different categories identified in the literature and from real cases. According to the location on the building envelope, a VFS can be located on rooftops, on the upper floor of the building or on façades – on the interior or exterior. Those on building façades can be located on balconies, windowsills or walls; behind windows; or in between a double-skin façade.

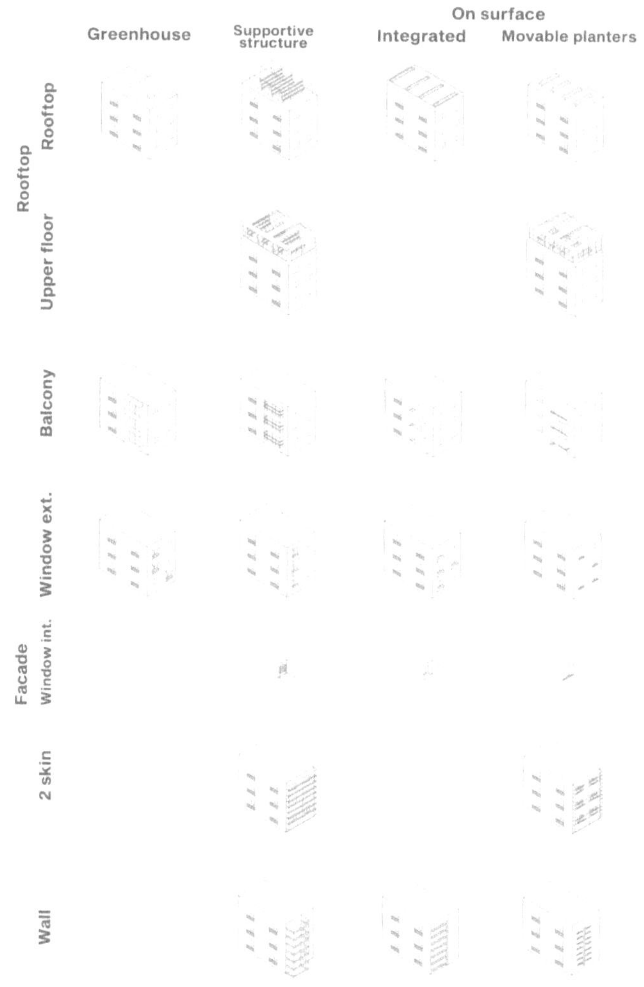

FIGURE 4.1 Classification of VFSs on building envelopes. (Image credit: Abel Tablada and Vesna Kosorić.)

According to the typology used, the VFS can be a greenhouse (most of the time on rooftops or balconies but also possibly on windows) or a supportive structure (the integrated or movable planters). The supportive structure refers to any added structure on the rooftop or façade that helps to support the hanging of the planters and containers and the irrigation system. Such structures are open and not part of the building structure. The integrated typology refers to those planters that are fixed and built together with building elements, e.g., a planter that is part of the balcony design or integrated into the rooftop floor and parapet. The movable category, on the contrary, refers to those planters and containers that can be placed on the rooftop, balcony or windowsills without the need to transform the original building elements. They may have fixing elements to resist strong winds but can be easily moved to another position or removed.

According to their technology, VFSs range from the most traditional ones, soil-based with manual irrigation, to the most advanced ones, fully automated hydroponic/aeroponic systems (Chatterjee et al., 2020; Despommier, 2009) or even aquaponic with fish cultivation.

Technology-wise, VFSs can be soil-based or hydroponic/aeroponic. In terms of irrigation and nutrient supply, they can be fully automated, semiautomated or manual. Those fully automated are, most of the time, hydroponics inside the greenhouses or in supportive structures. Most of the time, planters in windowsills and balconies are soil-based with hybrid or manual irrigation systems.

In terms of the arrangement, planters and containers can be placed vertically, horizontally or inclined. They can also be arranged in rows or scattered. A more detailed classification of VFSs on building façades can be found in Tablada and Kosorić (2022).

DESIGN CONSIDERATIONS FOR THE IMPLEMENTATION OF VFSs ON BUILDING ENVELOPES

Sunlight Availability on Building Envelopes

An important concern for wider dissemination of urban farming on other surfaces apart from rooftops, e.g., on façades, is whether the crops receive enough sunlight in different latitudes, orientations and contexts. The study of Tablada and Zhao (2016) introducing hypothetical residential buildings in Singapore with different urban densities and building typologies concluded that leafy vegetables would successfully grow at all façade orientations considering optimal required sunlight (ORS) of 10,000 lx. The factor of urban density was more influential than façade orientation and building typology for incident sunlight and, therefore, the potential for vegetable growth.

Song et al. (2018) also proved the feasibility of growing vegetables on façades and building corridors in Singapore. Based on the measured photosynthetically active radiation and the daily light integral (DLI), the study found that vegetables receiving half-day direct insolation can successfully grow. Palliwal et al. (2021) applied 3D city model simulations to identify residential building façades convenient for crop growing.

Using the same building model and installing experimental VFSs on building corridors, façades and rooftops, Song et al. (2022) compared the potential self-sufficiency of the three VFSs as retrofitting measures in Singapore's public housing buildings (Figure 4.2). Of the three systems, the concept of the productive façade (PF) system developed by Tablada et al. (2018), Kosorić et al. (2019) and Tablada et al. (2020) had the highest potential for food self-sufficiency (43%) due to the higher exposure to sunlight in comparison with planters placed on corridors (0.5%) and on rooftops (3%). The relatively low values obtained on rooftops compared to façade surfaces are explained by the small area available in high-rise rooftops relative to the number of residents.

Chaichana et al. (2022) developed a model for predicting the annual sunlight availability on vertical cultivation shelves by using the Grasshopper plugin in Rhinoceros 3D. The values of solar radiation are taken from Chiang Mai, Thailand, considering the shelve structure and orientation, and they were converted into photosynthetic photon flux density (PPFD) and DLI. The north–south orientations offer uniform PPFD and DLI to the shelves throughout the year.

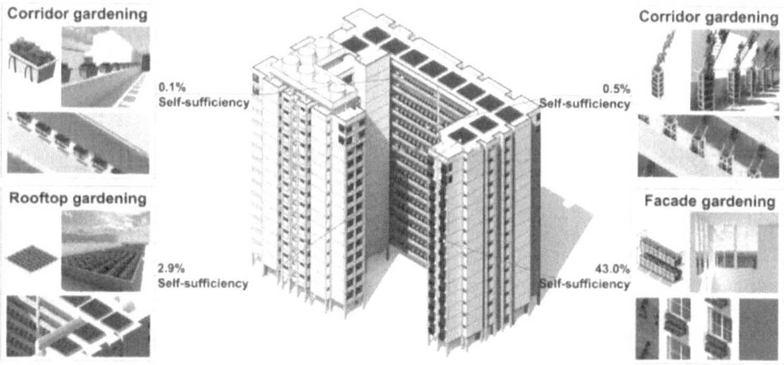

FIGURE 4.2 Model of the public housing in Singapore and the level of self-sufficiency achievable by different gardening methods. The illustration is not drawn to scale. In the calculation, it was assumed the façade gardens would be installed below windows and on windowless façades assuming accessibility via external scaffolds without substantial building structural alteration. (Image Credit: Song et al., 2022.)

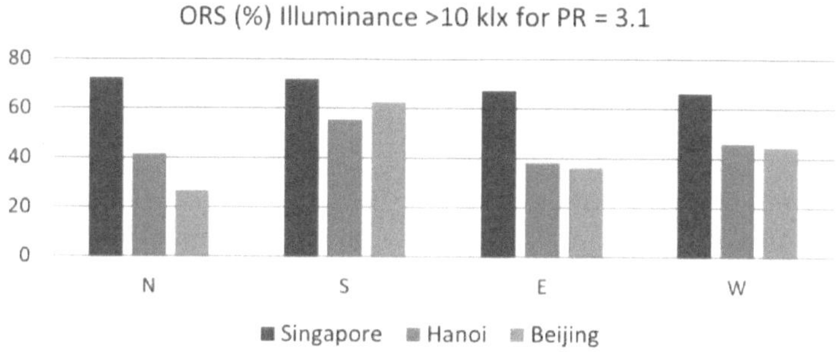

FIGURE 4.3 Comparison of the average optimal required sunlight (ORS) for the three typologies with plot ratio (PR) = 3.1 per façade between Singapore, Hanoi and Beijing. (Image Credit: Tablada (2015). Abel Tablada & Vesna Kosorić.)

Top shelves receive the highest values of DLI (36.80 mol × m^{-2} × day^{-1}), while middle and bottom shelves receive approximately 60% and 50% of the values at the top level, respectively, so specific crops can be placed according to incident sunlight. For east–west oriented shelves, a larger difference in DLI is observed between the top and bottom levels, but those on the bottom levels receiving 8–12 mol $m^{-2}d^{-1}$ can still be used from April to September to grow low-light crops such as microgreens, leafy vegetables, herbs and lettuces.

For higher latitudes (Hanoi: 21°N and Beijing: 39.9°N), the impact of urban density and façade orientation is more evident than for equatorial latitudes (Tablada, 2015). However, the reduction of food self-sufficiency due to higher latitudes in leafy vegetable potential production was less than 10% for the three building typologies and for the higher urban density of plot ratio (PR) = 5. For the lower urban density of PR = 1.3, the reduction is less than 6%. Regarding the orientation of façades, Figure 4.3 shows the percentage of ORS > 10 klx for the three typologies with the PR = 3.1 per façade between Singapore, Hanoi and Beijing. While sunlight incidence is quite similar for all façades in Singapore, the higher sunlight incidence on the southern façades compensates for reducing sunlight on the northern façades as latitude increases. However, this advantage is less evident in high-density urban contexts due to the shading effect of neighboring buildings. Therefore, the study concludes that the sunlight availability and, therefore, the potential for food self-sufficiency is more affected by the factor of density than by typology and latitude, especially if the southern latitudes are cloudier (Tablada, 2015).

For latitudes higher than 40°N, although sunlight may be scarce, VFSs on building envelopes can also be a suitable option if varieties of local vegetables are grown in low-to-middle urban density, especially on rooftops (Haberman et al., 2014).

Types of Crops and Protection Elements

When crops are exposed to exterior climatic conditions, they can be damaged by strong winds, heavy rains, excessive solar radiation and pollution. Therefore, a careful selection of crops should be conducted under the advice of agronomists and other horticulture experts, depending first on the climatic region, then on the location and dimensions of the planter boxes or containers and on their accessibility. Several measures can be applied, such as partially or fully covering the crops by a system of nets or placing them inside greenhouses to provide a more controlled environment. Those on façades may have a covering net against heavy rains and winds.

In general, those crops originally from the same region and planted according to their natural season require fewer additional measures for them to grow healthy. Higher investment and operational costs are needed if out-of-season or exotic crops are planted.

Supporting Structure

The structure where VFS is to be installed should fulfill several specific conditions related to their position on the rooftop or façades. Resistance to stormy winds is probably the most demanding requirement to be fulfilled. In tropical and subtropical regions where hurricanes, cyclones or typhoons are relatively frequent, having a strong structure is crucial for security and economic reasons. If the structure does not resist very strong winds, a demountable system should be designed considering the easiness of operation and a close and secure storage place.

Other requirements are related to accessibility for irrigation and other agricultural activities. In the case of façades, if the access is not directly from indoors, additional stairs and walkways should be considered, which may also add costs and weight to the structure.

Growing Media

The quality and adequacy of growing media or substrate is a crucial element for a successful implementation of VFSs, both indoors and on building envelopes. In the case of exposed-to-the-sun containers or planters, the substrate, whether it is natural soil or soilless media-engineered, should also be able to reduce water evaporation. However, in hydroponic and aeroponic systems, the chemical composition of water assures the right balance of nutrients according to the type of crops.

For soilless media, Shukla et al. (2021) suggest different combinations of soil constituents to satisfy the requirements of each vegetable for residential gardens on rooftops and balconies. In an ideal substrate, the content of solids, water and air should be around 40%, 40% and 20% of the greenery volume, respectively. The mineral composition can range from 80% to 100% and the organic matter from 0% to 20%. A wide variety of substrates can be used for the preparation of engineered soil or soilless media.

Irrigation System

Irrigation on VFSs located on rooftops and building façades can use a variety of systems: from manual to the most advanced ones. Manual irrigation is preferred for planters located on balconies or in simple indoor structures, especially for elderly people who enjoy gardening activities (Kosorić et al., 2019). However, for more complex VFSs on rooftops and in the exterior of façades, automated irrigation systems, such as the drip micro-irrigation system (Friedman, 2023), are preferred to assure the required amount of water and nutrients depending on the type of crops, substrate and evaporation levels.

Accessibility and Security

Accessibility and security are very important considerations for VFSs in building envelopes. Planter boxes and other supporting structures on rooftops and open corridors can be easily accessed and may not require additional structures. Clearly, a parapet around the edge of the roof is needed to limit the circulation within the rooftop. Those located indoors next to semitransparent windows are designed for easy access, so both accessibility and security are assured. However, accessibility is a challenge for those planters located on the exterior of building façades. A scaffolding or other lightweight structure with stairs is required if a VFS is installed against a blind wall. This may increase the final cost of the installation, but exposed surfaces can be fully utilized. Exterior access structures are not needed if planter boxes are placed under the windowsills. In those cases, the width of planter boxes should not be larger than 500 mm, so an average person can reach and work on the outer extreme.

EXAMPLES OF VFSs ON FAÇADES AND ROOFTOPS

This section shows several cases of the integration of farming systems on façades and rooftops. Some cases are experimental studies, while others are VFSs with commercial or social purposes.

Building Façades

CDL-NUS Productive Façades for Food and Energy Generation

The PF is a façade concept developed by Tablada et al. (2018), which integrates photovoltaic (PV) modules as shading devices and VF systems on balconies and windows. The first prototypes were built and tested at the Tropical Technologies Laboratory (T^2 Lab) at the School of Design and Environment, National University of Singapore (NUS) (Figure 4.4).

The idea is to expand the range of functions of building envelopes to include energy and food production, utilizing solar radiation. Hence, building design aims toward higher "outer" productivity, in parallel with increased indoor comfort for tenants (e.g., enhanced thermal conditions and aesthetic appeal).

FIGURE 4.4 Productive façades – north and east – at the NUS-CDL T^2 Lab in January 2019 at 9 a.m. (Image Credit: Abel Tablada.)

One of the tests carried out at the T^2 Lab was the assessment of the efficiency and productivity of the PF integrating PV and VFSs (Tablada et al., 2018). Eight PFs were designed and installed at four cardinal orientations representing eight testing cells. Two types of façades were selected as typical Housing Development Board (HDB) building façades and assessed for each orientation: balcony façade and window façade (Figure 4.4). The façades were designed using 3D simulation algorithms and a multi-criteria decision-making process. The VIKOR optimization method (Opricovic and Tzeng, 2004, 2007) was applied to find the best trade-off among the five performance indicators: (1) food production potential, (2) potential electricity generation, (3) indoor daylight, (4) solar heat gain and (5) view angles from the interior. The design was also assessed from the standpoint of cost, accessibility and aesthetics.

The design variants were further optimized using the input from an online survey among 100 PV experts and architects (Tablada et al., 2020) and an in-person social acceptance survey among 391 HDB residents (Kosorić et al., 2019). Accounting for weather conditions in Singapore (wind, solar, radiation, humidity, etc.), copper indium gallium selenide PV modules, convenient for non-clear skies, were located on the north and south façades, while monocrystalline silicon PV modules were installed on the east and west façades, as they respond well to direct solar radiation.

As for the crop selection, a tropical-adapted variety of lettuce was selected among several commonly cultivated leafy vegetables in Singapore, thanks to its short harvest cycles and moderate light requirement. Crop yields were measured in six rounds from December 2018 until June 2019.

Measurement data were collected on the PFs and inside the eight cells using 78 sensors for air and surface temperatures, relative humidity, solar irradiance, light intensity, water consumption and electric power production.

As for electricity generation from PV modules considering a typical HDB household with both north and south façades (20 m in total), around 30% of energy demand can be supplied by using PV modules as shading devices (Figure 4.5). The main façades oriented closer to the east and west could generate more electricity, thus meeting higher energy demand.

Regarding vegetable yield for the 6 months, the east and north façades had the greatest performance, producing 902 and 828 g of lettuce, followed by the south and west façades with 763 and 550 g, respectively (Figure 4.6). The amount produced represents 55–103% of the average leafy vegetable consumption of a four-member household in Singapore (ca. 16 kg per year). Soil quality turned out to be a factor with a great impact on vegetable yield.

Experimental Productive Double-Skin Façades in Tianjin

Experimental productive double-skin façades were built by researchers from the School of Architecture, Tianjin University (Zhang et al., 2022). The main objective of the study was to evaluate the

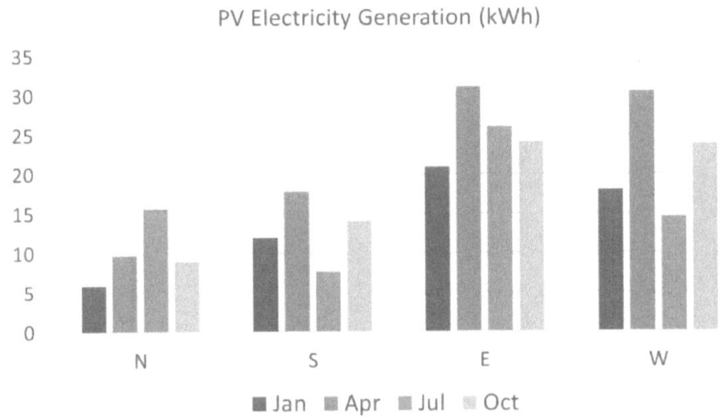

FIGURE 4.5 Electricity generation from PV modules per façade orientation for the months of January (2020), April, July and October (2019). (Image Credit: Tablada, A., in: Lau et al., 2021.)

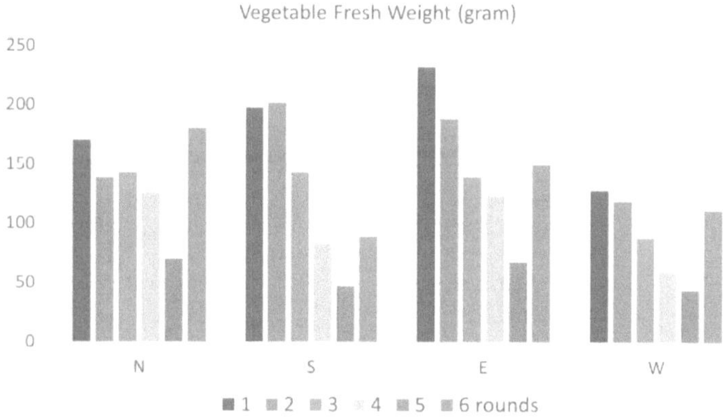

FIGURE 4.6 Weight of fresh lettuces cultivated at the T2 lab in each of the six rounds for the four façade orientations. (Image Credit: Data provided by Song Shuang, Department of Biological Sciences, NUS, in: Lau et al., 2021.)

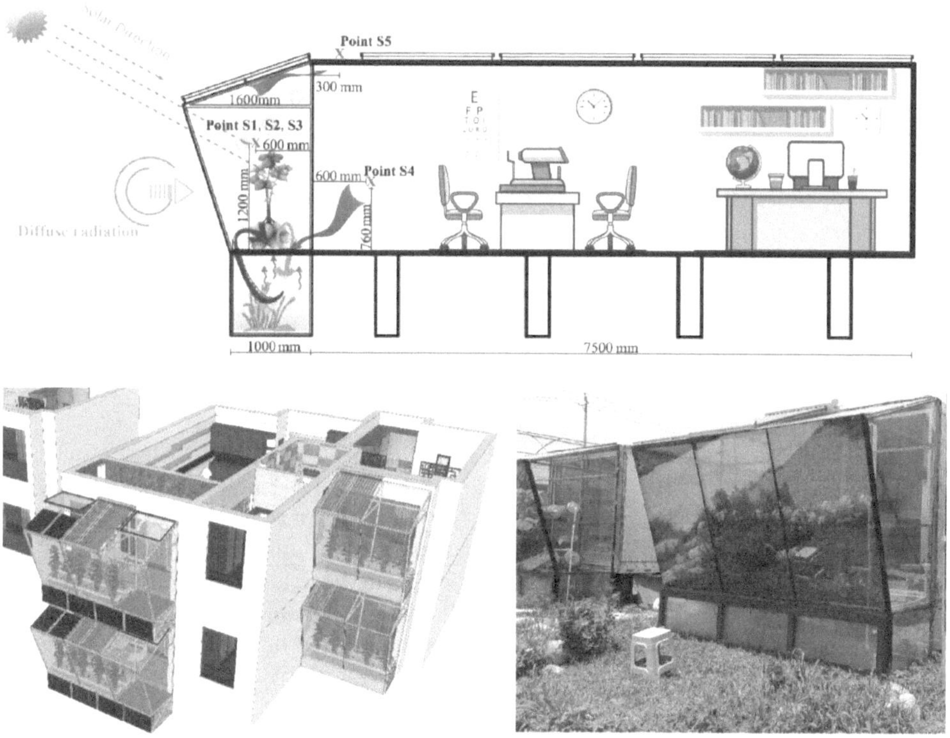

FIGURE 4.7 Proposed experimental double-skin façade. Top: section of the concept (top); bottom left: artistic impression of possible integration of DSF in a building; and bottom right: photo of the actual experimental setup at Tianjin University. (Image Credit: Zhang et al., 2022.)

daylighting comfort in rooms with this kind of façades. As shown in Figure 4.7, the double-skin façade has three compartments: a top triangular one with a tilted PV module acting as a shading device and a warmer space connected to the interior; a middle one where vegetables, herbs or ornamental plants are cultivated with an external inclined glazed surface from which daylight

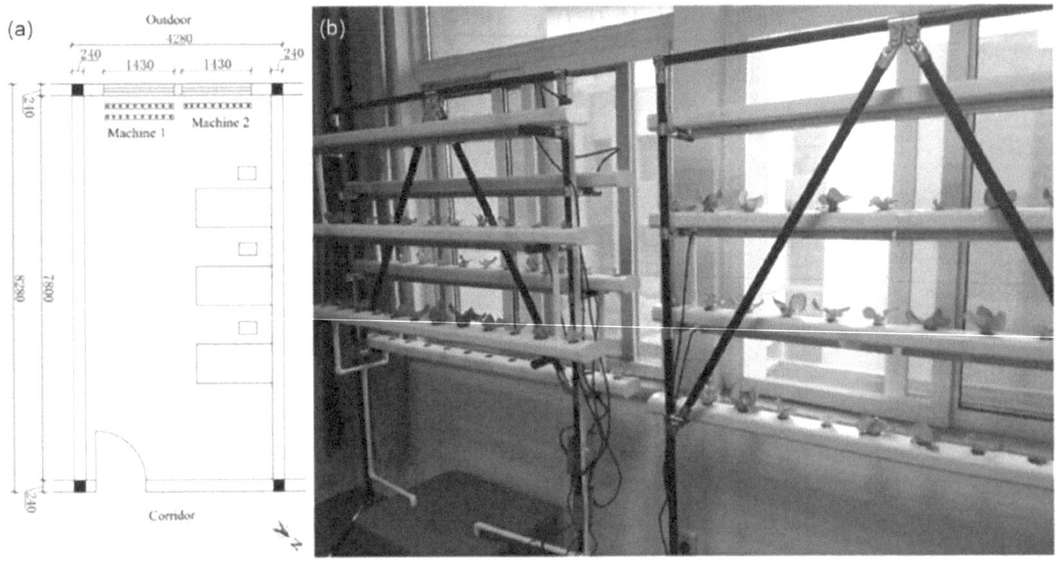

FIGURE 4.8 Laboratory layout of the urban family vertical farm. Left: floor plan of the laboratory; right: VF framework. (Image Credit: Shao et al., 2022.)

is obtained; and the bottom one as a square deposit of water for the irrigation of the plants above. The study concluded that the addition of the double-skin PF with the PV module as a shading device improved the overall daylight indicators by increasing the area with useful daylight intensity while reducing the direct sunlight being reflected onto the ground and the risk of glare from the interior. The models have the potential to be further analyzed and developed as a thermal buffer, especially in temperate and cold climates or in fully air-conditioned spaces in warmer regions. As the authors highlighted, the integration of production and building renewal at the building skin level has a high potential for research and development.

Experimental Hydroponic System behind Windows in Shanghai

Another interesting experiment is the proposal of an urban family VF by Shao et al. (2022) in Shanghai (Figure 4.8). It consists of a hydroponic system that may have a single or a double layer of containers for leafy vegetables to be implemented in the interior of residential buildings' façades behind windows. For the double layer, the alternate position of containers assures better daylight access. The study calculated the degree of self-sufficiency of vegetables and fruits according to the average demand of typical apartments. For a single household, it can reach between 15% and 20%. The economic aspect is also evaluated as having a payback period of only 3.5 years.

Rooftops

Organic Rooftop Farm at Thammasat University in Bangkok

The building of Thammasat University in Bangkok contains one of the largest organic rooftop farms in Asia (Holmes, 2020). The landscape architecture office LANDPROCESS (Landprocess, 2023) created a green roof of 22,000 m^2 by repurposing the unused roof area, of which even 7,000 m^2 (World Landscape Architecture, 2023) is intended for farming, first of all, rice and also various fruits and vegetables (Figure 4.9). Inspired by the form of traditional terraced rice fields, the landscape architects applied this motif and created an urban farm invoking, not only by its form but also by analogy with the use of rainwater, traditional forms of crop cultivation.

FIGURE 4.9 Thammasat University urban rooftop farm. Left: the largest urban farming green roof in Asia; right: outdoor education among the urban farm. (Image Credit: Panoramic studio/LANDPROCESS (Landprocess, 2023).)

Considering that the manufactured goods are used by the campus and that the crops are maintained by university students, the complex can be seen as a self-sustaining urban enclave. Also, the roof farm, apart from its basic activity, is a place of education and socialization for young people. This way, urban farming "creates space" for people to come together and interact with each other, which makes it highly socially valuable and, therefore, acceptable.

Rooftop Farm at Khoo Teck Puat Hospital

Through the architecture of the Khoo Teck Puat Hospital in Singapore (Khoo Teck Puat Hospital, 2022), the design team of CPG Consultants Pte Ltd (CPG Consultants Pte Ltd, 2022) reconsiders the traditional principles of treatment, bringing nature, greenery and urban gardens closer to patients and employees in the hospital sector, neighbors who are involved in the maintenance of gardens and many other visitors. Thanks to the biophilic design of the complex itself, nature therapy is introduced in the very treatment of patients (Kishnani, 2017; International Living Future Institute, 2023), in the correlation between two determinants: "hospital in the garden" or "garden in the hospital" (President's Design Award (P'DA), 2023). The hospital kitchen is supplied with fruits, vegetables and other plants from the gardens on the roof, whereby food transport and, therefore, pollution of the environment is reduced. Different gardening methods are used to grow a variety of edible fruits and vegetables, including hydroponics farming which was started by senior volunteers in 2019 (Figure 4.10).

Rooftop Gardens at Khoo Kampung Admiralty Housing

The Kampung Admiralty housing (Housing & Development Board, 2023), the first-of-its-kind development in Singapore, which integrates housing for the elderly with a wide range of social, healthcare, communal, commercial and retail facilities, was designed by the work of WOHA (WOHA, 2023), while the landscape elements were handled by Ramboll Studio Dreiseitl (Ramboll Group A/S, 2023). This design principle is inspired by the meaning of the word "Kampung" (Indesignlive, 2023), which in Malaysia, Singapore and the surrounding area denotes a traditional small village (Singapura Stories, 2023), also implying a more direct contact with nature and greenery, which in this project occupies a larger plot area than the building (Greenroofs, 2023). In accordance with the theme of urban gardens, small "plots" within the terraces are intended to cultivate various crops for the needs of the home for the elderly, while home residents participate in crop cultivation (Dezeen, 2023). The surrounding green area serves as a community park where the elderly live and socialize (Figure 4.11). The biotope formed by plants on the terraces and roofs has also contributed to biodiversity with the appearance of many species of insects and birds (Landezine International Landscape Award, 2023).

FIGURE 4.10 Farming with different technologies on rooftop gardens of Kho Teck Puat Hospital in Singapore. (Image Credit: Abel Tablada.)

FIGURE 4.11 Kampung Admiralty. Left: view of a courtyard and a community park with multi-layered sky terraces; right: view of raised planters with several bushes, herbs and ornamental plants. (Image Credit: Abel Tablada.)

Aquaponic on a Rooftop of Singapore Central Business District (CBD) Comcrop

Comcrop is a commercial firm that has installed an aquaponic system on the rooftop of a public building in the CBD of Singapore (Figure 4.12). The building is dedicated to supporting young entrepreneurship and innovation and hosts this company which, apart from producing vegetables and fish for commercial purposes, also uses the installation to educate the youth. Lettuces and other local leafy vegetables are produced in the A-shape hydroponic structure.

BENEFITS VS LIMITATIONS AND LIFE CYCLE ENVIRONMENTAL IMPACT

Benefits vs Limitations of VFSs on Building Envelopes

VFSs on building envelopes may add benefits to the urban environment, as in the case of vertical greenery systems (VGSs). Apart from helping the urban man reconnect with nature (Wong et al., 2010a), VGS has considerable functional (Cooper-Marcus & Barnes, 1999; Roehr & Laurenz, 2008) and formal-aesthetic values (Sutton, 2014).

Depending on the type of crops and plants and the arrangement of planter boxes, both VFSs and VGSs may contribute to indoor privacy by shielding building façades from external views.

FIGURE 4.12 Comcrop: Aquaponic system on the rooftop of a public building in Singapore. (Image Credit: Abel Tablada.)

They may also contribute to the aesthetics of buildings (Tan et al., 2009). In the case of VFSs, this is more assured if combined with flower plants acting as a natural insect repellent.

As for environmental benefits, VGSs contribute to reducing buildings' external surface temperature by 10–15°C compared to the buildings without the green façade (Pérez et al., 2017). For VFSs, this reduction may be much lower if the crops are in their early growing stage or if they do not cover large portions of the façade. When VFS includes fruit trees and bushes or has a similar arrangement as the ornamental VGS, the ambient temperature may also be reduced by up to 3.3°C at 0.15 m distance from the greenery surfaces (Tan et al., 2009). VFSs can also improve rainwater treatment and retention to safeguard from flooding during extreme rainy periods. They can create natural habitats for plant and animal species contributing to urban biodiversity (Brenneisen, 2004). The green layer on façades can create a sound-absorbing buffer, therefore improving the sound environment indoors (Wong et al., 2010b). Regarding air pollution, particulate matter (PM) 2.5 concentration decreases with height; thus, relative to street level, vegetables cultivated on rooftops and, to a certain degree, at higher levels of building façades, are less exposed to air pollutants, including heavy metals (Tong et al., 2016).

VGSs have economic benefits due to their improvement of building aesthetics and hence the increase in property value. They also contribute to reducing energy use in residential and office buildings because of their shading capacity, which can reduce cooling loads from 10% to as much as 74%, according to a Singapore study (Wong et al., 2009; Wong et al., 2010c). Depending on the design quality, the surface covered and maintenance, VFSs may also yield these economic benefits.

Compared with indoor VFSs, those located on building façades and rooftops may generate significant savings in rental costs which can be extremely high in high-income cities as reported in Benis and Ferrão (2018) for the case of the Netherlands. These exposed-to-the-sun systems can also save a significant amount of energy otherwise used by electric light systems. In Chapter 8, Ng and Mahkeswaran (2024) reported the results of Ong et al.'s (2019) study on aquaponic systems. Systems located outdoors save about 43% of energy using complementary grow lights compared to those located indoors, which need electric light for 14–16 hours.

In addition to shorter supply chains and reduced logistic costs, according to Benis and Ferrão (2018), VFSs promote rural–urban links, as they integrate productive and lifegiving aspects into urban spaces and serve as a vehicle to promote cultural values and social cohesion Zambrano-Prado et al. (2021b). Regarding market advantages, produce from VFS is fresher, which gives them an added value. Other social advantages of VFSs in building envelopes are discussed in Chapter 5 (Kosorić, 2024).

Rooftop productivity can also be relatively high according to the urban morphology. In non-residential morphologies, where roof surfaces are more extended, as in the case of industrial areas, the food output can be very generous. For example, for the studied case in Zaragoza, Spain, it reached 2.51 kg of tomatoes/m^2 in comparison with 0.31 kg of tomatoes/m^2 in single-family terraced morphology (Montealegre et al., 2022).

Some of the disadvantages that have been reported in relation to VGS, which can also be applicable to VFSs on building envelopes, are the following:

- High installation costs depending on the system, with direct green façades being the cheapest and the living wall system the most expensive (Perini and Rosasco, 2013). Living wall systems can also be used for VFSs or for a hybrid wall with ornamental and edible plants. However, more conventional systems, such as a supportive structure with planter boxes or containers attached to building walls, windowsills or on the rooftop, are less expensive than the living walls.
- Installation costs are much higher if greenhouse systems are installed on rooftops. Muñoz-Liesa et al. (2021) identified the conventional steel structure and polycarbonate material as the main factors affecting the environment and systems' costs in their life cycle due to their greater sophistication in comparison with conventional agriculture practices and with ground greenhouses. However, design improvement scenarios may decrease environmental impacts by about 24%;
- High operational and maintenance costs are also reported due to regular activities such as water pipe substitution, plant substitution and pruning. In the case of VFSs, operational costs can even be higher than for VGS because typical agricultural activities have to be considered and accessibility and security should be prioritized in the design. For more advanced systems such as high-tech greenhouses, apart from the initial high investments, running costs are high due to the demand for technical skills and the high degree of automation which still limits their applications even in industrialized countries (Roth et al., 2019);
- Appearance of unwanted pests and animals, which must be addressed by a pest control professional. Using plants that repel pests is a common practice in UA, which can also be applied to VFSs.

Other constraints on BIA were identified in a survey by Zambrano-Prado et al. (2021a, 2021b). The cost of water and pollution stood out as significant contextual and environmental constraining factors. Architectural and technical barriers include limits on building heights, historical buildings, the lack of specific building codes, building design and roof accessibility. On the contrary, on the environmental and economic side, the high cost of urban land is the primary factor in favoring VFSs on building envelopes. However, there is a long way to go to overcome economic constraints in the form of infrastructure costs and urban policies. The high cost of energy and water was also reported by Benis and Ferrão (2018) since local authorities do not subsidize them.

Life Cycle Environmental Impact of VFSs on Building Envelopes

VFSs are in the stage of exploration and development. Despite the bulk of commercial and research experiences, there is not yet robust evidence on their life cycle assessment and impact on GHG emissions.

However, some studies have already measured or estimated the effect of various VFSs compared with conventional food chains of certain produce. Benis et al. (2017) studied three hypothetical hi-tech urban farming scenarios in Lisbon, Portugal: a polycarbonate rooftop greenhouse, a vertical farm with windows and skylights on the top floor of a reinforced-concrete building, as well as a completely opaque VF with no penetration of natural light on the ground floor of a reinforced-concrete building. When compared with the existing supply chain for tomatoes, the experiments on the rooftop greenhouse and the top floor VF were able to reduce in half and in three GHG emissions, respectively. However, the indoor VFS has the highest GHG emissions due to the electricity needed for lighting (Figure 4.13). For Mediterranean climates and, therefore, for those regions with plenty of sunlight throughout the year, the study recommended those installations with plenty of incident sunlight instead of an indoor VFS, even though more efficient lighting systems can be used to reduce emissions.

Muñoz-Liesa et al. (2020) studied the impact of BIA on reducing GHG emissions, specifically the integration of rooftop greenhouses into unused spaces of urban fabric, which can be reused for offices and labs. By conducting energy simulations, the study demonstrated the decarbonized effect and the improvement of the energy performance of the symbiosis of energy and water systems between the host building and the added greenhouse. The combined effect of cooling and heating energy reduction achieved an overall 128 kWh/m^2 of net energy savings and 45.6 kg CO_2 eq/m^2 of

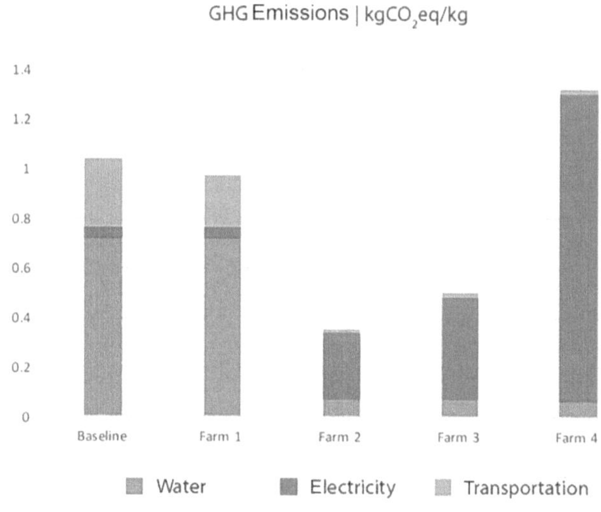

FIGURE 4.13 Baseline: The currently existing supply chain for tomatoes. Farm 1: a low-tech unconditioned urban rooftop greenhouse; farm 2: a polycarbonate rooftop greenhouse (RG); farm 3: a vertical farm with windows and skylights on the top floor of a reinforced-concrete building; farm 4: a completely opaque VF with no penetration of natural light on the ground floor of a reinforced-concrete building. (Image Credit: Benis et al., 2017.)

emissions reduction. The study considered a Mediterranean climate; therefore, different results can be obtained in other climatic regions, contexts and types of buildings. Similarly, the study of Khan et al. (2018), using a BIA information modeling tool in integration with building information modeling, demonstrated the benefits of BIA in optimizing water, nutrients and energy resources as well as the potential to increase productivity by simulating and monitoring desired conditions according to the type of produce.

Shao et al. (2022) compared their proposed model of the VFS in new residential buildings in Shanghai with other forms of UA, such as rooftop gardens, greenhouses, indoor VFS and allotment gardens. Their model to be installed in south-facing balconies and windows relying on natural light can provide around 24 kg/m^2/year of vegetables and fruits, which is much higher than some of the indoor VFS (2.8 and 10.3 kg/m^2/year) and the rooftop options (0.78 kg/m^2/year). In terms of self-sufficiency, the proposed VFS can provide from around 15% to 20% of the vegetables and fruits of a single family. However, considering the current public acceptance and preference, only 1% of self-sufficiency can be reached at the neighborhood level. Regarding cost and revenues, the fixed cost per household is from 330.7 to 433.0 GBP, the payback period is 3.5 years and the return on investment is about 30%.

Gumisiriza et al. (2022) investigated the economic viability of a VFS to produce lettuce at a low cost in Uganda. Several indicators related to profitability and affordability were analyzed for hydroponic systems outside greenhouses, hence, directly exposed to the exterior conditions. For annual crop production of six cycles, implementing the low-cost VFS was considered a cost-effective venture with substantial profits that, for local conditions, could contribute to food security in low-income communities. The VFS, although not designed to be implemented on building envelopes, can be an option for rooftops of denser urban settlements in developing countries.

FUTURE DIRECTIONS AND POLICIES

Besides traditional low-tech practices on rooftops and balconies, VFS on building envelopes is a relatively recent practice. Therefore, further technological innovation and public support are needed to reduce current limitations and costs, better synergize with building systems and increase public acceptance. Local policies and fiscal incentives may promote the achievement of these goals and adapt themselves to the continuous evolution of UA-related technologies from traditional to advanced practices using digital technology (Yuan et al., 2022).

Among social-related policies, D'Ostuni et al. (2022) recommend reconnecting people with food through healthy diet educational projects, making food production visible and tangible within the city boundaries. Together with technical innovation, municipal policies adapted to local conditions may drastically change the modern food system, contribute to the implementation of a circular food economy and reduce the overall carbon footprint in cities.

Promoting policies that integrate food and energy self-sufficiency in urban areas is another promising direction in which building envelopes can play a crucial role. This can be achieved by promoting the implementation of "green" certificates in new buildings and in building renovation works. Studies on the feasibility of applying current and improved technologies to generate renewable energy in VFSs (Kobayashi et al., 2022) also help convince local authorities to apply policies and incentives to combine food production and energy generation in urban areas.

CONCLUSIONS

The integration of VFSs on rooftops and façades is one of the available possibilities of UA. Considering the urgent need to increase food availability and food security in urban areas as part of international agendas on carbon footprint reduction, poverty alleviation and healthy food accessibility, building envelopes to produce food becomes a necessary step, despite the associated challenges. Owing to the availability of exposed-to-the-sun surfaces, growing vegetables and fruits on rooftops and façades is an effective alternative to producing food indoors, thus avoiding the use of additional energy or searching for expensive, vacant plots inside the cities.

This chapter has explored the potential of the building envelope to integrate VFSs by first classifying various design and implementation strategies and possibilities for rooftop and façade integrations according to their performance and relationship with interior and building envelope functions. Design considerations such as sunlight availability and structural and accessibility issues have been discussed. Several case studies involving experiments and community and commercial farms have been presented, showing a wide range of solutions and possibilities for the integration of VFSs into rooftops and façades. The advantages and potential threats of VFSs on building envelopes were also discussed, as well as the future trends and policies for counteracting limitations and incentivizing implementation.

There is a growing enthusiasm from business and public sectors for incorporating agricultural activities in urban areas. Although challenges are not a few, especially for VFSs on building envelopes, the incorporation of advanced technology and the integration of knowledge from multiple disciplines covering architectural design, agronomy, building engineering and information technology, among others, may gradually reduce current financial and technological drawbacks. Apart from the conventional functions of building envelopes, such as acting as a buffer between outdoors and indoors, the enormous number of exposed surfaces in most urban environments can provide greenery, food and energy to cities. The examples shown in this chapter of food growing on rooftops and façades have highlighted this yet neither appreciated nor exploited potential which may have multiple benefits to society from social, environmental, cultural and economic viewpoints.

REFERENCES

Benis, K., Ferrão, P., (2018), Commercial farming within the urban built environment – Taking stock of an evolving field in northern countries. *Global Food Security*, 17, 30–37. https://doi.org/10.1016/j.gfs.2018.03.005.

Benis, K., Reinhart, C., Ferrão, P., (2017), Development of a simulation-based decision support workflow for the implementation of building-integrated agriculture (BIA) in urban contexts. *Journal of Cleaner Production*, 147, 589–602. https://dx.doi.org/10.1016/j.jclepro.2017.01.130.

Brenneisen, S., (2004), The benefits of biodiversity from green roofs – Key design consequences, *Proceedings of 1st North American Green Roof Conference: "Greening Rooftops for Sustainable Communities"*, 29–30 May 2004, Chicago and Toronto, The Cardinal Group.

Chaichana, C., Man, A., Wicharuck, S., Mona, Y., Rinchumphu, D., (2022), Modelling of annual sunlight availability on vertical shelves: A case study in Thailand. *Energy Reports*, 8, 1136–1143. https://doi.org/10.1016/j.egyr.2022.07.077.

Chatterjee, A., Debnath, S., Pal, H., (2020), Implication of urban agriculture and vertical farming for future sustainability. In S. Shekhar, et al. (Eds.), *Urban Horticulture-Necessity of the Future*. Intechopen. https://doi.org/10.5772/intechopen.91133.

Chen, S., Chen, B., Feng, K. et al., (2020), Physical and virtual carbon metabolism of global cities. *National Communication*, 11, 182. https://doi.org/10.1038/s41467-019-13757-3.

Cooper-Marcus, C., Barnes, M., (1999), *Healing Gardens: Therapeutic Benefits and Design Recommendations*. New York: John Wiley.

CPG Consultants Pte Ltd., (2022), https://www.cpgconsultants.com.sg/. Accessed 27.12.2022.

Despommier, D., (2009), The rise of vertical farms. *Scientific American*, 301(5), 3239. https://doi.org/10.1038/scientificamerican1109-80.

D'Ostuni, M., Zaffi, L., Appolloni, E., Orsini, F., (2022), Understanding the complexities of building-integrated agriculture. Can food shape the future built environment? *Futures*, 144, 103061. https://doi.org/10.1016/j.futures.2022.103061.

Dezeen, (2023). https://www.dezeen.com/2018/12/07/kampung-admiralty-woha-singapore-world-building-year/. Accessed 17.01.2023.

Friedman, P. S., (2023), *Irrigation Methods, Reference Module in Earth Systems and Environmental Sciences*, Elsevier. https://doi.org/10.1016/B978-0-12-822974-3.00138-5.

Gazzotti, P., Emmerling, J., Marangoni, G. et al., (2021), Persistent inequality in economically optimal climate policies. *National Communication*, 12, 3421. https://doi.org/10.1038/s41467-021-23613-y.

Greenroofs, (2023), https://www.greenroofs.com/projects/kampung-admiralty/. Accessed 14.01.2023.

Gumisiriza, M.S., Ndakidemi, P., Nalunga, A., Mbega, E.R., (2022), Building sustainable societies through vertical soilless farming: A cost-effectiveness analysis on a small-scale non-greenhouse hydroponic system. Sustainable Cities and Society, 83, 103923. https://doi.org/10.1016/j.scs.2022.103923.

Haberman, D., Gillies, L., Canter, A., Rinner, V., Pancrazi, L., Martellozzo, F., (2014), The potential of urban agriculture in Montréal: A quantitative assessment. *ISPRS International Journal of Geo-Information*, 3(3), 1101–1117. https://doi.org/10.3390/ijgi3031101.

Holmes, D., (2020), Thammmasat University – The largest urban rooftop farm in Asia. Retrieved from WLA: https://worldlandscapearchitect.com/thammasat-university-the-largest-urban-rooftop-farm-in-asia/. Accessed 10.01.2023.

Housing & Development Board (HDB), (2023), https://www.hdb.gov.sg/residential/where2shop/explore/woodlands/kampung-admiralty. Accessed 14.01.2023.

Indesignlive, (2023), https://www.indesignlive.sg/segments/kampung-admiralty-design-architecture-history-more. Accessed 12.01.2023.

International Living Future Institute, (2023), https://living-future.org/case-studies/award-winner-khoo-teck-puat-hospital. Accessed 07.01.2023.

Khan, R., Aziz, Z., Ahmed, V., (2018), Building integrated agriculture information modelling (BIAIM): An integrated approach towards urban agriculture. *Sustainable Cities and Society*, 37, 594–607. https://doi.org/10.1016/j.scs.2017.10.027.

Kho Teck Puat Hospital, (2022), https://www.ktph.com.sg/. Accessed 27.12.2022.

Kishnani, N., (2017), Singapore's Khoo Teck Puat Hospital: Biophilic Design in Action. Retrieved from Interface: https://blog.interface.com/khoo-teck-puat-hospital-singapore-biophilic-design/.

Kobayashi, Y., Kotilainen, T., Carmona-García, G., Leip, A., Tuomisto, H. L., (2022), Vertical farming: A trade-off between land area need for crops and for renewable energy production. *Journal of Cleaner Production*, 379, 134507. https://doi.org/10.1016/j.jclepro.2022.134507.

Kosorić, V., (2024). Social acceptance of urban farming in and on buildings. In K. Al-Kodmany, A. K. Ng, A. Tablada, & C. Kole (Eds.),. *The Vertical Farm: Scientific Advances and Technological Developments*, Taylor & Francis Group, Boca Raton, FL, 57–81.

Kosorić, V., Huajing, H., Tablada, A., Lau, S.K., Lau, S.S.Y., (2019), Survey on the social acceptance of the productive facade concept integrating photovoltaic and farming systems in public high-rise residential buildings in Singapore. *Renewable & Sustainable Energy Reviews*, 111, 197–214. https://doi.org/10.1016/j.rser.2019.04.056.

Landezine International Landscape Award, (2023), https://landezine-award.com/kampung-admiralty/. Accessed: 27.01.2023.

Landprocess, (2023), https://landprocessdesign.wixsite.com/. Accessed: 07.12.2023.

Lau, S.S-Y., Tablada, A., Lau, S.K, Yuan, C., (2021), Vital signs revisited in the tropics – through the NUS-CDL tropical technologies laboratory pp. 95–110. In S. S.-Y. Lau, et al. (Eds.), *Design and Applications in Sustainable Architecture*, Springer Nature Switzerland AG, pages 95–110.

Montealegre, A.L., García-Pérez, S., Guillén-Lambea, S., Monzón-Chavarrías, M., Sierra-Pérez, J., (2022), GIS-based assessment for the potential of implementation of food-energy-water systems on building rooftops at the urban level. *Science of the Total Environment*, 803, 149963. https://doi.org/10.1016/j.scitotenv.2021.149963.

Muñoz-Liesa, J., Toboso-Chavero, S., Beltran, A.M., Cuerva, E., Gallo, E., Gasso-Domingo, S., Josa, A., (2021), Building-integrated agriculture: Are we shifting environmental impacts? An environmental assessment and structural improvement of urban greenhouses. *Resources, Conservation & Recycling*, 169, 105526. https://doi.org/10.1016/j.resconrec.2021.105526.

Muñoz-Liesa, J., Royapoor, M., Lopez-Capel, E., Cuerva, E., Rufí-Salís, M., Gasso-Domingo, S., Josa, A., (2020), Quantifying energy symbiosis of building-integrated agriculture in a mediterranean rooftop greenhouse. *Renewable Energy*, 156, 696–709. https://doi.org/10.1016/j.renene.2020.04.098.

Ng, A. K., & Mahkeswaran, R., (2024), A review on technological advances and challenges in aquaponics systems. In K. Al-Kodmany, A. K. Ng, A. Tablada, & C. Kole (Eds.). *The Vertical Farm: Scientific Advances and Technological Developments*, Taylor & Francis Group, Boca Raton, FL, 120–139.

Ong, Z.J., Ng, A.K., Kyaw, T.Y., (2019), Intelligent outdoor aquaponics with automated grow lights and Internet of Things. In IEEE International Conference on Mechatronics and Automation, Tianjin, 1778–1783.

Opricovic, S., Tzeng, G.H., (2004), Compromise solution by MCDM methods: A comparative analysis of VIKOR and TOPSIS. *European Journal Operational Research*, 156(2), 445455. https://doi.org/10.1016/S0377-2217(03)00020-1.

Opricovic, S., Tzeng, G.H., (2007), Extended VIKOR method in comparison with outranking methods. *European Journal of Operational Research*, 178(2), 514529. https://doi.org/10.1016/j.ejor.2006.01.020.

Palliwal, A., Song, S., Tan, H.T.W., Biljecki, F., (2021), 3D city models for urban farming site identification in buildings. Computers, Environment and Urban Systems, 86, 101584. https://doi.org/10.1016/j.compenvurbsys.2020.101584.

Pérez, G., Coma, J., Sol, S., Cabeza, L.F., (2017), Green facade for energy savings in buildings: The influence of leaf area index and facade orientation on the shadow effect. *Applied Energy*, 187(1), 424–437. https://doi.org/10.1016/j.apenergy.2016.11.055.

Perini, K., Rosasco, P., (2013), Cost-benefit analysis for green facades and living wall systems. *Building and Environment*, 70, 110–121. https://doi.org/10.1016/j.buildenv.2013.08.012.

President's Design Award (P'DA), (2023), https://pda.designsingapore.org/presidents-design-award/award-recipients/2011/khoo-teck-puat-hospital.html. Accessed 14.01.2023.

Ramboll Group A/S, (2023). https://ramboll.com/projects/singapore/kampung-admiralty. Accessed 07.01.2023.

Roehr, D., Laurenz, J., (2008), Living skins: Environmental benefits of green envelopes in the city context. In Proceeding of Eco Architecture II, WIT Press. Southampton, England: 149–158.

Roth, M., Frixen, M., Tobisch, C., Scholle, T., (2019), Finding spaces for urban food production – Matching spatial and stakeholder analysis with urban agriculture approaches in the urban renewal area of dortmund-Horde, Germany. Future of Food: Journal on Food, Agriculture and Society, 3(1), 79–88. Retrieved from https://www.thefutureoffoodjournal.com/index.php/FOFJ/article/view/123.

Samaniego, J. et al., (2023), Nature-based solutions and carbon dioxide removal, Project Documents (LC/TS.2022/224), Santiago, Economic Commission for Latin America and the Caribbean (ECLAC). https://repositorio.cepal.org/bitstream/handle/11362/48691/1/S2201263_en.pdf.

Shao, Y., Zhou, Z., Chen, H., Zhang, F., Cui, Y., Zhou, Z., (2022), The potential of urban family vertical farming: A pilot study of Shanghai. *Sustainable Production and Consumption*, 34, 586–599. https://doi.org/10.1016/j.spc.2022.10.011.

Shukla, K., Mishra, R., Sarkar, P., (2021), Understanding soilless engineered soil as a sustainable growing material for food production in a green roof. Materials Today: Proceedings, 43, 3054–3060. https://doi.org/10.1016/j.matpr.2021.01.397.

Singapura Stories, (2023), https://singapurastories.com/kampungcompound-houses/kampungcampongcompound/. Accessed 10.01.2023.

Song, S., Cheong, J.C., Lee, J.S.H., Tan, J.K.N., Chiam, Z., Arora, S., Png, K.J.Q., Seow, J.W.C., Leong, F.W.S., Palliwal, A., Biljecki, F., Tablada, A., Tan, H.T.W., (2022), Home gardening in Singapore: A feasibility study on the utilization of the vertical space of retrofitted high-rise public housing apartment buildings to increase urban vegetable self-sufficiency. Urban Forestry & Urban Greening, 78, 127755. https://doi.org/10.1016/j.ufug.2022.127755.

Song, X.P., Tan, H.T.W., Tan, P.Y., (2018), Assessment of light adequacy for vertical farming in a tropical city. Urban Forestry & Urban Greening, 29, 49–57. https://doi.org/10.1016/j.ufug.2017.11.004.

Sutton, R., (2014), Aesthetics for green roofs and green walls. Landscape Architecture Program: Faculty Scholarly and Creative Activity, 19. https://digitalcommons.unl.edu/arch_land_facultyschol/19.

Tablada, A., (2015), Impact of urban form on sunlight availability for urban farming in Asian cities at different latitudes. *9th International Conference of Urban Climate*, Toulouse. Available at: http://www.meteo.fr/cic/meetings/2015/ICUC9/LongAbstracts/ccma3-4-8831695_a.pdf.

Tablada, A., Kosorić, V., (2022), Vertical farming on facades: Transforming building skins for urban food security, In E. Gasparri, A. Brambilla, G. Lobaccaro, F. Goia, A. Andaloro, A. Sangiorgio (Eds.), *Rethinking Building Skins*, Woodhead Publishing, 285–311, ISBN 9780128224779. https://doi.org/10.1016/B978-0-12-822477-9.00015-2.

Tablada, A., Kosorić, V., Huajing, H., Chaplin, I., Lau, S.K., Yuan, C., Lau, S.S.Y., (2018), Design optimization of productive façades: Integrating photovoltaic and farming systems at the tropical technologies laboratory. *Sustainability*, 10(10), 3762. https://doi.org/10.3390/su10103762.

Tablada, A., Kosorić, V., Huajing, H., Lau, S.S.Y., Shabunko, V., (2020), Architectural quality of the productive façades integrating photovoltaic and vertical farming systems: Survey among experts in Singapore. *Frontiers of Architectural Research*. 9(2), 301–318. https://doi.org/10.1016/j.foar.2019.12.005.

Tablada, A., Zhao, X., (2016), Sunlight availability and potential food and energy self-sufficiency in tropical generic residential districts. *Solar Energy*, 139, 757–769. https://doi.org/10.1016/j.solener.2016.10.041.

Tan, P., Chiang, K., Chan, D., Wong, N., Chen, Y., Tan, A., (2009), *Vertical greenery for the tropics*. National Parks Board, National Parks Board Headquarters, Singapore.

Thomaier, S., Specht, K., Henckel, D., Dierich, A., Siebert, R., Freisinger, U., Sawicka, M., (2014), Farming in and on urban buildings: Present practice and specific novelties of Zero-Acreage Farming (ZFarming). Renewable Agriculture and Food Systems, 30, 43–54. https://doi.org/10.1017/S1742170514000143.

Tong, Z., Whitlow, T., Landers, A., Flanner, B., (2016), A case study of air quality above an urban roof top vegetable farm. *Environmental Pollution*, 208, 256–260. https://dx.doi.org/10.1016/j.envpol.2015.07.006.

van den Bergh, J.C., (2023), Contribution of global cities to climate change mitigation overrated. In Villamayor-Tomas, S., Muradian, R., (Eds.), *The Barcelona School of Ecological Economics and Political Ecology: A Companion in Honour of Joan Martinez-Alier*, Springer International Publishing, Cham, 335–346.

WOHA, (2023), https://woha.net/project/kampung-admiralty/. Accessed 07.01.2023.

Wong, N.H., Tan, A.Y.K., Tan, P.Y., Wong, N.C., (2009), Energy simulation of vertical greenery systems. *Energy and Buildings*, 41(12), 1401–1408. https://doi.org/10.1016/j.enbuild.2009.08.010.

Wong, N.H., Tan, A.Y.K., Tan, P., Sia, A., Wong, N.C., (2010a). Perception studies of vertical greenery systems in Singapore. Journal of Urban Planning and Development, 136(4), 330–338. https://doi.org/10.1061/(ASCE)UP.1943-5444.0000034.

Wong, N.H., Tan, A.Y.K., Tan, P.Y., Chiang, K., Wong, N.C., (2010b). Acoustics evaluation of vertical greenery systems for building walls. *Building and Environment*, 45(2), 411–420. https://doi.org/10.1016/j.buildenv.2009.06.017.

Wong, N.H., Tan, A.Y.K., Chen, Y., Sekar, K., Tan, P.Y., Chan, D., Chaing, K., Wong, N.C., (2010c), Thermal evaluation of vertical greenery systems for building walls. *Building and Environment*, 45(3), 663–672. https://doi.org/10.1016/j.buildenv.2009.08.005.

World Landscape Architecture (WLA), (2023), https://worldlandscapearchitect.com/thammasat-university-the-largest-urban-rooftop-farm-in-asia/. Accessed 17.01.2023.

Yuan, G.N., Marquez, G.P.B., Deng, H., Lu, A., Fabella, M., Salonga, R.B., Ashardiono, F., Cartagena, J.A., (2022), A review on urban agriculture: Technology, socio-economy, and policy. *Heliyon* 8, e11583.

Zambrano-Prado, P., Orsini, F., Rieradevall, J., Josa, A., Gabarrell, X., (2021a), Potential key factors, policies, and barriers for rooftop agriculture in EU cities: Barcelona, Berlin, Bologna, and Paris. Frontiers in Sustainable Food Systems, 5. https://doi.org/10.3389/fsufs.2021.733040.

Zambrano-Prado, P., Pons-Gumí, D., Toboso-Chavero, S., Parada, F., Josa, A., Gabarrell, X., Rieradevall, J., (2021b), Perceptions on barriers and opportunities for integrating urban agri-green roofs: A European Mediterranean compact city case, *Cities*. 114, 103196. https://doi.org/10.1016/j.cities.2021.103196.

Zhang, Y., Zhang, Y., Li, Z.A., (2022), Novel productive double skin façades for residential buildings: Concept, design and daylighting performance investigation. *Building and Environment*, 212, 108817. https://doi.org/10.1016/j.buildenv.2022.108817.

5 Social Acceptance of Urban Farming in and on Buildings

Vesna Kosorić

INTRODUCTION

In the context of challenging and deeply needed transitions related to the urban environment in the face of increasing urban population and climate change (Grebitus et al., 2017), the construction industry has resorted to a practice that has existed since ancient times across all geographies and communities – biological production in a spatial context (Lohrberg et al., 2016; Sroka et al., 2021). Playing many different roles for communities in urban areas, from meeting subsistence needs to large-scale commercial production (Yuan et al., 2022, urban agriculture (UA) is gaining more and more importance, attention, and justification. Lending itself to numerous definitions, interpretations, and classifications (Mougeot, 2006; Veenhuizen, 2006), today, it comprises various systems in cities using synergies between buildings and agriculture (Tablada and Kosorić, 2022).

In this chapter, farming in and on buildings refers to all types of soil-based and soilless farming, i.e. the growing of leafy vegetables, herbs, and other plant species on and in buildings. Therefore, it encompasses a wider range of systems than initially envisaged by building-integrated agriculture definitions of several authors (Marzouk et al., 2022; Gould and Caplow, 2012) and can be understood as a subtype of urban farming, representing a multifunctional land-use, combining different uses and functions within one building that Specht et al. (2013) named "zero-acreage farming" (ZFarming). Therefore, this chapter refers to farming on a building envelope – various types of rooftop urban agriculture (RUA) (rooftop greenhouses, rooftop open-air gardens, container farms) and various types of farming on façades (productive façades – PFs) (Tablada et al., 2018), vertically integrated greenhouses (VIGs) (Adams and Caplow, 2012) – and inside buildings, including various indoor farming types without the direct use of solar light (controlled-environment agriculture (CEA) systems, vertical farming (VF), container farms, in-store farms, appliance farms (Butturini and Marcelis, 2020)). Thus, it encompasses all technological farming types from the simplest Do-It-Yourself to the most high-tech systems.

Moreover, seen both as built structures and as building technology including various synergies with and complements to existing and newly designed buildings (Tablada and Kosorić, 2022), and also involving social activity, farming on and in buildings is a multi-dimensional (Rogus and Dimitri, 2015) and multifunctional (McClintock, 2010) concept, closely connected with sustainable development and sustainability and affecting all its three pillars: economy, society, and environment. Owing to its social and cultural impacts, benefits, and potentials, it is important to understand it as a concept intensively intertwined with social sustainability across a range of categories such as social equity and justice, social capital, social cohesion, social exclusion, environmental justice, quality of life, and urban liveability (Shirazi and Keivani, 2019). This is all supported by a large set of social activities and benefits involving volunteer and social work, leisure services, children and youth educational programs, job training, etc., encouraging the creation of powerful community ties and giving urbanites a chance to reconnect with the source of their food, resulting in the sense of belonging and feeling of accomplishment (Proksch, 2011; Eigenbrod and Gruda, 2015; Benis and Ferrão, 2018).

Surely, due to the symbiotic relationship with technology, in various settings, the social aspects are difficult to quantify (Sommerville et al., 2014), and benefits are most often measured by

DOI: 10.1201/b23309-5

modeling potential outcomes assuming the use of specific technology under different developmental, human, and climatic conditions (Rizal et al., 2018). However, despite the numerous benefits, which apart from food include a whole range of non-food related products and services, the UA's road to success is paved with numerous obstacles, such as legal barriers, high costs, the lack of space, conflicts with other urban functions, and health risks related to food produced on urban farms (van Tuijl et al., 2018).

As is usually the case with innovative technologies, technical infeasibility prevents many solutions from progressing beyond the prototype level (Specht et al., 2016a), and this also holds true for food production. UA, therefore, relies heavily on models (Banerjee and Adenaeuer, 2014; Al-Chalabi, 2015; Forchino et al., 2017), where the main focus is on life cycle analysis, production optimization, and technical feasibility, and these types of analyses are typically conducted in studies on aquaponics, rooftop greenhouses, and vertical and indoor farming (Benis et al., 2018; Cohen et al., 2018; Al-Chalabi, 2015; Sanyé-Mengual et al., 2015). However, whether new technologies will take root does not depend only on their technical feasibility and cost-effectiveness. Equally important is the social aspect, i.e. whether people accept them and want to integrate them into their living and working environment (Specht et al., 2019). Common social obstacles are that people in the city have weaned themselves from food production or have traditional ideas about agriculture that are unfeasible in an urban environment. Most often, the general public and decision-makers do not have sufficient information about new approaches in UA (Specht et al., 2015, 2016a; Sanyé-Mengual et al., 2015). Nothing designed and built, including farming systems in buildings, can be considered sustainable in the complete sense if they are not acceptable to people as pleasant places to live, work, meet, play, and interact. Economic and environmental assessments do not guarantee that people's needs and expectations will be adequately met (Kalantari et al., 2018). The downsides of technological progress have long taught us that technical and societal advances must go hand in hand to ensure sustainability (Schweizer-Ries, 2008; van Rijnsoever and Farla, 2014). Technological advancement must not be separated from people as it deeply impacts their lives (Kalantari et al., 2018). New technologies, including the ones in UA, call for a socio-technical system approach in which technological progress and people are seen as mutually impacting each other and contributing to symbiotic growth (Shamshiri et al., 2018). According to Germer et al. (2011), only such an approach will lead to global food safety and meeting the international sustainability criteria of environmental compatibility and public acceptability (Germer et al., 2011).

Social acceptance of projects and policies related to farming on and in buildings (Kosorić et al., 2019), relying on positive decisions and positive perceptions of various decision-makers and users, is the key prerequisite not only for successful planning, designing, constructing, and utilizing urban and building-integrated farming systems but also for fully exploiting their potential. Importantly, the acceptance of potential users should be carefully gauged and tested from very early planning stages since insights into potential social acceptance scenarios could make planning and design teams aware of things crucial for successful development. More delicate and even critical in terms of social acceptance may be more innovative forms of farming systems with connotations of intensive or high-tech agriculture, not consistent with the conventional picture of horticultural production (Benis and Ferrão, 2018; Butturini and Marcelis, 2020; Sanyé-Mengual et al., 2016; Specht et al., 2016b, 2016c). With its dynamics and complexity, social acceptance should be timely and appropriately considered in developing urban integrated sustainable concepts since negative responses could adversely affect the support for sustainable living environments (Kosorić et al., 2019).

This chapter focuses on the social acceptance of farming in and on buildings and has several aims. First, it taps into a body of literature to cast light on the concept of farming in and on buildings through various relevant domains and provide a holistic understanding of social acceptance. Mutually interrelated determining factors behind these two concepts are systematically clarified, suggesting possible links between farming facilities, food products, users, and social acceptance. The second aim is to give a summary overview of the latest studies, practices, and built projects closely related to social acceptance of various systems including all the main forms of farming systems on and in buildings: rooftop UA, façade farming, and indoor farming conducted in diverse latitudes and employing different technological solutions ranging from traditional open-air rooftop gardens, through technologically advanced VF in a controlled environment using, for example, hydroponics, to a sophisticated multifunctional solution of PFs (Tablada et al., 2018). Various ways of measuring, testing, and quantifying user attitudes are considered, including online surveys, a mixed-method approach comprising a survey with consumers and interviews with experts, a deconstructed theory of planned behavior model, and a novel methodology of text highlighting (Ares et al., 2021).

In reviewing these studies, the main findings related to consumers' and other stakeholders' inclination toward farming on and in buildings, attitudes, perceived benefits, and values, as well as risks and problems related to the farming systems, are highlighted. This chapter emphasizes the importance of the built projects as a tangible creative gift to users and society and a crucial evidence of farming systems' feasibility, values, benefits, and message to the society and young generations. It presents socially valuable and inspiring built farming systems integrated into NY public schools. Since the success of the concept of farming in and on buildings is closely related to social sustainability and acceptance, the third aim of this chapter is to give a rough overview, without going into detail, of hindering factors and strategies to address them, paving a path for the widespread adoption of building implemented agriculture systems and facilitating their social acceptance across the main relevant categories. The final aim is to emphasize the key factors for success, including governmental support starting as early as the policy-making stage and social acceptance underpinned by sensitivity to user needs from the design stage onwards, all of which should help to exploit rich architectural potentials of the concept of farming in and on buildings and successfully widespread it.

SOCIAL ACCEPTANCE OF FARMING IN AND ON BUILDINGS AND THE DETERMINING FACTORS

The key issue underlying the sustainability of anthropogenic systems (Takahashi and Sato, 2015) is social (sometimes called community or public) acceptance of urban farming on/in buildings. However, social acceptance is a multi-dimensional concept that is difficult to grasp using a single definition (González et al., 2016). It is a very complex and dynamic category including numerous mutually interrelated factors, both objective, such as a particular country and characteristics of a farming project, institutional environment and food prices, and subjective ones, such as the perception of risks and benefits of urban farming, socio-economic characteristics of users, their trust and beliefs. Figure 5.1 illustrates the factors determining the acceptance of farming in and on buildings: (1) stakeholder groups and domains, (2) socio-demographic characteristics, (3)socio-psychological characteristics, (4) characteristics of farming system/facility, (5) food product quality, (6) contextual factors, (7) perceived values and benefits and fulfillment of users' needs and expectations, and (8) perceived risks and problems. As can be seen, these factors are mutually related and interdependent, which calls for their simultaneous assessment.

According to Wüstenhagen et al. (2007), social acceptance is reflected in three interdependent main dimensions: (1) socio-political acceptance by the public, key stakeholders, and policymakers; (2) market acceptance by consumers and investors; and (3) acceptance by the community linked with procedural justice, distributional justice, and trust. These three aspects emphasize different dimensions of social acceptance from the perspective of different stakeholders (block 1, Figure 5.1). Some stakeholders involved are not directly connected with UA but strongly influence its development, such as policymakers, public administrators, and urban planners. For example, in market acceptance, the emphasis is on consumers and investors, i.e. financial supporters without whom no projects could exist (Kalantari et al., 2018). The key factor for successful urban farming project implementation is the fact that various stakeholder groups perceive and influence UA development differently, coupled with the various degrees of compatibility of UA with urban stakeholders' attitudes. This invites a holistic approach to bridge the potential gaps, prevent potential conflicts between various users, and ensure inequitable access to UA. Additionally, since UA is a context-related and multi-sectorial activity, its effective management requires a multi-stakeholder approach to achieve high engagement and participation (Cabannes and Marocchino, 2018). Di Fiore et al. (2021) have proposed a holistic multi-stakeholder framework to support UA harmonization with the contextual factors (Di Fiore et al., 2021).

An important group of variables includes stakeholders' socio-demographic characteristics (block 2, Figure 5.1). The significance of the socio-economic characteristics of residents, including their age, sex, education, and financial situation, with respect to social acceptance of urban farming, is acknowledged in literature (Wachenheim and Rathge, 2000; Malazewska and Was 2015). Furthermore, according to Sroka et al. (2021), variables showing the social distance to agriculture are understood as the intensity of contact between residents and farmers, relations between them, the sharing of certain values (e.g. love of growing plants), and origin (agriculture is

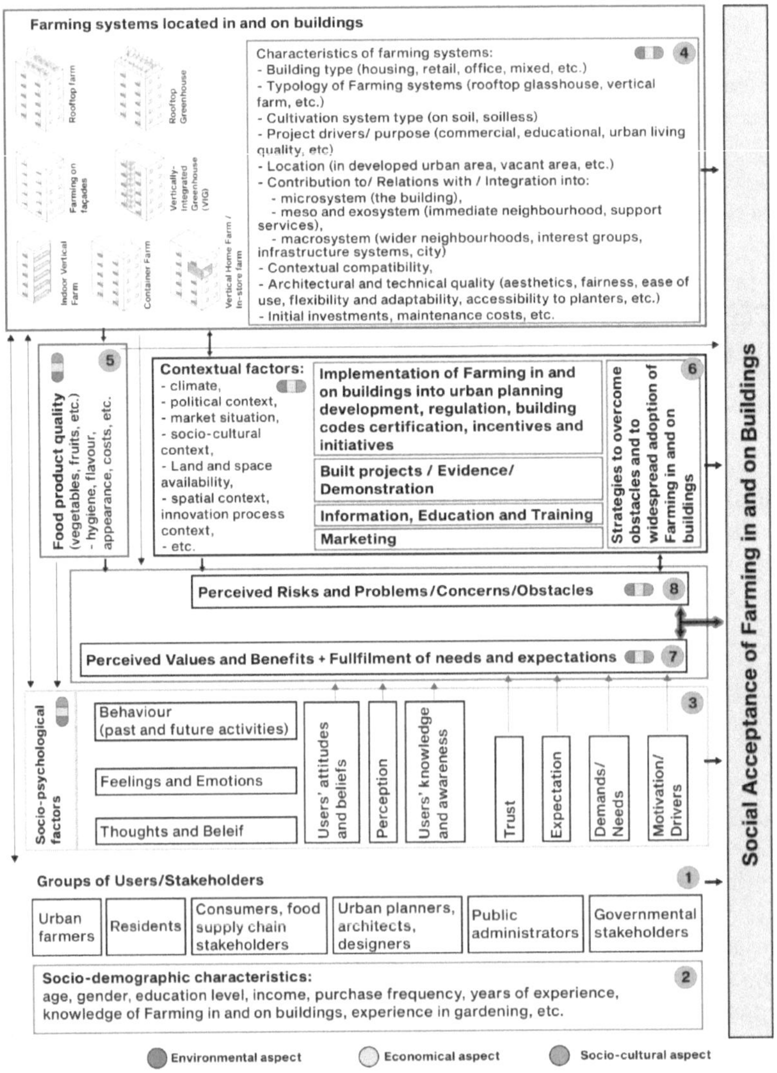

FIGURE 5.1 The holistic conceptual framework for analyzing social acceptance of farming systems located in and on buildings. (Image Credit: Vesna Kosorić.)

perceived differently by "a born city dweller" than by a person from the countryside), as well as the knowledge about the agriculture and awareness of the agricultural context. These are some of the main types of factors of importance in the acceptance of urban farming. Accordingly, better knowledge of agriculture and a greater awareness of the agricultural context contribute to increased acceptance of UA (Frick et al., 1995; Wachenheim and Rathge, 2000; Carr et al., 2014; Geneletti et al., 2018).

Socio-psychological characteristics of users (block 3, Figure 5.1) play an important role in the social acceptance of farming systems implemented in and on buildings. Knowledge awareness (Huijts et al., 2012), – the level of recognition of certain technology and how it functions (Yang et al., 2016) – and users' attitudes, perceptions, values, and beliefs are major covert factors in social acceptability evaluations (Hemström et al., 2014), largely affecting public acceptance. For example, the empirical results of Grebitus et al. (2017) demonstrated the importance of psychological and personal factors underlying consumer intention to participate in UA. With the increase in subjective knowledge, users develop a more affirmative attitude and become more prone to buy and grow produce at urban farms (Grebitus et al., 2017). Furthermore, according to Wolsink (2012), Hall et al. (2013), and Hanger et al. (2016), trust is a key factor and predictor of the effect on acceptance and later development. Therefore, to ensure acceptance in the development of farming projects located in and on buildings, users' expectations, needs, demands, and motivations should be taken into account, taking care that the developed farming systems and the food products meet them.

Various environmental, social, and economic characteristics of farming systems and facilities located in and on buildings (block 4, Figure 5.1) including (1) farming systems typology (rooftop farms, VF, etc.), (2) building type (housing, office, retail, etc.), (3) cultivation system type (on soil, soilless), (4) project drivers, i.e. purpose (commercial, educational, etc.), (5) location, (6) contribution to microsystem (the building), meso- and exosystem (immediate neighborhood, support services) and macrosystem (wider neighborhoods, city), (7) contextual compatibility, (8) architectural and technical quality (aesthetics, fairness, ease of use, flexibility and adaptability, accessibility to planters, etc.), and (9) costs play an important role in shaping community acceptance. For example, differently driven farming projects located in and on buildings are most likely to be differently accepted, as confirmed by the study of Specht et al. (2016c) conducted with potential consumers in Berlin, suggesting that multifunctional UA projects shaped around a mix of commercial, environmental, and social goals generally win the highest social acceptance, while those that were purely commercial and technologically intensive frequently experience rejection (Specht et al., 2016c). Social acceptability and perception are, of course, dependent on local and cultural factors. For example, in their study on consumer attitudes to VF in China, Singapore, the UK, and the USA, Ares et al. (2021) found that attitudes toward VF were largely positive in the four countries. Thus, indoor cultivation with artificial lighting under fully controlled conditions, plants growing in nutrient-rich substrates instead of soil, and being vertically stacked for optimal space – were generally perceived more positively than negatively by the participants (Ares et al., 2021).

Food product quality (block 5, Figure 5.1) is important for public acceptance of farming systems. Research studies should focus on food product quality since it is a common finding that consumers are concerned about the safety of food produced in cities (Specht et al., 2016a). In the study of Specht et al. (2019), the experts report the lack of consumer trust in product quality related to new approaches in urban food production since urban settings are generally viewed as environmentally unsafe and polluted. There are, for example, studies showing that rooftop crops receive a lower pollution load, including heavy metals, than vegetables grown in the ground near roads in New York (Tong et al., 2016) and that rooftop-grown vegetables in Guangzhou are competitive in both quality and cost in comparison to high-end vegetables sold on the market (Liu et al., 2016), more focus on the research in the domain of health risks and food safety would be very beneficial for acceptance of farming in and on buildings. So Buscaroli et al. (2021) developed a food safety assessment framework that supports decision-making and empowers stakeholders from an initiative to the policy-making level. In promoting novel food production technologies, informing users is particularly important, so they can make an informed decision and favor a new product over an already familiar one (Frewer, 1998; Frewer et al., 2011). For example, in their study related to aquaponics products in Europe, they found that few people (17%) were willing to pay higher prices for food produced in aquaponics, Miličić et al. (2017), urged to invest in educating consumers about aquaponics to raise their awareness about this new technology, as well as to ensure the organic certification of aquaponics products.

The success of farming in and on buildings is inseparable from the context and contextual factors (block 6, Figure 5.1). According to Li et al. (2016), emotional and cultural relations with a place are significant factors in defining public acceptability. Moreover, "place attachment" focuses on individual feelings and experiences (Rezaee and Kalantari, 2019). According to Kalantari and Akhyani (2021), social acceptance is correlated with local conditions and cultural values. Implementation of the concept of farming in and on buildings into local urban planning development, regulations, and certification systems, together with information, education, and training, are the key factors for spreading these farming practices. In addition to all mentioned, marketing and built projects providing evidence have an important role in shaping social acceptance of farming in and on buildings.

Finally, summarized and interrelated with all previously mentioned categories and factors determining farming projects' quality and their acceptance, the key factors affecting and directly determining the level of social acceptance are user-perceived benefits and values (Bearth and Siegrist 2016; Khorsand et al., 2015; Schumann 2015; Specht et al., 2016b, Wang and Li 2016; Yazdanpanah et al. 2015), the fulfillment of their needs and expectations (block 7, Figure 5.1) and user-perceived risks and problems (Bearth and Siegrist 2016; Schumann 2015; Specht et al., 2016a, Wang and Li 2016; Yazdanpanah et al., 2015), (block 8, Figure 5.1). Although it is evident that urban farming brings multiple benefits for all pillars of sustainable development (Brinkley, 2012; Specht et al., 2016a; Sroka et al., 2021, Van Tuijl et al., 2018; Ayambire et al., 2019; Azunre et al., 2019) – environmental, economic and socio-cultural (engaged and cohesive communities, health and wellbeing, economic opportunities, and education (Ilieva et al., 2022)), the achieved advantages and perceived benefits are always specific for each case. For example, according to Parkes et al. (2022), the advantages of specific types of CEA depend on the choice of ag-tech (environment type and growing system), the location, the product, and the degree of integration with built structures or existing systems. Together with positive perceptions related to benefits and values, negative perceptions about risks and problems including health risks, risks related to soilless growing techniques, departure from traditional conceptions of agriculture, etc., also shape social acceptance, and both should be carefully treated starting from early design stages to project finalization and exploitation. In addition to all mentioned, to achieve full progress in the domain of social acceptance, besides considering all the defined individual determining factors and categories, longitudinal studies should be encouraged, including a holistic approach considering also inter-category connections and the mutual causes and consequences of social acceptance.

SOCIAL ACCEPTANCE OF FARMING PRACTICES IN AND ON BUILDINGS

Although still limited and all the more precious, farming projects located in and on buildings and related studies are gaining ground, both in the Western world and in Asian megacities. The present sub-chapter covers research studies conducted on the topic in focus – the social acceptability of farming systems implemented in and on buildings.

Social Acceptance of Rooftop Farming

Rooftop Farming in European Cities: Barcelona, Berlin, Bologna, and Paris

In European cities, the lack of space for agricultural use on one side and the multiple benefits of UA on the other led to new ways of urban farming, including RUA. In Barcelona, the 2021 World Capital of Sustainable Food (Zambrano-Prado et al., 2021a), different RUA projects, such as a pilot rooftop greenhouse, were launched in the ICTA-ICP building of the Universitat Autonòma de Barcelona in 2014 (Fertilecity, 2023), followed by various programs for integrated production of vegetables (Zambrano-Prado et al., 2021a) and dedicated to large stakeholder groups, including socially vulnerable groups in the city (Barcelona Laboratory for Urban Environmental Justice and Sustainability, 2023). Furthermore, Barcelona's Climate Plan 2018–2030 considers RUA implementation to mitigate climate change and improve the quality of life in the city (Zambrano-Prado et al., 2021a). In Berlin, commercial urban farming enterprises have developed different prototypes and technologies for food production on buildings (Specht et al., 2016b), and RUA projects are already running, including, for example, two open-air rooftop gardens and one rooftop greenhouse located on the Humboldt University building (Tao et al., 2020). In Bologna, one of the first cities in Italy to adopt a local plan for adaptation to climate change, different projects have been developed and realized including three temporary pilot community rooftop gardens installed on the 10th floor of social housing buildings (Orsini et al., 2014) and an educational rooftop greenhouse at the multifunctional space SALUS (Pennisi et al., 2020). In Paris, according to the Paris Climate Action Plan, the city promotes RUA on municipal buildings (City of Paris, 2018), and accordingly, the Parisculteurs program was launched in 2016 for installing UA on buildings (Collé et al., 2018).

In Barcelona, stakeholder perceptions of urban rooftop farming have been studied by Sanyé-Mengualet al. (2016). Semi-structured interviews with 25 core stakeholders showed that UA is largely perceived as a social activity rather than a food production initiative in light of the traditional leisure and cultural purpose of Barcelona Gardens.

In Berlin and Barcelona, relying on qualitative expert interviews, Specht K and Sanyé-Mengual E have studied stakeholder perceived risks in rooftop agriculture. Five main categories were revealed: (1) risks associated with urban integration (e.g. conflicts with images of "agriculture"), (2) risks associated with the production system, (3) risks associated with food products (e.g. soilless growing techniques are "unnatural"), (4) environmental risks, and (5) economic risks (Specht and Sanyé-Mengual 2017). The authors found that the perceived risks are primarily related to the lack of knowledge and awareness and nonintegrative policy-making. An interesting specificity of Berlin was that the stakeholders were afraid of a "copy-paste" process from other cities and insisted on locally created mechanisms (Sanyé-Mengual et al., 2016; Zambrano-Prado et al., 2021a).

In the mentioned four cities, through a workshop and a survey of stakeholders involved in RUA, Zambrano-Prado, Orsini, Rieradevall, Josa, and Gabarrell have studied potential key factors, policies, and barriers associated with the integration of RUA. They found that education, environmental research, technological innovation, food production, and social factors play an important role in implementing RUA. Further, productive spaces, cultural values, social cohesion, social rural–urban links, and the high cost of urban land are the factors that strongly speak in favor of RUA (Zambrano-Prado et al., 2021a). According to their study, the cost of water and pollution are the major contextual factors that constrain RUA, and policies related to food trade and urban planning are those that mainly limit RUA development. The major architectural and technical barriers are related to the limits on building heights, historical buildings, the lack of specific building codes, building design, and roof accessibility. Regarding social barriers, exclusive access to rooftop food and projects and lack of interest in the society were identified as "rarely" present; resistance of residents as a barrier that "sometimes" appears. The top two economic constraints were infrastructure costs and urban policies hindering the sale of RUA products. The four cities differed in the respondents' perceived relevance of economic and policy barriers and factors related to pollution.

Social Value of Urban Rooftop Farming in Hong Kong

Hong Kong, an extreme example of a high-rise, high-density urban settlement in which urban lots are extremely costly and in high demand, has spontaneously emerged over 60 rooftop farms on industrial and institutional buildings (Hui, 2011).

Within HKU's broad-based "edible roof" initiative, Wang and Pryor M examined eight such farms to determine the nature and scale of social values that urban rooftop farms could generate

(Wang and Pryor, 2019). Their researchers surveyed user perspectives on intangible social values promoted by UA, categorized them under six factors: health, education, community recreation, urban improvement, social empowerment, and social group integration, and tried to quantify them. Most respondents (77%) perceived social values to be the most important benefit of urban rooftop farming, followed by environmental (58%) and economic benefits (10%) (Wang and Pryor, 2019). A typical rooftop farm user is female and middle-aged (30–50) and earn middle to high income. The benefit of personal socialization was valued most strongly among the six factors, followed by health and education, while less importance was attached to planning social welfare, social group integration, community recreation, and social empowerment (Wang and Pryor, 2019).

The appreciation of social values derived from urban rooftop farming made many respondents willing to participate financially. Willingness to pay increased with education, income level, and prior experience. The study of Wang and Pryor suggests that against the background of galloping urban densification, UA re-interpreted as an activity imbued with complex social values and responding to diverse user interests, can become a public good for cultural exchange and social coherence. According to the authors, this changing perception should be used to attract greater stakeholder support and advocate for legislative changes toward the integration of UA with urban planning and building control processes (Wang and Pryor, 2019).

Social Acceptance of Façade Farming – PFs in Singapore

PFs, a concept developed by Tablada et al. (2018), integrate photovoltaic (PV) modules and VF planters. Originating from Singapore – a country with at the same time exceptional potential (solar energy, robust real estate sector, strong economy fundamentals, government commitment (Kosorić et al., 2021)) and facing challenges such as high dependence on imported energy and food, scarcity of land, planned reduction of greenhouse gases (GHGs), and increase of high-rise greenery coverage – PFs are designed to turn building envelopes into producers of energy and food (Tablada and Kosorić, 2022). Testing social acceptance was a part of the comprehensive PF design development (see Figure 5.2). After developing and finding optimal design solutions including both the balcony and window façades to be installed, monitored, and assessed at the Tropical Technologies Laboratory (T2 Lab) at the campus of the National University of Singapore (NUS), two surveys were performed by the research team: one with potential users and the second one with experts.

An in-person survey in Singapore including 391 English- and Mandarin-speaking tenants in public housing examined their attitudes toward PF, covering both aesthetic and maintenance concerns and respondents' willingness to participate in VF gardening. The attitudes were generally positive – 80% of respondents agreed or strongly agreed that VF and PV modules positively impact residents (only 3% disagreed, while nobody strongly disagreed). However, 25% of the respondents were uninterested in home gardening in high-rise apartment buildings, mostly younger adults, while interest increased with age (Kosorić et al., 2019) (the same age-related preferences are reported in Møller 2005).

The second survey, an online questionnaire, included 97 professionals and experts in horticulture, agronomy, PV systems, and architecture (Tablada et al., 2020). The results enabled more detailed feedback regarding aesthetic, formal, and functional aspects of façade components, including PV modules, planters, and safety grills, as well as the operation and accessibility of farming systems. For example, once asked about the ease of operation and functionality, experts prefer the designs where users access bottom planters through an opening on balcony façades or directly on window façades, which is consistent with the preference of residents in the first survey. The findings from the two surveys helped formulate general design recommendations for PF to enhance productivity, aesthetics, and other architectural qualities (Tablada et al., 2020). The main takeaway of these studies is that incorporating user feedback is a crucial step toward the future scalability of this innovative concept. As for productivity enhancement, Song et al. (2022) systematically estimated the food production potential of a high-rise public housing apartment building in Singapore,

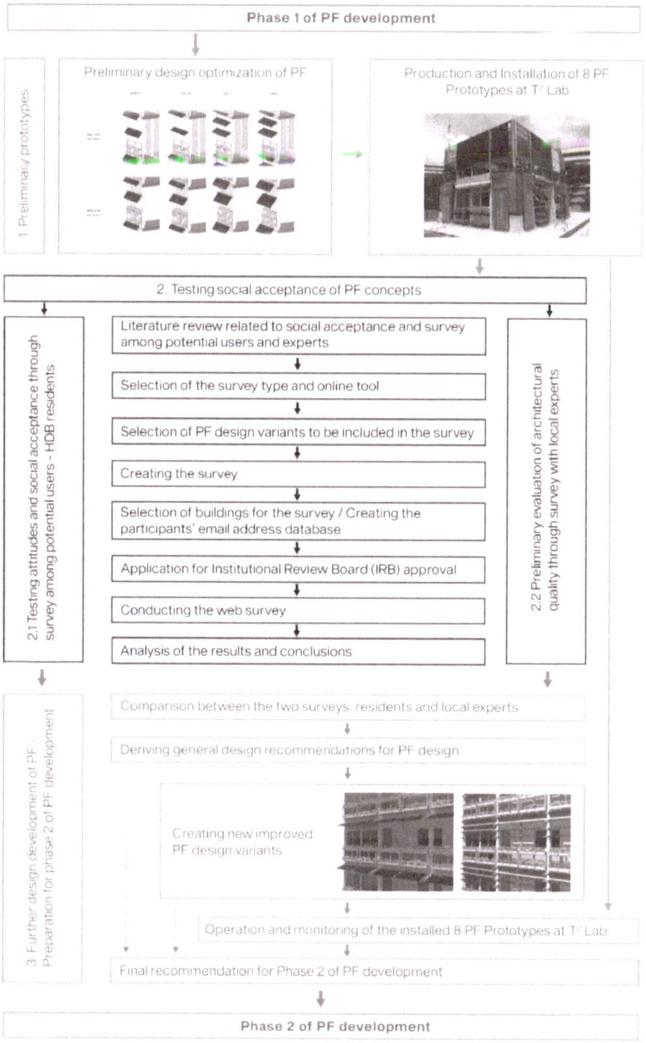

FIGURE 5.2 Graphical presentation of the methodology applied in the development of productive façades installed at the tropical technologies laboratory (T2 Lab), at the NUS campus. (Adapted from Kosorić et al., 2019; Tablada et al., 2020.)

including corridor, rooftop, and façade gardening using an experimental approach and showed that home gardening can potentially meet 46.4% of residents' need for vegetables. Building façade was seen as very promising but faced some downsides, such as poorer accessibility compared to home gardens (Song et al., 2022). It was concluded that the design of vertical space gardening systems for high-rise buildings needs to be improved, in parallel with raising awareness and encouraging residents to participate in home gardening.

Social Acceptance of Indoor Farming

Consumer Acceptance of Three VF Systems in Germany

The study of Jürkenbeck et al. (2019) provides insight into consumers' acceptance of three different VF systems with a fully controlled environment: (1) vertical home farm, fully controllable by the consumer via smartphone; (2) in-store vertical farm in grocery stores or restaurants, where consumers can watch the growth and put the product in their shopping carts; and (3) indoor vertical farm, the system without any direct consumer contact. Consistent with prior research (Vecchio and Annunziata, 2015; Hüttel et al. 2018; Grunert et al., 2014; Grunert 2011) and environmentalists' position in the debate on agricultural systems (Grunert, 2011), the study reveals that the perceived sustainability is the main driver of the acceptance of VF (Jürkenbeck et al., 2019). The indoor vertical farm was assessed as the most promising when it comes to sustainable food production and as the greatest contributor to regional food production. Therefore, the comparison of the three VF systems suggested that the consumers perceive the larger the system, the more value toward sustainability (Jürkenbeck et al., 2019).

Consumer Attitudes to VF in the USA, UK, Singapore, and China

By using a novel research methodology of text highlighting in online surveys (637–683 participants per country with matched gender and age group distributions), Ares et al. (2021) conducted a study on consumer attitudes toward VF in four countries (USA, UK, Singapore, and China). The text was about an indoor plant factory with artificial lighting, and the advantage of the text highlighting methodology was that it collected participants' attitudes about specific, targeted aspects of a particular food production system.

The results of the study showed 63% of participants (n=1,559) showed positive attitudes toward VF (Ares et al. 2021), consistent with the findings in the relevant literature (Grebitus et al., 2020; Jaeger et al. 2022; Jürkenbeck et al., 2019; Specht et al., 2019; Yano et al., 2021; Coyle & Ellison 2017; Broad et al., 2021). The characteristics of VF that aligned with the United Nations Sustainable Development Goals (Jensen, 2021) were identified as key drivers of positive attitudes (i.e. reduction of carbon emissions and securing access to food). For example, the researchers pointed out the connection between securing access to food as a selected key driver and the widespread concerns about food insecurity (Kirshenbaum and Buhler, 2018), and this topic gained prominence during the COVID-19 pandemic, at both individual and national levels (Crush and Si, 2020; Niles et al., 2020). On the contrary, negative attitudes toward VF were fueled by galloping energy prices, given the high energy consumption in this type of farming. While participants' reactions to the text were generally positive, in every country, there was also a small group with a negative or neutral/ambivalent attitude (17% showed a neutral sentiment (n=411) and 20% showed a negative sentiment (n=499) (Ares et al., 2021)). Differences between an affirmative and disapproving group were greater than those across countries. As for country differences, China and Singapore were the most affirmative toward VF. There were also some interesting aspects, such as Chinese participants being the least negative about using robots in planting and harvesting.

Consumers and Stakeholders' Acceptance of VF in Shanghai, China

Zhou et al. (2022) conducted a study on the social, ecological, and economic dimensions of VF in the densely populated metropolitan area of Shanghai. The survey among potential consumers (713 persons) showed a high general acceptance of VF technology (84%). The experts (20 persons) raised more doubts about the economic dimension, which remains a controversial aspect, as pointed out by Specht et al. (2013) and Al-Kodmany (2018). In contrast, the social and ecological dimensions and the contextual framework of indoor farming were considered to be positive. The experts dwelt extensively on hindering factors about productivity when compared to conventional agriculture. Consumers' acceptance was reflected in a high share of quotes on promoting factors related to multi-functionality, social awareness, value, and food. Experts were concerned with China's current food safety issues, which could be a driving force for future indoor farming development. Generally, interviewees considered the restoration of consumer trust an essential mission. A high preference for governmental investment (59%) highlighted the importance of ensuring political support because, unlike in Western market mechanisms, the government in China has a major impact on promoting or hindering new concepts and technologies, and hence also on the development of farming projects in buildings (Zhou et al., 2022). Despite the controversial debate about indoor farming, the first two-story vertical farm was constructed in 2016 in Shanghai, demonstrating its great potential and market attractiveness.

Also in Shanghai, but focusing on a micro-VF system in residential buildings, Shao et al. (2022a) investigated the influence of public acceptance based on the decomposed theory of planned behavior (DTPB) DTPB models which use behavioral intention have–as of recently, become an important part of studies on user acceptance (Compeau et al., 1999). The key finding is that the general public's overall intention to use VF is generally above "neutral" but slightly below "positive". The most critical factors are the cost-effectiveness of VF and the consumption of time and money required for the planting process. The biggest motivation for the public to engage in micro-VF is the possible economic benefits, while the biggest obstacle is the general lack of knowledge about VF and unfamiliarity with the growing process and other expertise. In their further study, Shao et al. (2022b) calculated that urban family VF could at least increase the self-sufficiency rate of vegetables and fruits per household by 6.3–8.3% and vegetable self-sufficiency rate of the entire city by 1%.

Consumers' Attitudes toward VF in Germany and Singapore

Jaeger et al. (2023) conducted consumer research on VF by exploring how consumers associate different characteristics of VF with the technology's pros and cons and how these are linked to their personal values. The researchers revealed consumers' cognitive structures relating to VF using means-end chains suitable for representing the relationship between a product and a consumer and the laddering methodology (Kahle et al., 1986; Grunert and Grunert, 1995; Olson 1995) convenient for online surveys. The participants in Singapore (n=547) and Germany (n=537) read an informative text about VF and completed the questionnaires. Data analysis resulted in hierarchical value maps, a graphical representation of means-end structures aggregated across participants (Gengler et al., 1995), which were compared by country and groups of participants with different attitudes to VF. The survey uncovered both altruistic ("Feeding the world", "Protecting nature", "Caring for future generations") and self-centered ("Long and healthy life"; "Pleasure and enjoyment") underlying motives of VF acceptance, consistent with other studies which also identified these two groups of drivers for acceptance of innovative and sustainable food technologies and behaviors (Tobler et al., 2011; Biasini et al., 2021; Gonera et al., 2021). Some cross-linking features and benefits were found across both groups of drivers, such as "Out-of-season vegetables will be available all year round" to "There will be enough food for everyone", and energy use also emerged as a concern in both groups of motives. Regarding inter-country differences, German participants were more sensitive to energy consumption and less frequently connected vegetables from VF to pleasure and enjoyment and health and wellbeing (Jaeger et al., 2023).

Community Acceptance of VF in Kuala Lumpur, Malaysia

Kalantari and Akhyani (2021) chose Kuala Lumpur as their location for research on VF community acceptance. Based on interviews with experts as well as a questionnaire survey, the research identified the following eight factors affecting community acceptance of VF, namely a visionary type of VF, a multi-story structure (Al-Kodmany, 2018) in Kuala Lumpur: (1) Perceived Benefit (1.1 Landscape Greenery, 1.2 Environmental Benefit, 1.3 Economic Benefit, 1.4 Social Benefit), (2) Perceived Risk (2.1 Surrounding and Concern, 2.2 Economic Concern of VF Implementation, 2.3 Socio-Economical Concern), (3) Location (3.1 Ecocentric, 3.2 Apathetic, 3.3 Anthropocentric), (4) Demographic characteristics (4.1 Anywhere, 4.2 Developed Area, 4.3 Vacant Area), (5) Value and Belief, (6) Trust, (7) Fairness, and (8) Knowledge (Kalantari and Akhyani, 2021). The results of the exploratory factor analysis revealed an underlying structure of variables and factors different from what could be found in the related literature. This indicated that local conditions, culture, and emotional factors are highly correlated with social acceptance (Li et al., 2016), and, therefore, what is practiced in one environment may not work in other contexts (Brunsting et al., 2015; Jaeger et al., 2015). Given that relevant drivers in diverse political and cultural contexts cannot be assumed a priori (Hanger et al., 2016), public participation methods must be fostered (Aitken, 2010), and the results cross-checked via comparison with other geographical and cultural settings.

Consumers' Attitudes to VF in the UK

Similarly, as Ares et al. (2021) did in their study, Jaeger et al. (2022) used text highlighting to measure explicit attitudes in a case study on indoor VF with 837 UK consumers. The text highlighting responses was summarized using word clouds, frequency tables, and sentiment scores, enabling consumer segmentation (Hollywood et al., 2007; Shepherd and Raats, 1996) on attitudinal variables to reveal an overall positive attitude to VF among participants.

Real Case Study of Urban Indoor Farm in Högdalen, Sweden

One of the rare real case studies was an indoor farm in Högdalen, a southern suburb of Stockholm, located in a cellar under a shopping and community center. The project was jointly launched by commercial, civil society, and municipal actors. The aim was to utilize vacant urban space for food production and generate new jobs. The study focused on GHG emissions, while the second component was a qualitative study of social factors to assess the broader issues related to the sustainability of the farm, including (1) interviews with key actors, combined with (2) field visits and observations (Milestad et al., 2020). While the Högdalen farm did not contribute much to food security (due to its small size and a relatively insignificant food item – lettuce), it contributed to urban development – by an effort to create jobs (though only part-time) and increase the attractiveness of a suburban area through reactivating the dormant urban space and joining the forces of local commercial, public and civil society actors toward the common goal (Milestad et al., 2020). It should, however, be noted that at the heart of these achievements was the idea of sustainable urban food production, i.e. growing lettuce.

Built Projects Fostering Socio-cultural Values and, Therefore, Social Acceptance

Projects, buildings, and facilities successful in terms of function, aesthetics, and educational purpose enrich people and spaces and contribute to the social acceptance of farming practices in and on buildings. This sub-chapter presents one such example, while several other successful and inspiring built projects covering residential and institutional buildings are presented within the chapter "Productive building envelopes".

The three rooftop greenhouse classroom/science labs in New York City public schools: PS 333, The Manhattan School for Children (PS 333, 2023); PS 84 Josê de Diego in Williamsburg, Brooklyn (PS 84 Josê de Diego in Williamsburg, 2023); PS/IS 686, Brooklyn School of Inquiry (PS/IS 686, 2023)) (see Figure 5.3), which are part of the Greenhouse Project, have been done by Kiss+Cathcart – Gregory Kiss, Colin Cathcart, and their team (Kiss+Cathcart, 2022), well known for great dedication to productive architecture, in collaboration with New York Sun Works (New York Sun Works, 2023). These proved to be truly inspirational infrastructures, beneficial for the society.

Each greenhouse includes a VIG prototype and hydroponic classroom layout by the program provider, New York Sun Works. Two of these projects also include graphics integrated into the greenhouse walls, which are part of the curriculum: images of hydroponically grown plants and fish commonly used in aquaponics. Having been integrated into the school curricula (from kindergarten to 8th grade) and offering hands-on learning to students in a range of areas, from food growing to sustainable development, these rooftop greenhouse labs contributed to students' better test scores and instilled in them respect for the environment. The greenhouses intertwined nature with culture, also contributing to social studies.

Due to the success of the program, New York Sun Works developed the "classroom conversion" model, a cost-effective option to integrate hydroponic labs in most NYC public school buildings

FIGURE 5.3 Rooftop greenhouse classroom/science labs in New York City public schools. Top left and top right: The NY Sun Works Center at PS 333, the Manhattan School for Children; Center left and center right: The NY Sun Works Greenhouse at PS 84 Josê de Diego in Williamsburg, Brooklyn; Bottom left and bottom right: The NY Sun Works Greenhouse at PS/IS 686, Brooklyn School of Inquiry. (Image Credit: Kiss+Cathcart Architects, 2022.)

(see Figure 5.4). Students from kindergarten to 12th grade grow food from seed to harvest while meeting the mandated science standards and connecting with sustainability and climate concepts impacting the planet.

The Penzel Valier (2022) Swiss architectural office designed a vertical garden in the city in Zurich (2020–2024) (Figure 5.5). Multi-family housing is counterpoised to urban gardens, attempting to soften the border between private and public. Therefore, the urban vertical garden forms a zone of semi-public space that brings residents closer together, giving them the opportunity to meet and communicate with each other. The vertical garden effect was achieved with planters on the terraces according to the needs and affinities of tenants. The architects gave recommendations on how the greenery could be maintained, emphasizing the variety and natural irregularity that would give the façade a specific and unique appearance (Hochparterre, 2022; Penzel Valier, 2022).

FIGURE 5.4 Interior images of the NY Sun Works Hydroponic Farm Classrooms. (Image Credit: Top left and right: Ari Burling; Down left and right: Daphne Youree.)

STRATEGIES FOR WIDESPREAD ADOPTION OF FARMING SYSTEMS IN AND ON BUILDINGS AND FACILITATING THEIR SOCIAL ACCEPTANCE

The complexity of farming systems located in and on buildings, their integration into buildings and utilization, and the dynamics of the social acceptance dimension invite holistic and multi-level mechanisms and actions to overcome diverse obstacles from all eight (8) categories defining social acceptance (see Figure 5.1). Figure 5.6 schematically presents a holistic multi-level and multi-disciplinary mechanism framework consisting of six main categories: economic, regulatory/legal, professional, architectural/technological, market and social, psychological, and cultural obstacles and the strategies to address them. Each obstacle group lists barriers, concerns, and negative perceptions, while in each group of strategies, proper mechanisms, actions, guidelines, and recommendations are listed. Since the six mechanism categories of obstacles and strategies are mutually interdependent and interrelated, they should be simultaneously considered. Furthermore, due to distinctive correlations between environmental, social, and economic aspects of sustainability in UA practices (Sanyé-Mengual et al., 2019), all three sustainability pillars should be considered properly.

The multi-level mechanism is envisaged to include all stakeholders in decision-making through regulatory strategies (Sanyé-Mengual et al., 2019). Policy-making should be addressed from the building and farming standpoints (Appollini et al., 2021), covering a range of areas, such as food planning, water efficiency, energy efficiency, education, health, and biodiversity to guarantee food supply. Furthermore, policies should be adaptive, maximizing UA's potential to increase resiliency and sustainability and incentivize its organic integration in cities while equally serving social justice (Yuan et al., 2022). The importance of adjusting to the environment and stakeholders' needs and aspirations has also been emphasized by Gómez-Villarino and Ruiz-Garcia (2021).

The research project FEW-meter, building on the existing policies in France, Germany, Poland, the United Kingdom, and the United States, offers an UA roadmap until 2050 (Fox-Kämper et al.,

FIGURE 5.5 Part of the Façade/Section of a vertical garden in the city of Zurich (2020–2024). (Image Credit: Penzel Valier, 2022.)

2022). Policy recommendations are formulated under seven categories: (1) political frame conditions, (2) economic regime, (3) urban growth dynamics, (4) urban planning policies, (5) land-use policies, (6) climate change, and (7) technical trends. For example, regarding political frame conditions, UA integration is proposed as part of health care schemes, e.g. on dietary regime and mental health issues. As for urban growth dynamics, it is proposed to share the knowledge base about the existing UA initiatives, create and publish an open-access database of available land plots for UA, including indoor farming, provide information about the location and status of sites and facilities available to encourage the uptake of spaces for UA and develop new food hubs for producer–consumer interaction and alternative food networks (Fox-Kämper et al., 2022).

Planning and legislation must go hand in hand with applied research and innovation to roll out optimal solutions that ensure maximum resource utilization and sustainability (Appollini et al., 2021). Moreover, to optimize crop yields, expand product diversity, and maximize profitability while maintaining product quality and environmental sustainability (O'Sullivan et al., 2019), innovations and research support across a wide range of disciplines and demonstration/pilot projects are needed.

Having in mind that dominant barriers to the adoption of concepts of farming in and on buildings are related to awareness, apart from conducting research, it is also necessary to increase

FIGURE 5.6 Holistic and multi-level mechanism framework to overcome obstacles and help widespread adoption and social acceptance of farming systems located in and on buildings. (Vesna Kosorić.)

education, information, and knowledge dissemination about these farming systems' characteristics and benefits among different actors, particularly building owners and users (Zambrano-Prado et al., 2021b). Accordingly, targeted education programs for residents, investors, design teams, and city policymakers should be designed in cooperation with academics, research centers, or specialized companies involved in farming projects. Especially for new types of farming in buildings, consumer-oriented targeted marketing and communication strategies (Grebitus et al., 2017), as well as educational strategies, play a key role and should be launched through various channels, including traditional and digital media and platforms. Ecogradia (Ecogradia, 2023) is a good example of a sustainable architecture and urbanism platform, with a podcast and blog on green projects, products, and technologies. The strategies envisaged by the holistic framework (Figure 5.6) are numerous and also include various specific pieces of training for existing and future users, for example, in crop management skills (e.g. maintenance, bed preparation, organic practices) and communication skills as it is acknowledged in the study of Ochoa et al. (2019) who assessed training needs in community gardens.

Finally, built facilities, successful projects showing evidence of possibilities and the value of farming in and on buildings, and inspiring design teams, investors, users, etc., are all crucial for adopting systems. Therefore, the development of design guidelines and recommendations for planning and design of farming systems in and on buildings clearly invites an integrative user-oriented participatory process, utilization of multicriteria decision-making methods and concepts, taking into account contextual and cultural values, understanding farming technologies and their implementation into buildings while accounting for user needs, expectations, and preferences. Such approaches, maximizing contextual potentials and minimizing constraints, focused on multi-functionality, aesthetics, and users' present and future needs, are urgently needed. Multi-level spatial consideration taking into account micro- (building part, building), meso- and exo- (neighborhood), and macro level (city) is of vital importance. Therefore, long-term success in farming projects located in and on buildings requires multi-disciplinary teamwork, conscious simultaneous understanding of all eight categories of factors that determine social acceptance presented in Figure 5.1, an iterative design process with proper balancing of design goals (including social, educational and economic goals) and, most importantly, perseverance in searching for a truly optimal solution.

CONCLUSION

Farming systems located in and on buildings in which food production and – in the case of agrovoltaic systems and PFs – energy production facilities complement or even substitute the building envelope, spaces, or other building stock elements are a promising planning and design strategy that not only contributes to food and energy self-efficiency in buildings but also yields other environmental, economic and social benefits for the built environment and its inhabitants. These farming projects thus have the potential to serve as a cornerstone of positive urban transformations toward a more resilient, low-carbon, inclusive sustainable future. Civil society could play a key role in this, on condition that policymakers opt for the path of participatory urban governance and planning, working jointly with the private and civil sectors toward the successful development of UA (Sanyé-Mengual et al., 2019). A balanced multi-stakeholder involvement, including proper information dissemination, transparent knowledge sharing, and communication within planning and design, ensures an appropriate understanding of the characteristics of the concept of farming in and on buildings and potential long-term benefits and risks on multiple levels and is, therefore, key to successful development.

The importance of considering the social aspect and social acceptance of farming projects implemented in and on buildings is increasingly recognized, as it undoubtedly plays a crucial role in adopting and spreading various farming practices. Therefore, RUA, façade farming, or indoor farming projects should be implemented only after proper consideration of their social impact, i.e. timely and proper considering of the following key factors: (1) stakeholder groups and domains,

(2) their socio-demographic characteristics, (3) socio-psychological characteristics, (4) characteristics of farming system/facility, (5) food product quality, (6) contextual factors, (7) potential and user-perceived values and benefits and fulfillment of users' needs and expectations, and (8) potential and user-perceived risks and problems (Figure 5.1). A project's acceptance and, therefore, its success largely depends on whether or not these relevant factors have been addressed (Specht et al., 2016a). Further, timely recognition and proper dealing with project-related potential challenges and various environmental, technological, economic, and social obstacles, thus finding a compromise/balance between competing objectives – all in accordance with the changing societal demands and environmental conditions (Jastrzębska et al., 2022; Siebrecht, 2020), increase the chances of project success. Interestingly and sometimes surprisingly, cultural factors and barriers are often more deeply entrenched and profoundly affect personal beliefs and perceptions (Moser and Ekstrom, 2010), being even harder to overcome than material barriers (De Jalón et al., 2015; Giles et al., 2021). Due to the systems' close connection to consumers, farming operators need to properly specify the project's aims and philosophy, products, and technologies (Specht et al., 2016a) – all according and adaptive to users' present and future needs. Moreover, defining and creating the economic, social (social cohesion, education, health, well-being, etc.), and environmental values of farming projects in and on buildings serves as a rationale for architects and engineers to justify the integration of food production systems within their design, in both new and existing buildings (Thomaier et al., 2014; D'Ostuni et al., 2022) and can be the key for project success, as it proves the meaningfulness and multi-functionality of new systems.

Furthermore, by integrating food growing with living and working spaces, farming projects implemented in and on buildings can shape new architectural forms and urban looks, making building skins greener and more life-giving. By making food visible and livable for all citizens, they reconnect people with food and nature, motivating them to get in touch with nature, engage in food growing, or at least lean toward a healthy diet. Making food production visible and tangible within the city boundaries should be considered an important goal of farming projects implemented in and on buildings changing how citizens see food and inspiring unforeseen transformations in the modern food system (D'Ostuni et al., 2022). Successful projects, created with careful consideration and understanding of user and society needs context, and technology, can make a huge step toward these ambitious goals and further strengthen the social acceptance of farming in and on buildings.

ACKNOWLEDGMENTS

I am grateful to Professor Abel Tablada for arousing my interest in scientific research related to his concept of PFs, implemented at the T2 Lab in Singapore, and for providing excellent input and valuable suggestions that helped strengthen the prospective contribution of this book chapter. I want to express my sincere gratitude to Đorđe Ćebić for his great technical support and assistance and to Svetlana Mladenović for language editing and proofreading during the preparation of this book chapter.

REFERENCES

Adams, Z.W., & Caplow, T., (2012), Vertically-integrated greenhouse. US Patent 8,151,518 B2.

Aitken, M., (2010), Wind power and community benefits: Challenges and opportunities. *Energy Policy*, 38(10), 6066–6075. https://doi.org/10.1016/j.enpol.2010.05.062.

Al-Chalabi, M., (2015), Vertical farming: Skyscraper sustainability? Sustainable Cities and Society, 18, 74–77. https://doi.org/10.1016/j.scs.2015.06.003.

Al-Kodmany, K., (2018), The vertical farm: A review of developments and implications for the vertical city. *Buildings*, 8(2), 24. https://doi.org/10.3390/buildings8020024.

Appolloni, E., Orsini, F., Specht, K., Thomaier, S., Sanyé-Mengual, E., Pennisi, G., Gianquinto, G., (2021), The global rise of urban rooftop agriculture: A review of worldwide cases. Journal of Cleaner Production, 296, 126556. https://doi.org/10.1016/j.jclepro.2021.126556

Ares, G., Ha, B., Jaeger, S. R., (2021), Consumer attitudes to vertical farming (indoor plant factory with artificial lighting) in China, Singapore, UK, and USA: A multi-method study, *Food Research International*, 150(Part B), 110811. https://doi.org/10.1016/j.foodres.2021.110811

Ayambire, R. A., Amponsah, O., Peprah, C., & Takyi, S. A. (2019). A Review of Practices for Sustaining Urban and Peri-Urban Agriculture: Implications for Land Use Planning in Rapidly Urbanising Ghanaian Cities. Land Use Policy, 84, 260-277.
https://doi.org/10.1016/j.landusepol.2019.03.004

Azunre G.A., Amponsah O., Peprah C., Takyi S.A., Braimah I. A review of the role of urban agriculture in the sustainable city discourse. Cities. 2019;93:104–119. https://doi.org/10.1016/j.cities.2019.04.006.

Banerjee, C., Adenaeuer, L. (2014). Up, up and away! The economics of vertical farming. Journal of Agricultural Studies, 2(1), 40. https://doi.org/10.5296/jas.v2i1.4526.

Barcelona Laboratory for Urban Environmental Justice and Sustainability, (2023), https://www.bcnuej.org/. Accessed 05.01.2023.

Bearth, A., Siegrist, M., (2016). Are risk or benefit perceptions more important for public acceptance of innovative food technologies: A meta-analysis. Trends in Food Science & Technology, 49, 14–23. https://doi.org/10.1016/j.tifs.2016.01.003.

Benis, K, Ferrão, P., (2018), Commercial farming within the urban built environment - Taking stock of an evolving field in northern countries. *Global Food Security*. 17, 30–37. https://doi.org/10.1016/j.gfs.2018.03.005

Benis, K., Turan, I., Reinhart, C., Ferrão, P., (2018), Putting rooftops to use-A cost-benefit analysis of food production vs. energy generation under Mediterranean climates. *Cities*, 78, 166–179. https://doi.org/10.1016/j.cities.2018.02.011

Biasini, B., Rosi, A., Giopp, F., Turgut, R., Scazzina, F., Menozzi, D., (2021), Understanding, promoting and predicting sustainable diets: A systematic review. Trends in Food Science and *Technology*, 111, 191–207. https://doi.org/10.1016/J.TIFS.2021.02.062.

Brinkley, C. (2012). Evaluating the benefits of peri-urban agriculture. Journal of Planning literature, 27(3), 259-269. DOI:10.1177/0885412211435172.

Broad, G.M., Marschall, W., Ezzeddine, M., (2021), Perceptions of high-tech controlled environment agriculture among local food consumers: Using interviews to explore sense-making and connections to good food. *Agriculture and Human Values*, 39, 417–433. https://doi.org/10.1007/s10460-021-10261-7.

Brunsting, S., Mastop, J., Kaiser, M., Zimmer, R., Shackley, S., Mabon, L., Howell, R., (2015), CCS Acceptability: Social site characterization and advancing awareness at prospective storage sites in Poland and Scotland. *Oil & Gas Science and Technology - Revue d'IFP Energies Nouvelles*, 70(4), 767–784. https://doi.org/10.2516/ogst/2014024.

Buscaroli, E., Braschi, I., Cirillo, C., Fargue-Lelièvre, A., Modarelli, G. C., Pennisi, G., Righini, I., Specht, K., Orsini, F., (2021), Reviewing chemical and biological risks in urban agriculture: A comprehensive framework for a food safety assessment of city region food systems. *Food Control*, 126, 108085. https://doi.org/10.1016/j.foodcont.2021.108085.

Butturini, M., Marcelis, L.F.M.., (2020), Chapter 4 – Vertical farming in Europe: Present status and outlook. T. Kozai, G. Niu, M. Takagaki (Eds.). *Plant Factory* (2nd Ed.). Academic Press, pp. 77–91. https://doi.org/10.1016/B978-0-12-816691-8.00004-2.

Cabannes, Y., Marocchino, C. (Eds.), (2018), *Integrating Food into Urban Planning*. Rome: UCL Press, Food and Agriculture Organization of the United Nations.

Carr, A., Zylmans, A., Drozdzik, C., Vena, M., Finkelstein, N., (2014), Public perceptions of urban agriculture's positionality in Vancouver, Canada. Geography 371 – Research Strategies in Human Geography, 32. https://www.lmnhs.bc.ca/wp/wp-content/uploads/2014/02/UA-Final-Report.pdf.

City of Paris, (2018), Paris climate action plan towards a carbon neutral city and 100% renewable energies. Paris: City of Paris. Available online at: https://cdn.paris.fr/paris/2019/07/24/1a706797eac9982aec6b767c56449240.pdf. Accessed: 19.02.2023.

Cohen, A., Malone, S., Morris, Z., Weissburg, M., Bras, B., (2018), Combined fish and lettuce cultivation: An aquaponics life cycle assessment. *Procedia CIRP*, 69, 551–556. https://doi.org/10.1016/j.procir.2017.11.029.

Collé, M., Daniel, A.C., Aubry, C., (2018), Call for projects "parisculteurs": Catalyst for urban agriculture development on rooftops in Paris. *Acta Hortic*, 1215, 147–151. https://doi.org/10.17660/ActaHortic.2018.1215.28.

Compeau, D.R., Higgins, C.A., Huff, S.L., (1999), Social cognitive theory and individual reactions to computing technology. *MIS Q*, 23(2), 145–158. https://doi.org/10.2307/249749.

Coyle, B.D., Ellison, B., (2017), Will consumers find vertically farmed produce "out of reach"? Choices. The Magazine of Food, Farm, and Resources Issues, 32(1): 1–8.

Crush, J., Si, Z., (2020), COVID-19 containment and food security in the Global South. Journal of Agriculture, Food Systems, and Community Development, 9(4), 149–151. https://doi.org/10.5304/jafscd.2020.094.026.

De Jalón, S.G., Silvestri, S., Granados, A., Iglesias, A., (2015), Behavioural barriers in response to climate change in agricultural communities: An example from Kenya. *Regional Environmental Change*. 15, 851–865. https://doi.org/10.1007/s10113-014-0676-y.

Di Fiore, G., Specht, K., Zanasi, C., (2021), Assessing motivations and perceptions of stakeholders in urban agriculture: A review and analytical framework. International Journal of Urban Sustainable Development, 13(2), 351–367. https://doi.org/10.1080/19463138.2021.1904247.

D'Ostuni, M., Zaffi, L., Appolloni, E., Orsini, F., (2022), Understanding the complexities of building-integrated agriculture. Can food shape the future built environment? *Futures*, 144, 103061. https://doi.org/10.1016/j.futures.2022.103061.

Ecogradia, (2023). https://www.ecogradia.com. Accessed 07.02.2023.

Eigenbrod, C., Gruda, N. (2015), Urban vegetable for food security in cities. A review. Agron. Sustain. Dev. 35, 483–498. https://doi.org/10.1007/s13593-014-0273-y

Fertilecity, (2023), Available online at: https://www.fertilecity.com/. Accessed: 07.02.2023.

Forchino, A.A., Lourguioui, H., Brigolin, D., Pastres, R., (2017), Aquaponics and sustainability: The comparison of two different aquaponic techniques using the life cycle assessment (LCA). *Aquacultural Engineering*, 77, 80–88. https://doi.org/10.1016/j.aquaeng.2017.03.002.

Fox-Kämper, R., Specht, K., Caputo, S., Hawes, J. K., Lélièvre, A., Cohen, N., Poniży, L., with contributions by Béchet, B., Jean-Soro, L., Schoen, V., Stark, M., (2022). Roadmap to Resource Efficient Urban Agriculture. FEW-meter, project database and outputs. https://doi.org/10.5281/zenodo.6622125.

Frewer, L., (1998), Consumer perceptions and novel food acceptance. *Outlook on Agriculture*, 27(3), 153–156. https://doi.org/10.1177/003072709802700304.

Frewer, L.J., Bergmann, K., Brennan, M., Lion, R., Meertens, R., Rowe, G., Siegrist, M., Vereijken, C., (2011), Consumer response to novel agri-food technologies: Implications for predicting consumer acceptance of emerging food technologies. *Trends in Food Science & Technology*, 22(8), 442–456. https://doi.org/10.1016/j.tifs.2011.05.005.

Frick, M.J., Birkenholz, R.J., Machtmes, K., (1995), Rural and urban adult knowledge and perceptions of agriculture. *Journal of Agricultural Education*, 36(2), 44–53.

Geneletti, D,. Scolozzi, R., Adem Esmail, B., (2018), Assessing ecosystem services and biodiversity tradeoffs across agricultural landscapes in a mountain region. International Journal of Biodiversity Science, Ecosystem Services & Management, 14(1), 188–208. https://doi.org/10.1080/21513732.2018.1526214.

Gengler, C. E., Klenosky, D. B., Mulvey, M. S., (1995), Improving the graphic representation of means-end results. International Journal of Research in *M*arketing, 12(3), 245–256. https://doi.org/10.1016/0167-8116(95)00024-V.

Germer, J., Sauerborn, J., Asch, F., de Boer, J., Schreiber, J., Weber, G., Müller, J., (2011), Skyfarming anecological innovation to enhance global food security. Journal Für Verbraucherschutz und *Lebensmittelsicherheit*, 6(2), 237–251. https://doi.org/10.1007/s00003-011-0691-6.

Giles, J., Grosjean, G., Le Coq, J.F., Huber, B., Bui, V.L., Läderach, P., (2021), Barriers to implementing climate policies in agriculture: A case study from Vietnam. Frontiers in Sustainable Food Systems, 5. https://doi.org/10.3389/fsufs.2021.439881.

Gómez-Villarino, M.T., Ruiz-Garcia, L., (2021), Adaptive design model for the integration of urban agriculture in the sustainable development of cities. A case study in northern Spain. Sustainable Cities and Society, 65, 102595. https://doi.org/10.1016/j.scs.2020.102595.

Gonera, A., Svanes, E., Bugge, A.B., Hatlebakk, M.M., Prexl, K.M., Ueland, Ø., (2021), Moving consumers along the innovation adoption curve: A new approach to accelerate the shift toward a more sustainable diet. *Sustainability*, 13(8), 4477. https://doi.org/10.3390/su13084477.

González, A.M., Sandoval, H., Acosta, P., Henao, F., (2016), On the acceptance and sustainability of renewable energy projects-A systems thinking perspective. *Sustainability*, 8(11), 1171. https://doi.org/10.3390/su8111171.

Gould, D., Caplow, T., (2012), Chapter 8: 'Building-integrated agriculture: A new approach to food production'. In F. Zeman (Ed.). *Metropolitan Sustainability: Understanding and Improving the Urban Environment*. Cambridge: Woodhead Publishing Limited. https://doi.org/10.1533/9780857096463.2.147.

Grebitus, C., Chenarides, L., Muenich, R., Mahalov, A., (2020), Consumers' perception of urban farming-An exploratory study. Frontiers in Sustainable Food Systems, 4, 79. https://doi.org/10.3389/fsufs.2020.00079.

Grebitus, C., Printezis, I., Printezis, A., (2017), Relationship between consumer behavior and success of urban agriculture. *Ecological Economics*, 136, 189–200.

Grunert, K.G., (2011), Sustainablitity in the food sector: A consumer behaviour perspective. *International Journal on Food System Dynamics*, 2(3), 207–218. Available at SSRN: https://ssrn.com/abstract=2008735

Grunert, K. G., Grunert, S.C., (1995), Measuring subjective meaning structures by the laddering method: Theoretical considerations and methodological problems. International *J*ournal of *R*esearch in *M*arketing, 12(3), 209–225. https://doi.org/10.1016/0167-8116(95)00022-T.

Grunert, K.G., Hieke, S., Wills, J., (2014), Sustainability labels on food products: Consumer motivation, understanding and use. *Food Policy*, 44, 177–189. https://doi.org/10.1016/j.foodpol.2013.12.001.

Hall, N., Ashworth, P., Devine-Wright, P., (2013), Societal acceptance of wind farms: analysis of four common themes across Australian case studies. *Energy Policy Research*, 58, 200–208. https://doi.org/10.1016/j.enpol.2013.03.009.

Hanger, S., Komendantova, N., Schinke, B., Zejli, D., Ihlal, A., Patt, A., (2016), Community acceptance of large-scale solar energy installations in developing countries: Evidence from Morocco. *Energy Research & Social Science*, 14, 80–89. https://doi.org/10.1016/j.erss.2016.01.010.

Hemström, K., Mahapatra, K., Gustavsson, L., (2014), Public perceptions and acceptance of intensive forestry in Sweden. *AMBIO*. 43(2), 196–206. https://doi.org/10.1007/s13280-013-0411-9.

Hochparterre, (2022), Hochparterre - Die Gartenstadt wird vertikal. https://www.hochparterre.ch/nachrichten/wettbewerbe/die-gartenstadt-wird-vertikal/. Accessed 27.12.2022.

Hollywood, L.E., Armstrong, G.A., Durkin, M.G., (2007), Using behavioural and motivational thinking in food segmentation. International Journal of Retail & *Distribution Management*, 35(9), 691–702. https://doi.org/10.1108/09590550710773246.

Hui, S.C.M., (2011), Green roof urban farming for buildings in high-density urban cities. In *The 2011 Hainan China World Green Roof Conference*. Hainan, China: The University of Hong Kong, 18–21 March 2011. Available at: https://hub.hku.hk/bitstream/10722/140388/1/Content.pdf?accept=1.

Huijts, N. M. A., Molin, E. J. E., Steg, L., (2012). Psychological factors influencing sustainable energy technology acceptance: A review-based comprehensive framework. Renewable and Sustainable Energy Reviews, 16(1), 525-531. https://doi.org/10.1016/j.rser.2011.08.018

Hüttel, A., Ziesemer, F., Peyer, M., Balderjahn, I., (2018), To purchase or not? Why consumers make economically (non-) sustainable consumption choices. *Journal of Cleaner Production*. 174, 827–836. https://doi.org/10.1016/j.jclepro.2017.11.019.

Ilieva, R.T., Cohen, N., Israel, M., Specht, K., Fox-Kämper, R., Fargue-Lelièvre, A., Poniży, L., Schoen, V., Caputo, S., Kirby, C.K., Goldstein, B., Newell, J.P., Blythe, C., (2022), The socio-cultural benefits of urban agriculture: A review of the literature. *Land*, 11(5), 622. https://doi.org/10.3390/land11050622.

Jaeger, H., Knorr, D., Szabó, E., Hámori, J., Bánáti, D., (2015), Impact of terminology on consumer acceptance of emerging technologies through the example of PEF technology. Innovative Food Science & Emerging Technologies, 29, 87–93. https://doi.org/10.1016/j.ifset.2014.12.004.

Jaeger, S.R., Chheang, S.L., Ares, G., (2022), Text highlighting as a new way of measuring consumers' attitudes: A case study on vertical farming. *Food Quality and Preference*. 95, 104356.

Jaeger, S. R., Chheang, S. L., Bredahl, L., (2023), Means-end chain generation with online laddering: A study on vertical farming with consumers in Singapore and Germany. *Food Quality and Preference*. 106, 104794. https://doi.org/10.1016/j.foodqual.2022.104794.

Jastrzębska, M, Kostrzewska, M., Saeid, A., (2022), Chapter 2 – Sustainable agriculture: A challenge for the future. In K. Chojnacka, A. Saeid, *Smart Agrochemicals for Sustainable Agriculture*, Academic Press, pp. 29–56. https://doi.org/10.1016/B978-0-12-817036-6.00002-9.

Jensen, L (Ed.) (2021), *The Sustainable Development Goals Report 2021, UN Department of Economic and Social Affairs*, United Nations Publications, New York, pp. 28–56.

Jürkenbeck, K-, Heumann, A., Spiller, A., (2019), Sustainability matters: Consumer acceptance of different vertical farming systems. *Sustainability*, 11(15), 4052. https://doi.org/10.3390/su11154052.

Kahle, L. R., Beatty, S. E., Homer, P., (1986), Alternative measurement approaches to consumer values: The list of values (LOV) and values and life style (VALS). *Journal of Consumer Research*. 13(3), 405–409. https://doi.org/10.1086/209079.

Kalantari, F., Akhyani, N., (2021), Community acceptance studies in the field of vertical farming-A critical and systematic analysis to advance the conceptualisation of community acceptance in Kuala Lumpur. International Journal of Urban Sustainable Development, 13(3), 569–584. https://doi.org/10.1080/19463138.2021.2013849.

Kalantari, F., Tahir, O.M., Kbari Joni, R., Aminuldin, N.A., (2018), The importance of the public acceptance theory in determining the success of the vertical farming projects. *Management Research and Practice*, 10(1), 5–16.

Khorsand, I., Kormos, C., MacDonald, E.G., Crawford, C., (2015), Wind energy in the city: An interurban comparison of social acceptance of wind energy projects. *Energy Research & Social Science*. 8, 66–77. https://doi.org/10.1016/j.erss.2015.04.008.

Kirshenbaum, S., Buhler, D.., (2018), Americans are confused about food and unsure where to turn for answers, study shows. In *Alliance for Science*. Available at: https://allianceforscience.org/blog/2018/03/americans-confused-food-unsure-turn-answers/. Accessed at: 19.02.2023.

Kiss+Cathcart, Architects, (2022), https://kisscathcart.com/. Accessed 07.12.2022.

Kosorić, V., Huang, H., Tablada, A., Lau, S. K., Tan, H. T. W., (2019), Survey on the social acceptance of the productive façade concept integrating photovoltaic and farming systems in high-rise public housing blocks in Singapore. Renewable and Sustainable Energy *Reviews*, 111, 197–214. https://doi.org/10.1016/j.rser.2019.04.056.

Kosorić, V., Lau, S. K., Tablada, A., Bieri, M., Nobre, A. M., (2021). A holistic strategy for successful photovoltaic (PV) implementation into Singapore's built environment. *Sustainability*, 13, 6452. https://doi.org/10.3390/su13116452.

Li, X., Hijazi, I., Koenig, R., Lv, Z., Zhong, C., Schmitt, G., (2016), Assessing essential qualities of urban space with emotional and visual data based on GIS technique. ISPRS International Journal of Geo-Information, 5(11), 218. https://doi.org/10.3390/ijgi5110218.

Liu, T., Yang, M., Han, Z., Ow, D.W., (2016), Rooftop production of leafy vegetables can be profitable and less contaminated than farm-grown vegetables. Agronomy for Sustainable Development, 36, Article number: 41. https://doi.org/10.1007/s13593-016-0378-6.

Lohrberg, F., Lička, L., Scazzosi, L., Timpe, A., (2016), *Urban Agriculture Europe*. Berlin: Jovis.

Malazewska, S, Was, A. 2015. Determinanty wartości krajobrazu rolniczego jako dobra publicznego (in engl. Determinants of value of agricultural landscape as public good). *Roczniki Naukowe Ekonomii Rolnictwa i Rozwoju Obszarów Wiejskich*, 102(4), 26–40.

Marzouk, M.A., Salheen, M.A., Fischer, L.K., (2022), Functionalizing building envelopes for greening and solar energy: Between theory and the practice in Egypt. Frontiers in Environmental Science, 10. https://www.frontiersin.org/articles/10.3389/fenvs.2022.1056382.

McClintock, N., (2010), Why farm the city? Theorizing urban agriculture through a lens of metabolic rift. Cambridge Journal of Regions, Economy and Society, 3(2), 191–207. https://doi.org/10.1093/cjres/rsq005.

Milestad, R., Carlsson-Kanyama, A., Schaffer, C., (2020), The Högdalen urban farm: A real case assessment of sustainability attributes. *Food Security*. 12, 1461–1475. https://doi.org/10.1007/s12571-020-01045-8.

Miličić, V., Thorarinsdottir, R., Santos, M-D., Hančič, M.T., (2017), Commercial aquaponics approaching the European market: To consumers' perceptions of aquaponics products in Europe. *Water*, 9(2), 80. https://doi.org/10.3390/w9020080.

Møller, V. (2005), Attitudes to food gardening from a generational perspective. *Journal of Intergenerational Relationships*, 3(2), 63–80. https://doi.org/10.1300/J194v03n02_05.

Moser, S. C., Ekstrom, J.A., (2010), A framework to diagnose barriers to climate change adaptation. *Proceedings of the National Academy of Sciences of the United States of America*. 107, 22026–22031. https://doi.org/10.1073/pnas.1007887107.

Mougeot, L.J.A.., (2006), *Growing Better Cities: Urban Agriculture for Sustainable Development*. Ottawa, Canada: International Development Research Centre. ISBN: 1552502260.

New York Sun Works, (2023) https://nysunworks.org/. Accessed 14.01.2023.

Niles, M.T., Bertmann, F., Belarmino. E.H., Wentworth, T., Biehl, E., Neff, R. (2020), The early food insecurity impacts of COVID-19. *Nutrients*, 12(7), 2096. https://doi.org/10.3390/nu12072096.

Ochoa, J., Sanyé-Mengual, E., Specht, K., Fernández, J.A., Bañón, S., Orsini, F., Magrefi, F., Bazzocchi, G., Halder, S., Martens, D., Kappel, N., Gianquinto, G., (2019), Sustainable community gardens require social engagement and training: A users' needs analysis in Europe. *Sustainability*, 11(14), 3978. https://doi.org/10.3390/su11143978.

Olson, J.C. (1995), Introduction. Special issue on means-end chains analysis. International *J*ournal of *R*esearch in *M*arketing, 12(3), 189–191.

Orsini, F., Gasperi, D.,Marchetti, L., Piovene, C., Draghetti, S., Ramazzotti, S., Bazzocchi, G., Gianquinto, G., (2014), Exploring the production capacity of rooftop gardens (RTGs) in urban agriculture: The potential impact on food and nutrition security, biodiversity and other ecosystem services in the city of Bologna. *Food Security*, 6, 781–792. https://doi.org/10.1007/s12571-014-0389-6.

O'Sullivan, C.A., Bonnett, G.D., McIntyre, C.L., Hochman, Z., Wasson, A.P., (2019), Strategies to improve the productivity, product diversity and profitability of urban agriculture. *Agricultural Systems*, 174, 133–144. https://doi.org/10.1016/j.agsy.2019.05.007.

Parkes, M.G., Azevedo, D.L., Domingos, T., Teixeira, R.F.M., (2022), Narratives and benefits of agricultural technology in urban buildings: A review. *Atmosphere*, 13(8), 1250. https://doi.org/10.3390/atmos13081250.

Pennisi, G., Magrefi, F., Michelon, N., Bazzocchi, G., Maia, L., Orsini, F., Gianquinto, G., (2020), Promoting education and training in urban agriculture building on international projects at the research centre on urban environment for agriculture and biodiversity. *Acta Horticulturae*, 1279, 45–51. https://doi.org/10.17660/ActaHortic.2020.1279.7.

Penzel Valier, A.G., (2022), https://penzelvalier.ch. Accessed 14.11.2022.

Proksch, G., (2011), Urban Rooftops as Productive Resources. Rooftop Farming versus Conventional Green Roofs, in: ARCC 2011 Conference «Considering Research: Reflecting upon current themes in Architectural Research".

PS/IS 686, Brooklyn School of Inquiry, (2023) https://bsi686.org/stemlab/. Accessed 13.01.2023.

PS 333, The Manhattan School for Children, (2023), https://www.ps333.org/. Accessed 14.01.2023.

PS 84 Josê de Diego in Williamsburg, Brooklin, (2023), https://www.ps84k.org. Accessed 14.01.2023.

Rezaee, E.D., Kalantari, F., (2019), Proposal of an operational model to measure feelings and emotions in urban space. *Journal of Landscape Ecology*, 12(3), 34–52. https://doi.org/10.2478/jlecol-2019-0014.

Rizal, A., Dhahiyat, Y., Zahidah, Andriani, Y., Handaka, A.A., Sahidin, A., (2018), The economic and social benefits of an aquaponic system for the integrated production of fish and water plants. IOP Conference Series: Earth and Environmental Science, 137, 012098. https://doi.org/10.1088/1755-1315/137/1/012098.

Rogus, S., Dimitri, C., (2015), Agriculture in urban and peri-urban areas in the United States: Highlights from the census of agriculture. Renewable Agriculture and Food Systems, 30(1), 64–78. https://doi.org/10.1017/S1742170514000040.

Sanyé-Mengual, E., Anguelovski, I., Oliver-Solà, J., Montero, J.I., Rieradevall, J., (2016), Resolving differing stakeholder perceptions of urban rooftop farming in Mediterranean cities: Promoting food production as a driver for innovative forms of urban agriculture. *Agriculture and Human Values*, 33, 101–120. https://doi.org/10.1007/s10460-015-9594-y.

Sanyé-Mengual, E., Oliver-Solà, J., Montero, J.I., Rieradevall, J., (2015), An environmental and economic life cycle assessment of rooftop greenhouse (RTG) implementation in Barcelona, Spain. Assessing new forms of urban agriculture from the greenhouse structure to the final product level. *The International Journal of Life Cycle Assessment*, 20, 350–366. https://doi.org/10.1007/s11367-014-0836-9.

Sanyé-Mengual, E., Specht, K., Grapsa, E., Orsini, F., Gianquinto, G., (2019), How can innovation in urban agriculture contribute to sustainability? A characterization and evaluation study from five Western European Cities. *Sustainability*, 11(15), 4221. https://doi.org/10.3390/su11154221.

Schumann, D., (2015), Public acceptance. In *Carbon Capture, Storage and Use*, Germany: Springer International Publishing, pp. 221–251. https://doi.org/10.1007/978-3-319-11943-4_11.

Schweizer-Ries, P. 2008. Energy sustainable communities: Environmental psychological investigations. *Energy Policy* 36(11), 4126–4135.

Shamshiri, R.R., Kalantari, F., Ting, K.C., Thorp, K.R., Hameed, I.A., Weltzien, C., Ahmad, D., Shad, Z.M., (2018), Advances in greenhouse automation and controlled environment agriculture: A transition to plant factories and urban farming. International Journal of Agricultural and Biological Engineering, 11(1), 1–22. https://doi.org/10.25165/j.ijabe.20181101.3210.

Shao, Y., Wang, Z., Zhou, Z., Chen, H., Cui, Y., Zhou, Z., (2022a), Determinants affecting public intention to use micro-vertical farming: A survey investigation. *Sustainability*, 14, 9114. https://doi.org/10.3390/su14159114.

Shao, Y., Zhou, Z., Chen, H., Zhang, F., Cui, Y., Zhou, Z., (2022b), The potential of urban family vertical farming: A pilot study of Shanghai. *Sustainable Production and Consumption*, 34, 586–599. https://doi.org/10.1016/j.spc.2022.10.011.

Shepherd, R., Raats, M. M., (1996), Attitudes and beliefs in food habits. In H.L. Meiselman, H.J.H. MacFie (Eds.). *Food Choice, Acceptance and Consumption*, London: Blackie Academic & Professional, pp. 346–364.

Shirazi, M. R., Keivani, R., (2019), Social sustainability discourse – A critical revisit. In M.R. Shirazi R. Keivani (Eds.), *Urban Social Sustainability: Theory, Policy and Practice*. New York, NY: Routledge, pp. 1–26.

Siebrecht, N., (2020), Sustainable agriculture and its implementation gap - overcoming obstacles to implementation. *Sustainability*, 12(9), 3853. https://doi.org/10.3390/su12093853.

Sommerville, C., Moti, C., Edoardo, P., Austin, S., Alessandro, L., (2014), Small-scale aquaponic food production - Integrated fish and plant farming FAO. Italy: Fisheries and Aquaculture Technical Paper Rome, no. 589.

Song, S., Cheong, J. C., Lee, J. S. H., Tan, J. K. N., Chiam, Z., Arora, S., Png, K. J. Q., Seow, J. W. C., Leong, F. W. S., Palliwal, A., Biljecki, F., Tablada, A., Tan, H. T. W., (2022), Home gardening in Singapore: A feasibility study on the utilization of the vertical space of retrofitted high-rise public housing apartment buildings to increase urban vegetable self-sufficiency. Urban Forestry & Urban Greening, 78, 127755. https://doi.org/10.1016/j.ufug.2022.127755.

Specht, K., Siebert, R., Hartmann, I., Freisinger, U.B., Sawicka, M., Werner, A., Thomaier, S., Henckel, D., Walk, H., Dierich, A., (2013), Urban agriculture of the future: An overview of sustainability aspects of food production in and on buildings. Agriculture and Human Values, 31, 33–51. https://doi.org/10.1007/s10460-013-9448-4.

Specht, K., Siebert, R., Thomaier, S., Freisinger, U., Sawicka, M., Dierich, A., Henckel, D., Busse, M., (2015), Zero-acreage farming in the city of Berlin: An aggregated stakeholder perspective on potential benefits and challenges. *Sustainability*, 7, 4511–4523. https://doi.org/10.3390/su7044511.

Specht, K., Siebert, R., Thomaier, S., (2016a), Perception and acceptance of agricultural production in and on urban buildings (ZFarming): A qualitative study from Berlin, Germany. Agri*culture and* Human Values, 33(4), 753–769. https://doi.org/10.1007/s10460-015-9658-z.

Specht, K., Zoll, F., Siebert, R., (2016b), Application and evaluation of a participatory "open innovation" approach (ROIR): The case of introducing zero-acreage farming in Berlin. *Landscape and Urban Planning*, 151, 45–54. https://doi.org/10.1016/j.landurbplan.2016.03.003.

Specht, K., Weith, T., Swoboda, K., Siebert, R., (2016c), Socially acceptable urban agriculture businesses. *Agronomy for Sustainable Development*, 36, 17. https://doi.org/10.1007/s13593-016-0355-0.

Specht, K., Sanyé-Mengual, E., (2017), Risks in urban rooftop agriculture: Assessing stakeholders' perceptions to ensure efficient policy-making. *Environmental Science & Policy*, 69, 13–21. https://doi.org/10.1016/j.envsci.2016.12.001.

Specht, K., Zoll, F., Schümann, H., Bela, J., Kachel, J., Robischon, M., (2019), How will we eat and produce in the cities of the future? From edible insects to vertical farming-A study on the perception and acceptability of new approaches. *Sustainability*, 11(16), 4315. https://doi.org/10.3390/su11164315.

Sroka, W., Bojarszczuk, J., Łukasz, S., Szczepańska, B., Sulewski, P., Lisek, S., Luty, L., Zioło, M., (2021), Understanding residents' acceptance of professional urban and peri-urban farming: A socio-economic study in Polish metropolitan areas. *Land Use Policy*, 109(C). https://doi.org/10.1016/j.landusepol.2021.105599.

Tablada, A., Kosorić, V., Huajing, H., Chaplin, I. K., Lau, S. K., Yuan, C., Lau, S. S. Y., (2018), Design optimisation of productive facades: Integrating photovoltaic and farming systems at the Tropical Technologies Laboratory. *Sustainability*, 10(10), 3762. https://doi.org/10.3390/su10103762.

Tablada, A., Kosorić, V., Huang, H., Lau, S. S. Y., Shabunko, V., (2020), Architectural quality of the productive façades integrating photovoltaic and vertical farming systems: Survey among experts in Singapore. *Frontiers of Architectural Research*, 9(2), 301–318. https://doi.org/10.1016/j.foar.2019.12.005.

Tablada, A., Kosorić, V., (2022), Vertical farming on facades: Transforming building skins for urban food security. E. Gasparri, A. Brambilla, G. Lobaccaro, F. Goia, A. Andaloro, A. Sangiorgio. In *Woodhead Publishing Series in Civil and Structural Engineering, Rethinking Building Skins*. Woodhead Publishing, pp. 285–311, ISBN 9780128224779 https://doi.org/10.1016/B978-0-12-822477-9.00015-2.

Takahashi, T., Sato, T., (2015), Inclusive environmental impact assessment indices with consideration of public acceptance: Application to power generation technologies in Japan. *Applied Energy*, 144, 64–72. https://doi.org/10.1016/j.apenergy.2015.01.053.

Tao, P., Yi Meng, W., Diebel, C.,Wenye, G., Yi, H. C., Sternfeld, E., Antoni, C., Paul, M., (2020), *Urban Farming Incubator Guide to Urban Farming in Shanghai and Berlin*. Berlin: Citymakers Urban Farming Incubator.

Thomaier, S., Specht, K., Henckel, D., Dierich, A., Siebert, R., Freisinger, U., Sawicka, M., (2014), Farming in and on urban buildings: Present practice and specific novelties of zero-acreage farming (ZFarming). Renewable Agriculture and Food Systems, 30, 43–54. https://doi.org/10.1017/S1742170514000143.

Tobler, C., Visschers, V.H.M., Siegrist, M., (2011), Eating green. Consumers' willingness to adopt ecological food consumption behaviors. *Appetite*, 57(3), 674–682. https://doi.org/10.1016/j.appet.2011.08.010.

Tong, Z., Whitlow, T.H., Landers, A., Flanner, B., (2016), A case study of air quality above an urban roof top vegetable farm. *Environmental Pollution*, 208, 256–260. https://doi.org/10.1016/j.envpol.2015.07.006.

United Nations, (2015), The 17 goals. In Sustainable Development Goals (Vol. 2021): UN Department of Economic and Social Affairs.

van Rijnsoever, F.J., Farla, J.C.M., (2014), Identifying and explaining public preferences for the attributes of energy technologies. *Renewable and Sustainable Energy Reviews*, 31, 71–82. https://doi.org/10.1016/j.rser.2013.11.048.

Van Tuijl, E., Hospers, G.J., Van Den Berg, L., (2018), Opportunities and challenges of urban agriculture for sustainable city development. European *S*patial *R*esearch and *P*olicy, 25(2), 5–22. https://doi.org/10.18778/1231-1952.25.2.01.

Vecchio, R., Annunziata, A., (2015), Willingness-to-pay for sustainability-labelled chocolate: An experimental auction approach. *Journal of Cleaner Production*, 86, 335–342. https://doi.org/10.1016/j.jclepro.2014.08.006.

Veenhuizen, R.V. (ed.), (2006), *Cities Farming for the Future: Urban Agriculture for Green and Productive Cities*. RUAF Foundation, IDRC and IIRR Publishing, Philippines

Wachenheim, C., Rathge. R., (2000), Societal perceptions of agriculture. Agribusiness and applied economics report no. 449. Fargo, ND: Department of Agribusiness and Applied Economics Agricultural Experiment Station. North Dakota State University, p. 58105. https://doi.org/10.22004/ag.econ.23541.

Wang, Y., Li, J., (2016), A causal model explaining Chinese university students' acceptance of nuclear power. *Progress in Nuclear Energy*, 88, 165–174. https://doi.org/10.1016/j.pnucene.2016.01.002.

Wang, T., Pryor, M., (2019), Social value of urban rooftop farming: A Hong Kong case study. Agricultural Economics – Current Issues. https://doi.org/10.5772/intechopen.89279.

Wolsink, M., (2012), Undesired reinforcement of harmful 'selfevident truths' concerning the implementation of wind power. *Energy Policy*, 48, 83–87. https://doi.org/10.1016/j.enpol.2012.06.010.

Wüstenhagen, R, Wolsink, M, Bürer, M. J., (2007), Social acceptance of renewable energy innovation: An introduction to the concept. *Energy Policy*, 35(5), 2683–2691. https://doi.org/10.1016/j.enpol.2006.12.001/.

Yang L, Zhang X, McAlinden KJ., (2016), The effect of trust on people's acceptance of CCS (carbon capture and storage) technologies: evidence from a survey in the People's Republic of China. *Energy*. 96:69–79. https://doi.org/10.1016/j.energy.2015.12.044.

Yano, Y., Nakamura, T., Ishitsuka, S., Maruyama, A., (2021), Consumer attitudes toward vertically farmed produce in Russia: A study using ordered logit and co- occurrence network analysis. *Foods*, 10(3), 638. https://doi.org/10.3390/ foods10030638.

Yazdanpanah, M., Komendantova, N., Ardestani, R.S., (2015), Governance of energy transition in Iran: Investigating public acceptance and willingness to use renewable energy sources through socio-psychological model. *Renewable Sustainable Energy Reviews*, 45, 565–573. https://doi.org/10.1016/j.rser.2015.02.002.

Yuan, G. N., Marquez, G. P. B., Deng, H., Iu, A., Fabella, M., Salonga, R. B., Ashardiono, F., Cartagena, J. A., (2022), A review on urban agriculture: Technology, socio-economy, and policy. *Heliyon*, 8(11), E11583. https://doi.org/10.1016/j.heliyon.2022.e11583.

Zambrano-Prado, P., Orsini, F., Rieradevall, J., Josa, A., Gabarrell, X., (2021a), Potential key factors, policies, and barriers for rooftop agriculture in EU cities: Barcelona, Berlin, Bologna, and Paris. Frontiers in Sustainable Food Systems, 5. https://doi.org/10.3389/fsufs.2021.733040.

Zambrano-Prado, P., Pons-Gumí, D., Toboso-Chavero, S., Parada, F., Josa, A., Gabarrell, X., Rieradevall, J., (2021b), Perceptions on barriers and opportunities for integrating urban agri-green roofs: A European Mediterranean compact city case. *Cities*, 114, 103196. https://doi.org/10.1016/j.cities.2021.103196.

Zhou, H., Specht, K., Kirby, C.K., (2022), Consumers' and stakeholders' acceptance of indoor agritecture in Shanghai (China). *Sustainability*, 14, 2771. https://doi.org/10.3390/su14052771.

6 Digital Urban Agriculture
Implementing Digital Tools in the Design Phase

Szu-Cheng Chien, Hui An, and Chew Beng Soh

INTRODUCTION

Urban agriculture is increasingly recognized as a promising solution to address various challenges metropolitan cities face, such as food security, environmental sustainability, and climate change resilience (Orsini et al., 2013). As cities grow, the need for innovative and efficient urban farming systems becomes more pressing (Chatterjee et al., 2020). Compared with traditional urban agriculture, digital urban agriculture (DUA), as described in this context, involves farming practices within urban and peri-urban areas that integrate aspects of automation, software, and silicon-based hardware into their operations (Carolan, 2020). Digital tools and technologies have the potential to transform urban agriculture (WEF, 2022), enabling it to meet these challenges more effectively.

In the design phase, DUA involves various digital tools. Such tools include building information modeling (BIM) and computational performance simulations of the site, daylighting (Pacheco Diéguez et al., 2016; Eaton et al., 2021), computational fluid dynamics (CFD) (Bartzanas et al., 2013), heat load, etc. These tools can optimize the urban farming system design for optimal resource utilization, crop yields, and minimization of environmental impacts in urban farming systems (Berger, 2001; Pfister et al., 2005; Rendel et al., 2013). However, such applications in agriculture are relatively rare compared with their use in building design, specifically with a comprehensive assessment using these simulation tools simultaneously.

The driving force behind DUA is to address the challenges of limited space, resources, and environmental conditions in urban areas to promote sustainability. Using digital tools enables practitioners to optimize space and resources while tailoring farming systems to urban contexts. Furthermore, digital tools can enhance climate change resilience by allowing design professionals to consider future scenarios and adapt designs accordingly.

This chapter aims to demonstrate the potential of DUA through a Singapore case study, emphasizing the importance of integrating digital tools during an urban farming project's planning and design phases. This chapter offers valuable insights for other cities interested in leveraging digital technologies for urban agriculture initiatives by examining a successful implementation in a densely populated city-state like Singapore. The focus will be on the crucial design phase, which lays the groundwork for successful agriculture projects and promotes a sustainable, resilient urban future.

DIGITAL TOOLS IN URBAN AGRICULTURE

As urban agriculture continues to expand and evolve (Konjoian, 2014), integrating digital tools has become increasingly critical to optimize farming systems, enhance efficiency, and reduce the environmental footprint. This section provides an in-depth analysis of the key digital tools in urban agriculture, including the BIM modeling platform and computational performance simulation and diagnostics.

DOI: 10.1201/b23309-6

Building Information Modeling

BIM is an established digital technology that has become increasingly common in the design and construction industry. It enables the creation of virtual, three-dimensional models of urban farming systems, facilitating better communication and coordination among stakeholders. BIM provides a comprehensive understanding of a project's design, construction, and operation, allowing for improved decision-making and enhanced performance (Azhar, 2011; Kubba, 2012; Fu, 2018).

In DUA, BIM could allow stakeholders to visualize the farm layout, assess different design variants, and optimize the configuration of multiple elements for urban farming systems (e.g., indoor vertical farming structures, hydroponic systems, and greenhouses). Also, BIM supports the integration of diverse components, ensuring that the urban farming system operates efficiently and sustainably. Thereby, utilizing BIM in the design phase offers a range of benefits, such as enhanced collaboration, reduced costs (involving accurate cost estimation and minimizing the risk of budget overruns), improved efficiency (e.g., material quantity take-offs and clash detection), and sustainability via the evaluation of multiple design variants and their environmental impacts.

Computational Performance Simulation and Diagnostics

Computational performance simulation and diagnostics tools play a crucial role in DUA by simulating and analyzing the performance of the urban farming system in terms of site context, daylight availability, heat load, ventilation, and air circulation capacity (Kozai et al., 2019). These tools allow design professionals to optimize the system's performance, achieving the required climate "recipe" for crop growth. Also, they can develop holistic solutions that address the complex challenges associated with urban agriculture. This subsection will discuss critical simulations, specifically horticulture lighting, heat load simulations, and CFD.

Horticulture Lighting Simulations

The success of urban farming is heavily reliant on adequate lighting (Wong et al., 2020). In the realm of DUA, horticulture lighting simulations enable professionals to analyze and evaluate various lighting scenarios and strategies, allowing them to make informed decisions when designing urban farming systems. In this approach, optimal growth and crop yields are ensured. This section will discuss horticulture lighting simulations' benefits, functions, and applications and review suitable simulation tools for use in urban agricultural projects.

Benefits – Horticulture lighting simulations can be conducted to evaluate the daylight and/or artificial lighting performance in greenhouses and indoor vertical farming in an operationally efficient and resource-manageable approach. It helps design professionals identify the solutions and make better-informed decisions about desired light recipes for crop growth and other essential physiological processes while minimizing energy consumption. This is particularly important in urban farming systems, where energy costs, healthy crop growth, yields, and enhanced crop quality can significantly impact overall profitability.

Functions – In horticulture lighting simulations, lighting distributions within the growing spaces are assessed to ensure uniform illumination and minimize the risk of over- or under-lit areas while considering parameters such as photosynthetically active radiation, photosynthetic photon flux density (PPFD), and daily light integral (DLI). For example, in greenhouse settings, simulations can help design professionals assess the impact of natural daylight on the overall lighting strategy based on the facility design. By considering both external factors (e.g., daylight availability at farm site and obstructions) and internal factors (involving greenhouse orientation, glazing, shading devices, and typology and layout of planters), design professionals can optimize the use of daylight, supplemented as required by artificial lighting in a temporal and spatial approach. On the contrary, in indoor vertical farming systems, where natural daylight is often limited or unavailable, horticulture lighting simulations are essential for determining the most effective artificial lighting strategies.

These simulations can help design professionals select the appropriate light fixtures (involving optimal placement, configuration, performance, wavelengths, and intensities), ensuring that light is delivered effectively to all crops within the growing space.

Suitable Lighting Simulation Tools – There are several lighting simulation tools available for evaluating the performance of horticulture lighting. Some of these tools are listed below.

DIALux (DIAL, n.d.): DIALux is a widely used lighting simulation software that can be adapted for use in horticultural lighting projects (Vizeu da Silva et al., 2016). It offers various features for analyzing light distribution, fixture placement, and energy consumption, enabling design professionals to create efficient and effective lighting strategies for urban farming systems.

SketchUp is a popular 3D modeling software with various plug-ins (e.g., Curic Sun, Sefaira, LightUp) for daylighting and shadow analysis (SketchUp, n.d.; Curic, n.d.; Trimble, n.d.; LightUp, n.d.). These plug-ins enable design professionals to simulate daylight availability, sun paths, and shadows within the modeled environment. By using SketchUp and its plug-ins, design professionals can optimize sunlight exposure for urban farming projects.

AGi32 and Elumtools: AGi32 (stand-alone) and Elumtools (Revit plugin) are versatile lighting simulation tools with similar functions, capable of modeling both daylight and artificial lighting scenarios (Lighting Analysts, n.d.a & n.d.b). Their features include daylighting, artificial lighting, and PPFD horticulture calculation.

Climate Studio, a plugin for Rhino 7, is a daylighting and artificial lighting simulation tool (Solemma, n.d.). Based on Rhino 7 platform, it allows users to assess daylight availability, illuminance levels, and distribution within growing spaces. It enables annual and point-in-time simulations. By easily modifying design variants and re-running simulations in Rhino 7/ClimateStudio, design professionals can evaluate the impact of design changes on daylight and artificial lighting performance, streamlining the design process.

Heat Load Simulations

Heat load management is critical for crop growth in urban agriculture. Heat load simulations assess the thermal performance of the urban farming system, calculating the amount of heat generated by various sources, such as solar radiation and equipment. These simulations enable design professionals to optimize the system's thermal performance via cover materials, shading, natural and mechanical ventilation, and evaporative cooling for crop growth (Teitel, 2020). Heat load simulations can also guide the selection of insulation materials, the design of passive cooling strategies, and the sizing of heating, ventilation, and air conditioning systems. This section will discuss heat load simulations' benefits, functions, and applications in urban agricultural systems and present an overview of suitable simulation tools.

Benefits and Functions – Heat load simulations enable design professionals to predict temperature variations and crop heat and mass transfer within the agricultural facility (Ledesma et al., 2022), ensuring that the optimal growing conditions for different crops are maintained throughout the year. This directly contributes to enhanced crop yield, quality, and consistency. By accurately estimating the heat gain and loss within the facility, design professionals can develop strategies to reduce energy consumption, such as optimizing insulation, implementing passive solar design, and selecting energy-efficient heating and cooling systems. It can significantly reduce operational costs, making urban agricultural systems more economically viable and sustainable.

In greenhouse farming, heat load simulations estimate the heat gain from solar radiation, internal heat sources (e.g., lights, equipment, and crops), and heat transfer through the building envelope (e.g., walls, roof, and glazing). These simulations can inform decisions about glazing materials, shading devices, insulation, and ventilation systems, ultimately improving the greenhouse's energy efficiency and thermal performance. On the contrary, heat load simulations in indoor vertical farming systems can help design professionals optimize artificial lighting, heating, and cooling systems to maintain the ideal temperature and Relative Humidity (RH) levels for crop growth.

By accurately predicting the heat generated by equipment, such as LED lights and HVAC systems, design professionals can develop strategies to minimize energy consumption and reduce thermal stress on crops (Mahan et al., 1995).

Suitable Heat Load Simulation Tools – Various software tools are available to perform heat load simulations in urban agricultural systems. Some options include:

EnergyPlus: Developed by the U.S. Department of Energy, EnergyPlus is a comprehensive energy simulation software that can model heat transfer, thermal comfort, and energy consumption in buildings, including greenhouses and indoor vertical farms (EnergyPlus, n.d.).

TRNSYS: TRNSYS is a flexible, modular software package for simulating energy systems, including greenhouses and controlled-environment agriculture (CEA) facilities. It can model various components, such as solar collectors, heating, and cooling systems, and building enclosures (TRNSYS, n.d.), to evaluate the overall performance of an agricultural facility.

IES VE: Integrated Environmental Solutions' Virtual Environment (IES VE) is a suite of building performance analysis tools that can model thermal, daylighting, and energy performance (IESVE, n.d.). IES VE offers a range of modules that can be used to simulate and optimize greenhouse and indoor vertical farming systems.

ClimateStudio: In addition to the lighting simulations, the ClimateStudio tool calculates the amount of solar radiation received by designated surfaces yearly and monthly. The simulation is based on sun and sky radiances that are generated from weather file data, which reflects typical meteorological conditions from the past (ClimateStudiodocs, n.d.).

Computational Fluid Dynamics

CFD is a powerful simulation tool that can model fluid flow, heat transfer, and other related phenomena in complex systems (Krühne et al., 2012). In urban agriculture, particularly for greenhouse and indoor vertical farming applications, CFD simulations can provide valuable insights into the microclimate and environmental conditions within the farming system, including flow, heat transfer, photosynthesis, and transpiration (Naranjani et al., 2022). This subsection will discuss the benefits and functions of CFD simulations in urban agriculture and present suitable CFD simulation tools.

Benefits and Functions – CFD simulations can accurately model the microclimate within an urban agricultural facility, including temperature, humidity, air velocity, and airflow patterns. By analyzing airflow patterns and velocity profiles, design professionals can optimize the placement of vents, fans, and other air circulation components (Wang et al., 2022). This information is critical for optimizing crop growth conditions and ensuring crop health, yields, and quality. It can also model heat and moisture transfer processes within the urban agricultural facility, allowing design professionals to evaluate different heating, cooling, and dehumidification systems. By identifying the most efficient and effective ventilation strategies, this approach can lead to more energy-efficient designs and reduced operational costs. Also, in CEA and some greenhouse settings, maintaining optimal carbon dioxide (CO_2) levels is essential for plant growth (Nurmalisa et al., 2022). CFD simulations can help design professionals assess the distribution and concentration of CO_2 within the farming facility, enabling them to design more effective CO_2 supplementation and management strategies. However, due to the advanced computational demand, the common practice is to evaluate the airflow performance with a static approach and based on selected time steps (at one single point in time) instead of an "annual based dynamic evaluation".

Suitable CFD Simulation Tools – Several software tools can be utilized for conducting CFD simulations in urban agricultural systems. Some of these options include:

ANSYS Fluent: ANSYS Fluent is a widely used CFD simulation software that offers advanced modeling capabilities for fluid flow, heat transfer, and other related phenomena (Ansys, n.d.; Wu et al., 2016). It is suitable for a wide range of applications, including greenhouse and indoor vertical farming system design.

SimScale: SimScale is a cloud-based CFD simulation platform that enables users to run complex simulations in a web browser without requiring specialized hardware or software installations

(SimScale, n.d.; Yilmaz and Hu, 2018). This makes it an accessible and convenient option for urban agricultural system design professionals.

OpenFOAM: OpenFOAM is an open-source CFD simulation software that offers a comprehensive range of modeling capabilities for fluid dynamics, heat transfer, and chemical reactions (Jasak, 2009). Its open-source nature and flexible licensing make it an attractive option for researchers and design professionals in the urban agriculture sector.

CASE STUDY: IMPLEMENTING DUA IN SINGAPORE

Singapore, a densely populated city-state with limited land resources, has been actively exploring innovative solutions to enhance food security, reduce import dependency, and promote sustainable urban development (Diehl et al., 2020). DUA has emerged as a promising approach to address these challenges, offering the potential to increase local food production while minimizing the environmental impact. This section presents a case study of an urban agriculture project in Singapore, highlighting the application of digital tools and their integration throughout the system's lifecycle, including the design, construction, and operation phases.

This project is funded by the Singapore Food Story (SFS) R&D Program grant call toward Singapore's "30 by 30" policy, a strategic initiative to enhance the nation's food security and resilience. The primary goal of the policy is to produce 30% of Singapore's nutritional needs locally by 2030, a significant increase from the current figure of less than 10% (SFA, n.d.). This ambitious target seeks to reduce Singapore's dependence on food imports and increase self-sufficiency in the face of potential global supply chain disruptions, climate change impacts, and rising food demand. The "30 by 30" policy reflects Singapore's commitment to tackling the challenges of food security and sustainability in an innovative and forward-looking manner. By investing in advanced agricultural technologies, supporting local food production, and fostering collaborations, Singapore aims to create a resilient and sustainable food system for its residents (Kok, 2020).

This research project aims to attain the 30 by 30 target by improving yield via filtered natural lighting via novel coated film to optimize the morphogenetic stimulus and photo-morphogenetic response of Asian greens grown in Singapore's urban farmscape. To enhance the utilization of space, an eco-resilient plug and growth facility, "Urban-metabolic Farming-module (UmFm)" system, has been built for growth with minimal utilization of energy resources (William et al., 2022). Data analytic competencies have been built into IoT systems for the evaluation of the effectiveness of crop protection technologies and pest management systems with tracking of post-harvesting cycles to ensure crops' freshness.

To better understand the context constraint, the research team conducted simulations to support the design and construction of UmFm – a modular farming unit. Also, the system is optimized in terms of natural ventilation with the supplement of mechanical ventilation and optimized system envelope to introduce appropriate sunlight and reduce the solar heat gain suitable for operation by a commercial farmer in tropical residential areas. In the below subsections, we will describe how we utilize digitals to realize the architectural design and fabrication of the UmFm unit for optimal performance of the urban farming module. The study aims to serve as a model for sustainable urban farming system design, leveraging digital tools to optimize performance and resource efficiency.

SITE CONTEXT

The selected site for the UmFm project is situated in the campus courtyard of the Singapore Institute of Technology (SIT) (see Figure 6.1), consisting of three consecutive descending terraced surfaces, which evoke a stepped appearance to emulate a vibrant and multifaceted urban setting. It presents unique opportunities and challenges for developing a sustainable and efficient urban agricultural system in the tropical urban context. The site is characterized by its strategic location, surrounding

FIGURE 6.1 Site location of UmFm (open field with receding steps).

land use, climate conditions, and topography, which collectively inform the design and implementation of the urban farming system.

Climate Conditions

A thorough analysis of the site's microclimate is essential for successfully implementing an urban farming project. Two key factors to consider are wind patterns and sun exposure, which can be assessed using a wind rose diagram and a sun path diagram. By considering the above-mentioned unique characteristics, the project can effectively harness the benefits of these natural resources, creating a sustainable and efficient agricultural system responsive to the local microclimate.

Wind Rose Diagram – A wind rose diagram is a graphical representation of the frequency and intensity of wind from different directions at a specific location (NOAA, n.d.). Reviewing the wind rose diagram for the site provides insights into prevailing wind directions and speeds, allowing agricultural facility planning and design professionals to strategically position structures, openings, and natural ventilation systems to harness the cooling effects of wind and promote air circulation. Wind primarily originates from the north to northeast during the Northeast Monsoon (December to March) and from the south to southeast during the Southwest Monsoon (June to September). The wind's intensity is generally more significant during the Northeast than during the Southwest Monsoon period (MSS, n.d.).

Sun Path Diagram – A sun path diagram illustrates the sun's path across the sky at different times of the day and year for a particular location (Autodesk, n.d.). By reviewing the sun path diagram, professionals can understand the site's solar exposure, enabling them to optimize building orientation, shading devices, and glazing strategies. For example, by analyzing the solar elevation angle at various times of the month and day, design professionals can identify their project's physical constraints and opportunities. One such factor is the selection of polycarbonate panels used for UmFm. The solar elevation angle affects the angle of incidence at which sunrays strike the shelter, influencing the amount of reflection and sunlight penetration into the urban farming unit. Different elevation angles can either increase or decrease sunlight penetration, directly impacting the growing conditions within the unit. This knowledge can be used to maximize daylight for crop growth and improve the microclimate of the urban farming system.

Topography – The site showcases a stair-like artificial terrain, presenting opportunities for tiered structures that optimize available space.

FIGURE 6.2 Perspectives of modular urban farming unit: (left) generated by Autodesk Revit; (right) taken by drone.

By carefully considering the site's strategic location, surrounding land use, climate conditions, and topography, the project can maximize its potential for creating a sustainable and efficient agricultural system that contributes positively to the urban environment.

Application of Digital Tools in the Design Phase of UmFm

A systematic approach was conducted to achieve the project's goal. The process began with a thorough site analysis to obtain necessary measurements and identify environmental challenges associated with the site location. Following this, multiple case studies and projects related to structures built on terraced terrain were investigated. Also, meetings with agriculture experts and industry partners were conducted to determine the most suitable materials for the modular urban farming unit. Subsequently, extensive studies were carried out to understand how individual Revit families could be modeled. The proposed 3D BIM model was built based on the most accurate dimensions available from the product suppliers' catalogs (see Figure 6.2). Subsequently, we utilized the model for further performance analysis based on the lighting, heat load, and airflow (CFD).

BIM Implementation in Design Phase

During the design phase, the project team utilized BIM software Autodesk Revit 2021 to create a comprehensive 3D model of UmFm. This model integrates various components such as structural elements, grow systems, water management systems, and facility envelopes. BIM facilitated seamless communication among stakeholders, such as research members, agriculture experts, and industry partners, enabling the team to identify and resolve potential issues early in the design process, ultimately reducing costs and improving overall system performance (Cumo et al., 2020).

The design concept for the project involves a modular approach utilizing a scaffolding system, which offers adaptability to the changes in terrain at the site. Based on the identified design requirements, the structural elements must be robust and versatile to withstand the environment. The modular design concept involves subdividing the product into smaller components that can be assembled. These products consist of multiple prefabricated scaffolding modules that can save time, cost, and resources (see Figure 6.3). The modular design was incorporated into the project as it allows for high customizability, portability, reusability, and adaptability within the current design context.

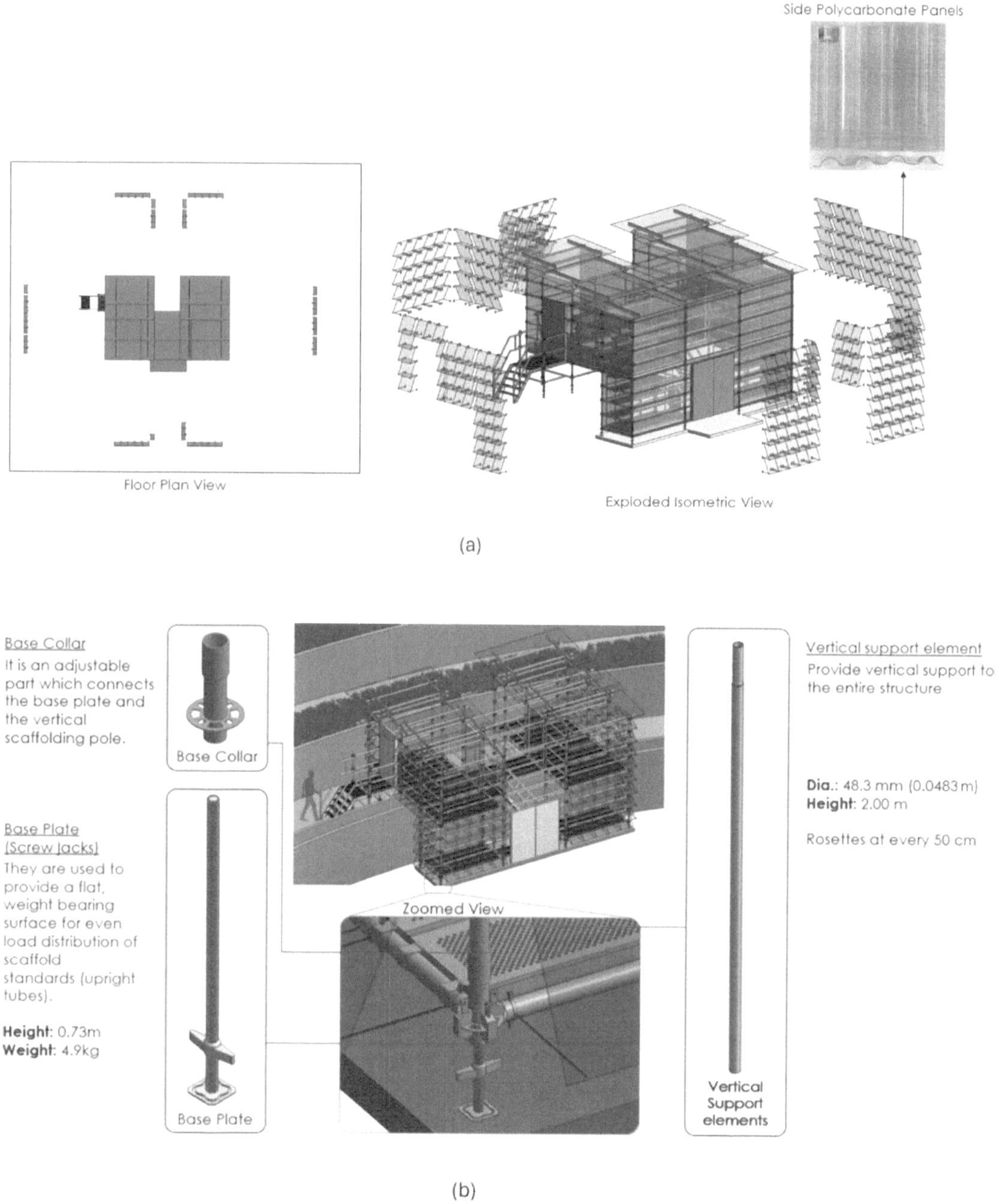

FIGURE 6.3 Component breakdown of (a) side polycarbonate panels; (b) base plate, base collar, and vertical support elements; (c) vertical (standards/uprights) and horizontal (ledger) support elements; (d) plan view of level 2 of UmFm; and (e) section view of UmFm.

(*Continued*)

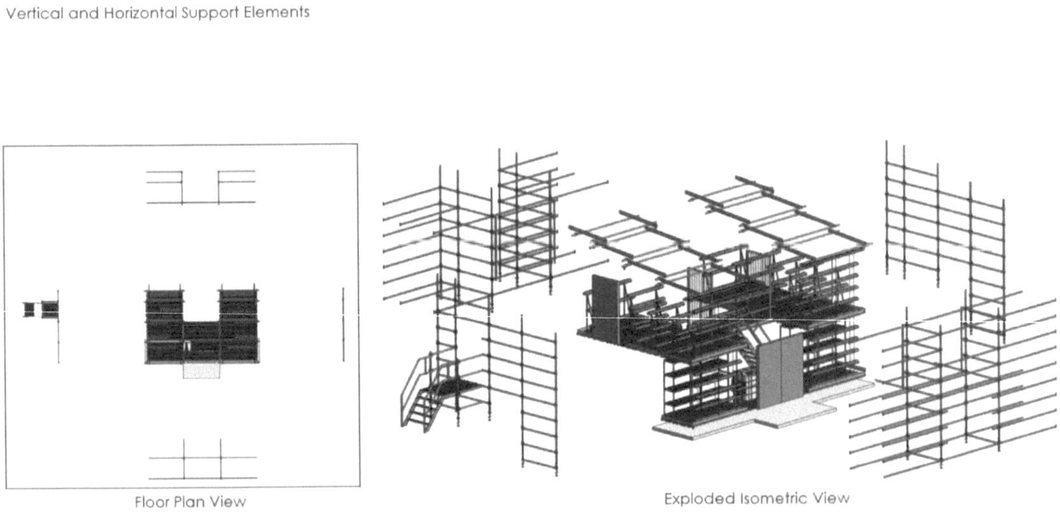

(c)

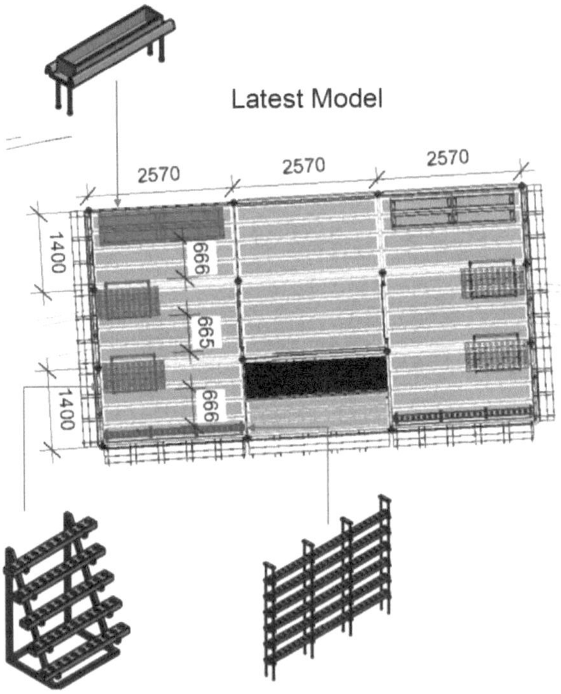

(d)

FIGURE 6.3 (*CONTINUED*) Component breakdown of (a) side polycarbonate panels; (b) base plate, base collar, and vertical support elements; (c) vertical (standards/uprights) and horizontal (ledger) support elements; (d) plan view of level 2 of UmFm; and (e) section view of UmFm.

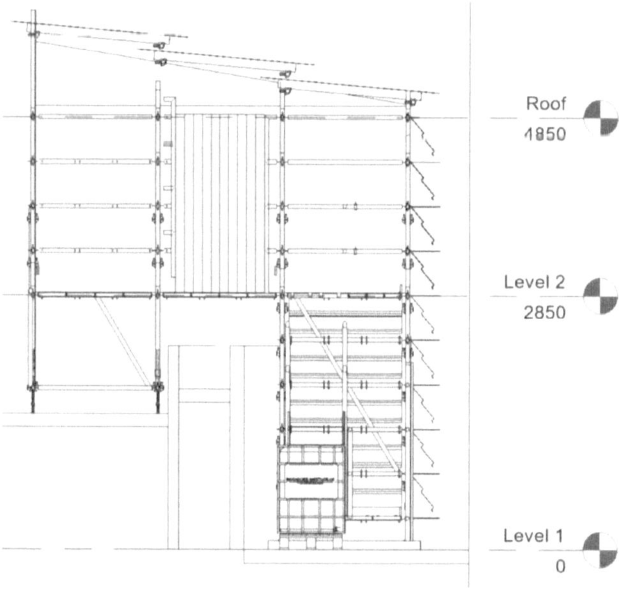

(e)

FIGURE 6.3 *(CONTINUED)* Component breakdown of (a) side polycarbonate panels; (b) base plate, base collar, and vertical support elements; (c) vertical (standards/uprights) and horizontal (ledger) support elements; (d) plan view of level 2 of UmFm; and (e) section view of UmFm.

Performance Simulations and Diagnostics

The project team employed computational performance simulations and diagnostics tools to optimize the UmFm design in terms of performance. Shadow analysis and daylighting simulations guided the system design for better daylight performance, ensuring optimal light distribution and intensity for crop growth. Heat load simulations were used to calculate the system's thermal load, informing the insulation materials selection and the enclosure's passive design strategies. CFD simulations were employed to analyze the airflow patterns and thermal behavior within UmFm. This allowed for identifying potential issues and optimizing the system's performance.

Shadow Analysis – This study aims to analyze and identify the most optimal location for building UmFm. The ideal location should enable the growing frames to absorb sunlight efficiently without compromising the performance of the photovoltaic panels situated on the grass patch to the south of the elevated garden. A shadow analysis has been carried out to examine the impacts of adjacent buildings on UmFm in a temporal approach. In contrast, the sunpath diagram, adjacent buildings, the site's topography, orientation, and available footprint were considered to determine the optimal layout and design of the facility to minimize heat gains and losses.

Two primary processes are involved in determining the farm location. First, shadow analysis was conducted on three locations to assess whether shadows impact the PV panels. Second, annual solar radiation analysis has been performed to determine which location offers the highest annual solar radiation for the facades. Following these analyses, a comparison will be conducted to finalize the site location. Figure 6.4 illustrates the three different locations under consideration.

The shadow analysis used SketchUp in the preliminary design phase, referencing conditions from URA's Privately-Owned Public Spaces Scheme (URA, n.d.). Figure 6.5 displays the sun path

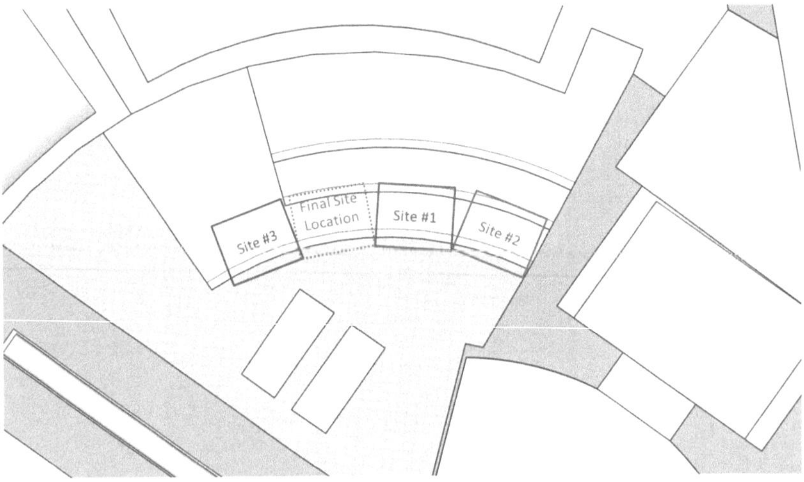

FIGURE 6.4 Three potential site locations, namely #1, #2, and #3, have been evaluated in the preliminary design stage.

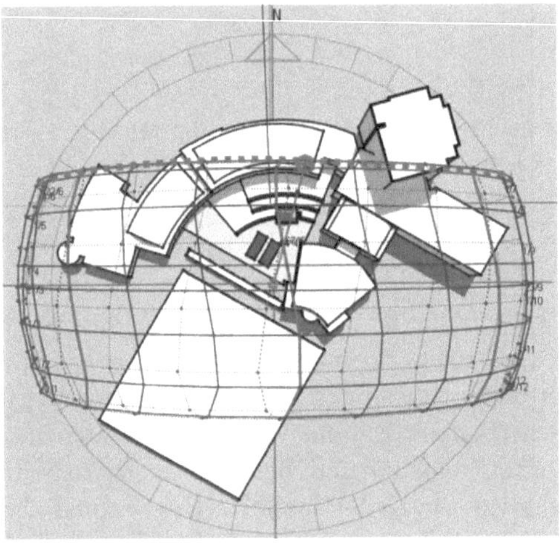

FIGURE 6.5 Sun path diagram of Singapore on 21 June.

diagram of Singapore on 21 June, highlighted in orange. For the sun shadow study, shadow diagrams must be examined for shadows cast on 21 June at three different times: 9:00 a.m., 12:00p.m., and 4:00 p.m. These conditions were employed to conduct the sun shadow study for the three locations.

Figure 6.6 illustrates the sun shadow study of the farm facility at positions #1, #2, and #3. Analyzing the results from the shadow analysis reveals that the adjacent buildings will entirely cover the farm as it moves along the northern portion of the stereographic diagram. Consequently, the solar radiation during the morning would likely be lower. This is probably due to the office building located in the northeastern portion of the site which prevents the sun from directly illuminating the vegetable farm in the morning. For timings at 12:00 p.m. and 4:00 p.m., the farm facility is barely affected by the surrounding buildings, as the buildings on the western side are of shorter heights, and the sun's angle does not permit the shadow of the buildings on the western side to impact the farm. Similarly, all three positions demonstrate that the farm facility would only be affected by the adjacent buildings during the morning, while the afternoon sun would still be able to shine directly onto the farm.

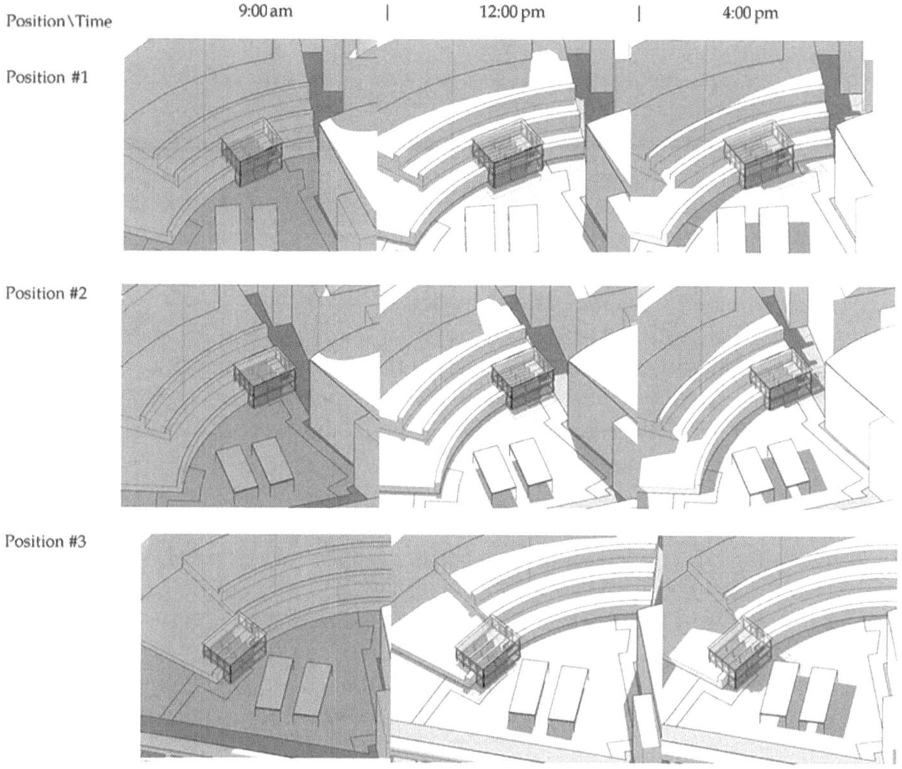

FIGURE 6.6 Selected sun shadow study for positions #1, #2, and #3 in Terms of 9 a.m., 12 p.m., and 4 p.m. on 21 June.

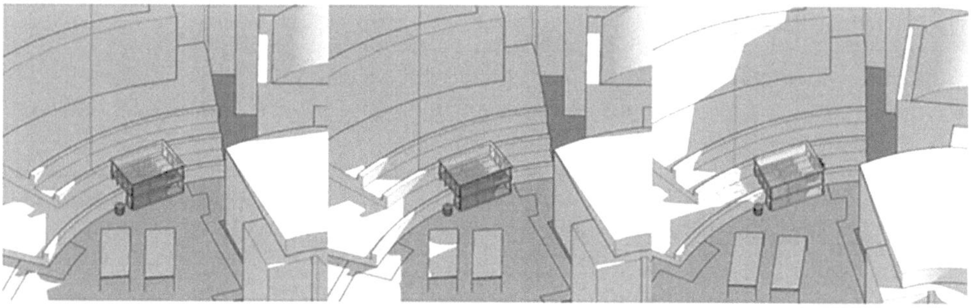

FIGURE 6.7 Sun shadow study for position 1 at 9 a.m. on 21 March (left), 21 September (middle), and 21 December (right).

As we analyze the site layout @SIT, we can understand that other critical dates shall also be considered apart from the sun path during the Summer Solstice (21 June). They are the Spring Equinox (21 March), the Autumn Equinox (21 September), and the Winter Solstice (21 December). Thus, additional sun shadow study should be conducted to understand the impact of the sun path across the entire year, as shown in Figure 6.7.

For this portion, the sun shadow study at 9:00 a.m. was selected for further analysis for the remaining three seasonal variations, as the taller buildings are mainly located at the eastern portion of the site. Based on the results, we can predict that throughout the year, buildings placed on the east would impact the solar access and the illuminance of the planter boxes during the morning. Position #1 is the most suitable site location, as it receives a more significant amount of sunlight than the other two positions. Specifically, Positions #1, #2, and #3 are covered by sunlight at 10:00 a.m., 11:00 a.m., and 9:15 a.m., respectively, on 21 June.

Daylighting Assessment – To further understand the daylight performance inside this urban farming facility, we conducted a simulation via Rhino 7 with its plugin ClimateStudio. This assessment enables us to better understand the PPFD (μmol/m^2/s) and DLI (mol/m^2/day) experienced by crops within the designed urban farming facility. This process involves several steps and considerations to ensure accurate results and practical design, namely geometry creation, identification of the areas, space usage patterns, and sky models (Chen et al., 2022).

Geometry Creation: To create a 3D model of UmFm, including all relevant architectural and structural elements. This model should accurately represent the dimensions, materials, and overall design of the facility, as well as the surrounding context, such as adjacent buildings or other shading elements.

Area of Interest: identify the specific areas within the facility where daylighting performance is crucial, such as crop growth zones or workspaces. These areas of interest will be the primary focus of the simulation and analysis.

Space Usage: It is important to define the usage of each space within the facility, as this information will inform the daylighting requirements and performance criteria. For example, zones dedicated to crop growth may require specific PPFD and DLI levels to ensure optimal growth, while germination chambers (i.e., Level 1 of UmFm) may have different daylighting needs for seedling production.

Sky Model: Select an appropriate sky model for the simulation, which represents the distribution of daylight in the sky based on location, time, and weather conditions. Commonly used sky models include the CIE Standard Overcast Sky, Clear Sky, and Intermediate Sky. The selection of the sky model will depend on the specific project requirements and the desired level of accuracy in the simulation.

Simulation Settings: Configure the simulation settings, including parameters such as the time step (e.g., sub-hourly or hourly), simulation period (e.g., a full year or specific seasons), and daylighting metrics (e.g., PPFD, DLI, or others). These settings will affect the granularity and scope of the results obtained from the simulation.

The simulation results show that the facility has an adequate distribution of daylight throughout most of the crop growth zones on level 2 (see Figure 6.8). Most of the areas achieved the desired PPFD and DLI levels required for optimal crop growth, aligning with the grow light requirements. However, there are two points worth addressing below.

Areas of Insufficient Daylight: A few areas within the facility were found to receive insufficient daylight, particularly those obstructed by adjacent structures or grow frames for a short period over a day. These areas may require additional passive design interventions, such as sunlight reflectors, to ensure that the daylighting needs of the space are met.

Areas of Excessive Daylight: On the contrary, some areas within the facility experienced excessive daylight, leading to potential issues such as overheating or crop stress. Design modifications, such as coated reflective roof panels or light-diffusing materials, may be necessary to mitigate these issues and improve the overall daylighting performance of the facility. It's worth noting that this research primarily focused on evaluating the performance of pre-designed planter systems provided by our industry partner. Consequently, parameters like dimensions, arrangement, spacing, and tilt angle of planter boxes remained constant throughout, marking a limitation. Future studies may explore the impact of altering these parameters on urban farming system performance.

Heat Load Simulation – In this greenhouse-type facility-UmFm, the heat load simulation primarily focuses on solar heat gain. Utilizing Rhino 7 and Climate Studio, the analysis was conducted using the previously constructed 3D model, which includes adjacent buildings. This simulation process involves a series of steps and considerations to guarantee precise results and efficient design. The following is a step-by-step overview of the process, emphasizing annual heat load data:

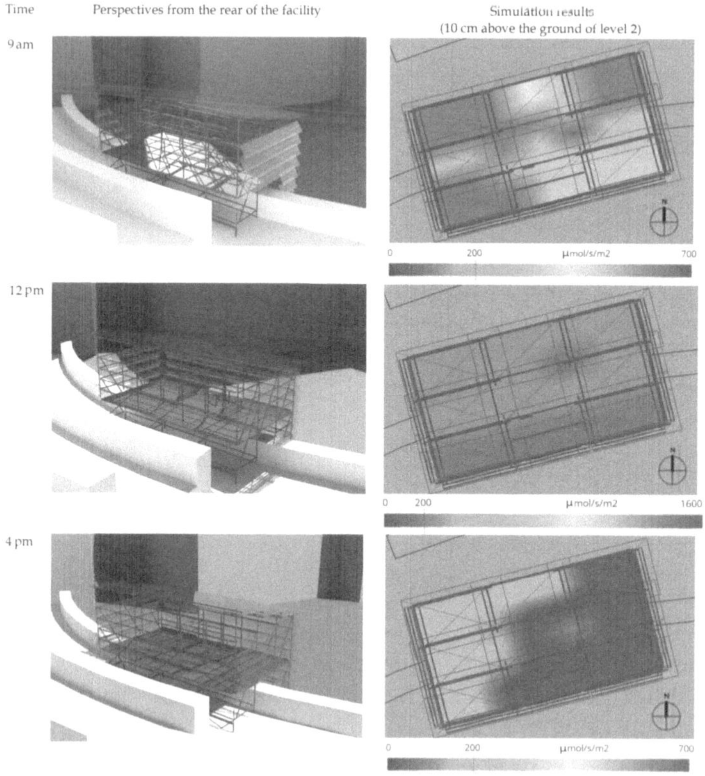

FIGURE 6.8 Selected daylighting simulation results after conversion to PPFD (Date: 21 January).

Space Usage: The use of each space within the facility was clearly defined, as different areas have varying heat load requirements depending on their function (e.g., diverse crop growth zones or systems, especially in terms of heights and horizontal extents).

Material Properties: Assign appropriate material properties to the 3D model, such as thermal conductivity, heat capacity, and reflectivity, to represent the building envelope's performance in the simulation accurately. The primary materials used for our facility consist of polycarbonate panels as the enclosure and a scaffolding framework for support.

Simulation Settings: Configure the simulation settings, including parameters such as the time step (e.g., day or month), simulation period (e.g., a full year or specific seasons), and heat load metrics (e.g., total heat gains and losses, or others). These settings will affect the granularity and scope of the results obtained from the simulation.

The heat load simulations are conducted with the model parameters set (including space usage, material properties, and simulation settings). The results, generated from point in time (e.g., 9 a.m., 12 p.m., and 4 p.m.) and annual irradiance exposure, provide an understanding of UmFm's heat load performance under conditions without mesh nets on the facades. This analysis identifies areas with high or low heat gains and losses within specific cells (i.e., high or low layers of left and right cells), assesses the effectiveness of insulation and shading strategies, and evaluates the overall thermal performance of the facility. Figure 6.9 presents an example of heat load simulation results, which assessed the heat exposure experienced by the grow layers within UmFm. The annual solar irradiance reaches 614 kWh/m^2, averaging 140 W/m^2 daily. As a result, heat load simulation data provides valuable insights into the heat load inside UmFm and the thermal conditions the plants are exposed to and making informed decisions regarding the design and placement of the growing layers and facility enclosure to optimize plant growth and overall system performance using passive design strategies and sustainable approaches. A comprehensive understanding of the local context is vital

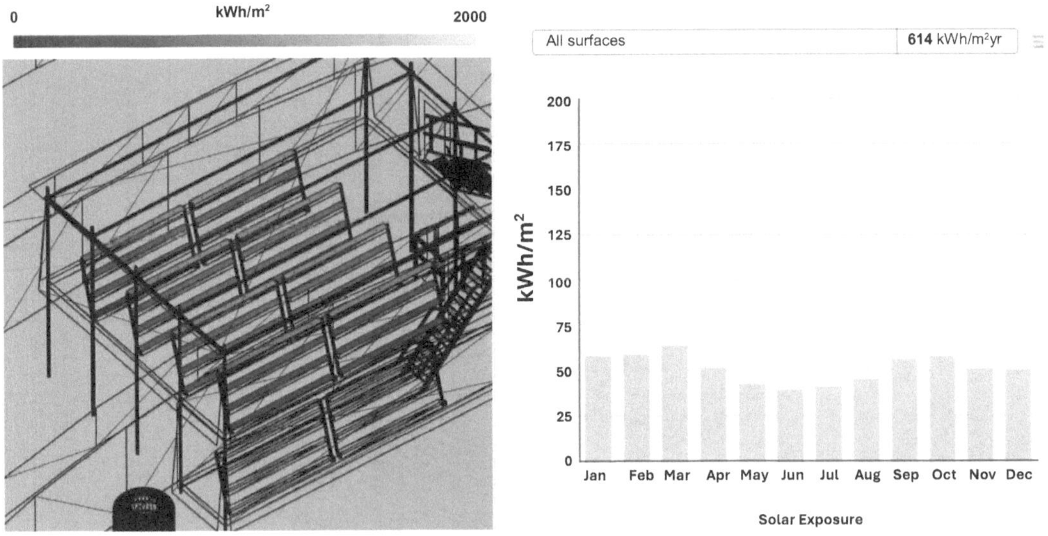

FIGURE 6.9 Annual solar irradiance experienced by the grow system at the second level of UmFm.

for developing urban agriculture projects in Singapore that function efficiently while fostering environmentally conscious practices. Given that Singapore experiences approximately 12 hours of daylight each day throughout the year, it is essential to consider this factor in urban farming initiatives' design and planning stages, ensuring that the projects are productive and sustainable.

CFD Simulation – To optimize the natural ventilation performance inside UmFm, a set of CFD simulations are conducted for the structure with consideration of the wind direction at different months of the year in Singapore. Following the consideration of concerns of bugs and pests, the Design Team decided to include mesh netting in the UmFm. The CFD models, with the inclusion of the mesh netting and the polycarbonate side façade with varying degrees of tilting, are established to understand the airflow in both chambers with prevalent wind directions. Figure 6.10 shows the computational domain of UmFm for CFD simulation (north wind scenario).

The selected results of the global airflow at the farming site are shown in Figure 6.11, with the farm's site location indicated by the red box. As demonstrated in s Figure 6.11a and b, the airflow

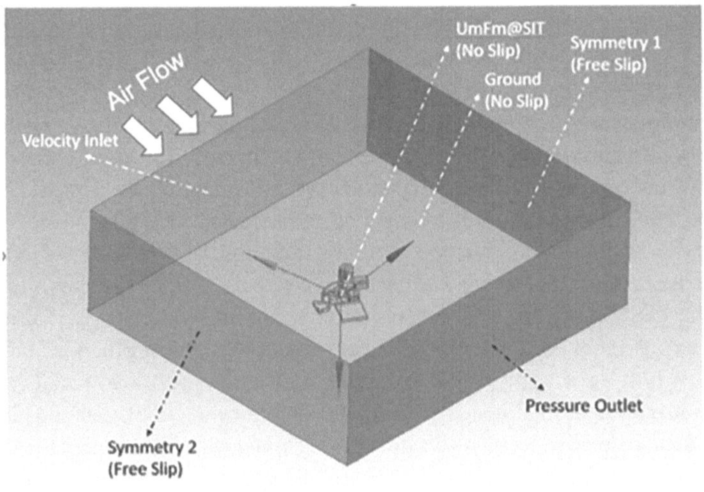

FIGURE 6.10 Computational domain of UmFm for CFD simulation (north wind scenario).

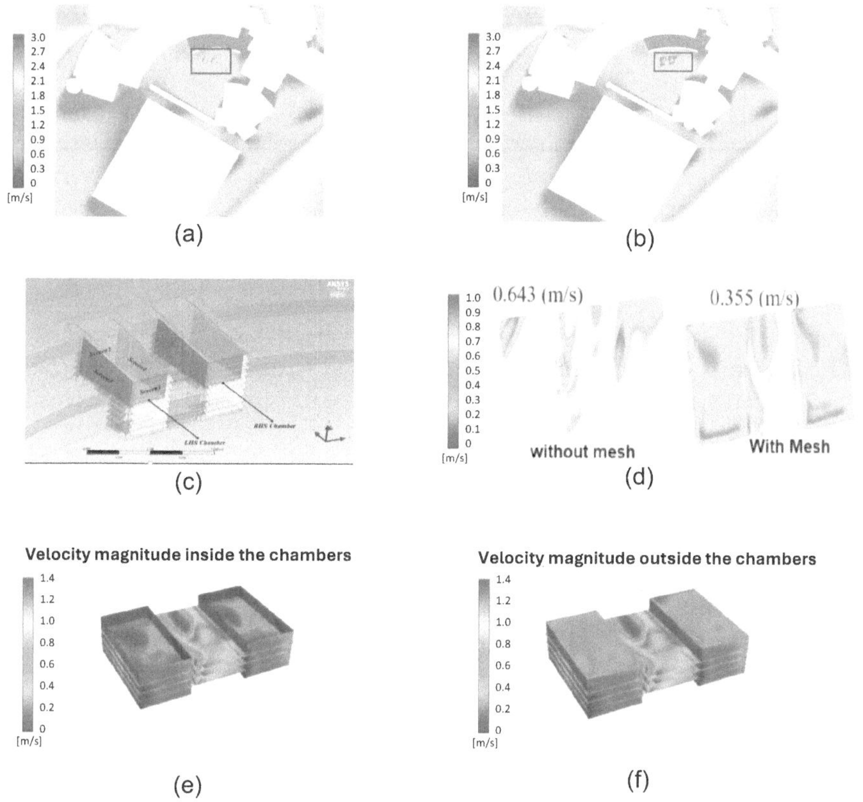

FIGURE 6.11 The global airflow around the farming site at a height of 3.5 m from ground level (a) without and (b) with mesh meeting of porosity of 36.5%. (c) A 3D model of a modular farm unit with screens 1–5 represents the mesh netting cover in one chamber. (d) UmFm setup without mesh netting has a significantly higher area-weighted average velocity (AWAV). (e) The airflow is less regulated inside the structure. (f) Better airflow regulation outside the chamber with mesh netting.

at 3.35 m above-ground level with mesh netting experiences a reduction in air velocity compared to the scenario without mesh netting.

Figure 6.11c depicts the inclusion of five screens representing the mesh netting in each chamber. The planar view of the CFD airflow modeling in Figure 6.11d illustrates that the UmFm setup without mesh netting has a significantly higher area-weighted average velocity of 0.643 m/s, as opposed to 0.355 m/s with mesh netting. The velocity contours at the chambers for the UmFm, as shown in Figure 6.11e, suggest that the airflow is less regulated (indicated by the patches of blue distribution) inside the structure. In contrast, Figure 6.11f demonstrates better airflow regulation outside the chamber with mesh netting, as represented by the more uniform patches of green.

The CFD simulation results in this study, which considered the ambient wind impact but not heat gain, allow us to evaluate the natural ventilation performance under various scenarios related to (1) the tilting angles of the polycarbonate panels mounted on the facades of UmFm and (2) the facility with and without the mesh netting. The CFD simulation was conducted isothermally, focusing primarily on the effect of natural or cross ventilation from the window façade. This is because we believe that internal heat gain has no significant impact on internal airflow when there is strong cross ventilation due to wind. By comparing the outcomes for different tilting angle configurations, we can determine the most effective panel angles that optimize natural ventilation in response to dynamic wind conditions.

Refinement and Iteration: Key Activities in the Design Phase – The design and evaluation of UmFm involves a dynamic, iterative process that relies on various digital tools for continuous improvement and assessment (Wachter et al., 2003). Two crucial activities within the design phase are refining the design and iterating through the process, which are essential components of this scientific support approach.

Design Refinement: Insights gained from simulations and analyses inform modifications to UmFm's design. These modifications may include changes to geometry, enclosure materials, insulation, shading devices, or coating technologies to optimize sunlight exposure and ventilation, minimize heat gains, and enhance the facility's overall performance.

Iteration: The simulation and analysis process are repeated as necessary to refine and optimize the performance of the urban farming facility. This iterative approach ensures that the final design is efficient and effective in meeting UmFm's specific requirements.

By integrating digital tools into the design process, professionals can analyze and model different scenarios, identify potential challenges, and develop simulated-data-driven solutions. This ultimately leads to better decision-making, reduced resource consumption, and the creation of more resilient and adaptable urban farming systems.

Based on the authors' knowledge, currently, no fully integrated simulation platform effectively streamlines this process. Consequently, professionals in the field must collaborate closely with various stakeholders, including design consultants, agricultural experts, farm owners/operators, engineers, and construction teams (Almeida et al., 2016). Such a collaborative design approach ensures that each project stage is comprehensively addressed, allowing for the successful implementation of urban farming initiatives that meet all parties' diverse needs.

CHALLENGES AND OPPORTUNITIES

In this section, we investigate the challenges and opportunities of implementing digitalization in urban agriculture. By implementing digital technologies during the design phase of our urban agriculture project, we can identify potential barriers and prospects in this area. However, it is worth noting that there may be further opportunities to explore during other phases of the building life cycle. We have outlined four key points below to elaborate further on the challenges and opportunities.

Continuous Optimization and Adaptation

Continuous optimization and adaptation in the operation and maintenance (O&M) phase present excellent opportunities to better inform the implementation of digital tools in the design phase. For instance, computational simulation results obtained during the design phase can be calibrated using real performance data from the O&M phase. This can help design professionals improve the future design of urban farming systems by refining the parameters used in simulation tools.

During the O&M phase, data can be collected on various aspects of the urban farming system, such as energy consumption, crop yield and quality, water usage, and growth conditions. By comparing this real-world data to the initial computational simulations, design professionals can identify discrepancies, gaps, and areas of improvement and/or modification. This feedback loop can fine-tune the digital tools and models used during the design phase, resulting in more accurate simulations and better-informed design decisions. Furthermore, the continuous optimization and adaptation process during the O&M phase can help identify new technologies, materials, and methods that can improve the overall performance of urban farming systems. This information can be integrated into the digital tools used in the design phase, allowing design professionals to stay updated with the latest advancements in the field.

In addition, involving key stakeholders such as farm owners, operators, and maintenance staff in the continuous optimization process can provide valuable insights into the practical aspects of urban farming system operation. Their input can help design professionals better understand the challenges faced during the O&M phase and incorporate this knowledge into the design requirements and process, resulting in more resilient, efficient, and adaptable urban farming systems.

Moreover, the previously built BIM model from the design phase could be further developed into a digital twin (Chien et al., 2017) that incorporates facility infrastructure, monitoring data collected from the microclimate of the urban farming facility, crop growth condition, the status of control devices, and even control commands. In this way, the digital twin can accurately reflect the real-world performance of the system (Chaux et al., 2021), with the potential for dynamic optimization and adaptation powered by artificial intelligence (AI) and machine learning.

Photogrammetry Solution

Ongoing technological advancements promise to further enhance the capabilities and performance of implementing digital tools in urban farming systems. One emerging tool will be the photogrammetry solution with the integration of both hardware (e.g., 3D laser scanner and drone) (Bouziani et al., 2021) and software (e.g., Reality Capture) (Berrett et al., 2021). It could support the rapid modeling of greenhouse and indoor vertical farming systems and offer precise, detailed 3D models through point clouds. For example, a preliminary photogrammetry study using the RealityCapture software was tested for its effectiveness in rapidly modeling urban farming systems. We took 476 photos using a DJI Mini 3 Pro drone, focusing on a farm facility, Agritech Farm, located near UmFm@SIT. This site was selected due to its relatively smaller size and simpler geometry. The drone photos were taken at noon to ensure optimal daylight conditions and minimize shadows on the ground, which could lead to incorrect identification of objects in the virtual environment. The drone-captured images at various heights and locations, following the flight trajectory shown in Figure 6.12.

Through this initial trial using drone-captured photogrammetry images, we identified areas that warrant further exploration. For example, smaller elements such as fencing and structural supports require the drone to fly closer to capture better images for the software to stitch together. Additionally, there were insufficient images for certain parts of Agritech Farm, as evidenced by holes in the rendered solar panels. Further testing and trials will be conducted to develop a more complete and accurate architectural model within the software.

Photogrammetry could assist in various stages of the urban agricultural process. Such examples include accurate site measurements (including dimensions, elevations, and topography) and contextual understanding (pertaining to the surrounding context of an urban agricultural facility, including nearby buildings, streets, and vegetation) in the planning stage for better-informing decisions related to site selection, orientation, and integration with the existing urban context. Also, photogrammetry could be integrated with BIM platforms, allowing design professionals to create

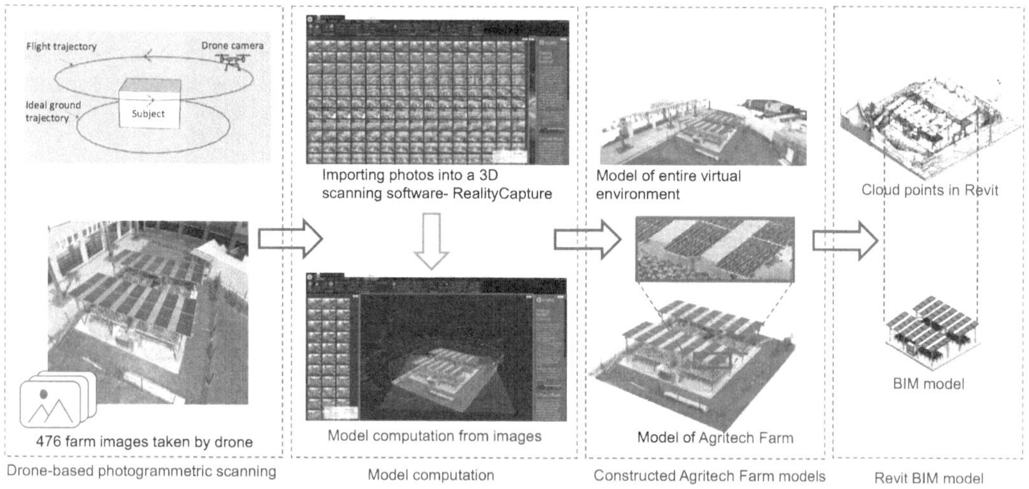

FIGURE 6.12 Photogrammetry workflow for farm scanning with the drone.

accurate and detailed 3D models of the agricultural facility. This can streamline the design and construction process by reducing errors and miscommunications between project stakeholders.

Barriers to Implementation of Digitalization

In our case study, we could conclude that implementing digitalization in the design phase of urban farming systems could bring along quite a few benefits. However, it is a complex process that involves some significant barriers. Some discussions of the barriers are listed below, namely high costs of implementation, lack of technical knowledge and skills, resistance to change, and limited availability.

High Costs of Implementation – The high cost associated with digitalization implementation (i.e., significant investment in terms of hardware, software, and upgrading employees' skills) make it difficult for small and medium enterprises in the fields of urban agriculture to afford these technologies, limiting their capability to adopt them.

Lack of Technical Knowledge and Skills – Urban farmers may face challenges related to technical knowledge and skills, which can result in steep learning curves when attempting to implement and effectively conduct digital technologies. This can limit their ability to take advantage of the benefits that digitalization can offer, such as increased efficiency, better farm design toward crop yields/quality, and sustainability.

Resistance to Change – Urban farmers may be hesitant to adopt new technologies, especially if they are "comfortable" with traditional/common practices and experiences and have concerns about the potential risks and uncertainties associated with new approaches.

Limited Availability of Relevant Data – Implementing digital technologies in urban agriculture often requires access to relevant data, such as climate and weather data. However, such data may not be readily available or may be difficult to access and interpret for them.

Usability – This study implemented advanced simulation tools to accurately evaluate critical parameters within the urban farming model, providing high-resolution scientific results. However, this approach can pose challenges to farmers with limited resources, technical expertise, or knowledge in utilizing such tools. There is a need for more holistic platforms that allow the evaluation of daylighting, CFD, sunpath, and solar heat gain parameters in a user-friendly manner. These platforms could be particularly useful for those applications with less stringent accuracy requirements.

To overcome these barriers, it is essential to collaborate and engage with key stakeholders (e.g., investors, grantors, and academic and industry experts). This includes investing in the development of more efficient workflow and user-friendly digital tools that can better cater to the needs of urban farmers. Furthermore, focusing on training and education programs on the interoperability of urban agriculture and digital technologies can help cultivate a workforce with the necessary skill sets and knowledge. By addressing these barriers, we can create a more supportive environment for the implementation of DUA, which can help to realize its full potential toward sustainable urban agriculture facility design.

Considering Climate Change in Computational Simulations

Addressing climate change is an emerging aspect of urban farm design, as it has significant implications for the sustainability, productivity, and resilience of these agricultural systems. Traditionally, computational simulations have relied on past weather data spanning 20–30 years to inform the design process (Herrera et al., 2017). However, this approach may fail to account for the potential impacts of climate change on future urban farming conditions, leading to less climate-resilient designs.

To better evaluate urban farms' performance in climate change, it is necessary to incorporate projected climate change data profiles into the computational simulation process (Jylhä et al., 2015). This approach can provide a more accurate representation of future environmental conditions, helping design professionals make informed decisions to enhance climate resilience in urban farm design. Thereby, the following points are discussed regarding the importance of considering climate change in computational simulations for urban farm design.

Future Climate Projections – Incorporating climate change data profiles, such as temperature/RH increase, shifts in precipitation patterns, and extreme weather events, can help design professionals and engineers better understand the potential challenges and risks associated with future climate conditions. This information can be considered to develop design strategies that enhance the resilience of urban farming systems against the changing climate.

Adaptable Microclimate Management – By considering projected climate change data, computational simulations can support the optimization of the microclimate within urban agricultural facilities to account for future environmental scenarios. This may involve the modification of the facility enclosure, adjusting ventilation methods and HVAC systems, or, furthermore, modifying the selection of crops and growth media to adapt to the anticipated climate changes (Raza et al., 2019).

Resource and Energy Efficiency – Climate change projections can inform decisions about resource and energy management in urban farms. Design professionals can utilize this information to develop effective strategies for minimizing water and energy consumption and managing renewable energy sources, contributing to greater climate resilience.

In recent years, several initiatives and research projects have focused on generating projected future weather files for simulation tools to evaluate the performance of buildings and farming systems under future climate conditions (Gaur and Lacasse, 2022). These files are typically derived from climate model outputs, which are downscaled and processed to create realistic, location-specific weather data. Such examples include (1) Future Weather Files for Building Performance Simulation (UK) based on UK Climate Projections (UKCP09) data (Watkins et al., 2011); (2) ClimateAP for the Asia-Pacific region (Wang et al., 2017); (3) morphed weather (Europe) developed by the Fraunhofer Institute for Solar Energy Systems ISE, Climate Change World Weather File Generator (CCWorldWeatherGen) and Weather Morph (an online version of CCWorldWeatherGen) (Rodrigues et al., 2023). These examples illustrate the increasing awareness and need to incorporate projected future weather data into simulation tools. By using these future weather files, design professionals, engineers, and planners can better understand the potential impacts of climate change and develop climate-resilient strategies for the future design of urban farming systems.

CONCLUSIONS

DUA has emerged as a promising solution to the challenges of urban food security, land scarcity, and environmental sustainability, specifically in metropolitan cities, such as Singapore. By implementing digital tools and technologies in the design phase of urban farming facilities, it is possible to optimize urban farm's performance, enhance resource efficiency, and minimize environmental impact appropriate to its local context. Also, these tools could support design professionals (architects, design consultants, and engineers) to create innovative and efficient urban farming solutions that integrate seamlessly with new and existing buildings and infrastructure in a variety of urban building typologies, such as rooftop gardens, vertical farming on building facades, or indoor farming systems within unused spaces. In contrast, a set of conditions (involving specific building types, sizes, and constraints) are considered.

As urbanization and climate change continue to intensify, the need for innovative and sustainable approaches to urban food production has never been more critical. DUA offers immense potential to address these challenges and contribute to developing resilient and sustainable urban food systems. By embracing the opportunities offered by technological advancements and developing strategies to overcome existing barriers, DUA can play a substantial role in shaping the future of food production in our cities, ensuring a healthier, more sustainable, and food-secure future for urban dwellers.

ACKNOWLEDGMENTS

This research is supported by the Singapore Food Agency, Singapore, under its Singapore Food Story Theme 1 program (Award No. SFS_RND_SUFP_001_09). We would like to express our sincere gratitude to Dr. Youhanna E. William, Mr. Jarrod Chin, Mr. Nicholas Tang, Mr. Desmond Ong, Mr. Kenneth Sin, and Mr. Arijit Saha for their invaluable research support and contributions to our project during their postdoc, postgraduate, master's, and undergraduate studies. We also wish to acknowledge the use of AI language models (ChatGPT 3.5 & 4), which provided helpful feedback on earlier versions of this chapter. The models assisted with enhancing the structure, clarity, presentation of perspectives, and reference formatting. However, the authors are fully responsible for the final edited content, analysis, and errors.

REFERENCES

Almeida, P., Solas, M., Renz, A., Bühler, M. M., Gerbert, P., Castagnino, S., & Rothballer, C. (2016). Shaping the future of construction: A breakthrough in mindset and technology. Technical Report. https://doi.org/10.13140/RG.2.2.21381.37605.

Ansys. (n.d.). Ansys fluent. Retrieved from https://www.ansys.com/products/fluids/ansys-fluent.

Autodesk. (n.d.). Reading sun path diagrams. Retrieved from https://www.autodesk.com/support/technical/article/caas/tsarticles/ts/2pGZ0xLAMCrBMy9xLtObJM.html.

Azhar, S. (2011). Building information modeling (BIM): Trends, benefits, risks, and challenges for the AEC industry. *Leadership and Management in Engineering*, 11(3), 241–252. https://doi.org/10.1061/(ASCE)LM.1943-5630.0000127.

Bartzanas, T., Kacira, M., Zhu, H., Karmakar, S., Tamimi, E., Katsoulas, N., In Bok Lee, & Kittas, C. (2013). Computational fluid dynamics applications to improve crop production systems. *Computers and Electronics in Agriculture*, 93, 151–167. https://doi.org/10.1016/j.compag.2012.05.012.

Berger, T. (2001). Agent-based spatial models applied to agriculture: A simulation tool for technology diffusion, resource use changes and policy analysis. *Agricultural Economics*, 25, 245–260.

Berrett, B. E., Vernon, C. A., Beckstrand, H., Pollei, M., Markert, K., Franke, K.W., & Hedengren, J.D. (2021) Large-scale reality modeling of a university campus using combined UAV and terrestrial photogrammetry for historical preservation and practical use. *Drones*. 5(4), 136. https://doi.org/10.3390/drones5040136.

Bouziani, M., Chaaba, H., & Ettarid, M. (2021). Evaluation of 3D building model using terrestrial laser scanning and drone photogrammetry. *The International Archives of the Photogrammetry, Remote Sensing and Spatial Information Sciences*, XLVI-4/W4-2021, 39–42. https://doi.org/10.5194/isprs-archives-XLVI-4-W4-2021-39-2021.

Carolan, M. (2020). Urban farming is going high tech: Digital urban agriculture's links to gentrification and land use. *Journal of the American Planning Association*, 86(1), 47–47.

Chatterjee, A., Debnath, S., & Pal, H. (2020). Implication of urban agriculture and vertical farming for future sustainability. *IntechOpen*. https://doi.org/10.5772/intechopen.91133.

Chaux, J. D., Sanchez-Londono, D., & Barbieri, G. (2021). A digital twin architecture to optimize productivity within controlled environment agriculture. *Applied Sciences*. 11(19), 8875. https://doi.org/10.3390/app11198875.

Chen, H. Y., Chien, S.-C., Yong, S. K. T., Yeh, I.-L., Fan, P. E. M., Qian, H. V., Leow, W. X. B., Nur, L., Ng, L. S., Aloweni, F., Leow, L. C., Phua, G. C., & Ang, S. Y. (2022). Simulation-based assessment of human-centric lighting performance in tropical hospitals. Plea Santiago 2022, 23–25 November 2022, Santiago, Chile.

Chien, S. C., Chuang, T. C., Yu, H. S., Han, Y., Soong, B. H., & Tseng, K. J. (2017). Implementation of cloud BIM-based platform towards high-performance building services. *Procedia Environmental Sciences*, 38, 436–444. https://doi.org/10.1016/j.proenv.2017.03.129.

ClimateStudiodocs. (n.d.). Radiation map [Web page]. https://climatestudiodocs.com/docs/radiationMap.html.

Cumo, F., Piras, G., Pennacchia, E., & Cinquepalmi, F. (2020). Optimization of design and management of a hydroponic greenhouse by using BIM application software. *International Journal of Sustainable Development and Planning*, 15, 157–163. https://doi.org/10.18280/ijsdp.150205.

Curic. (n.d.). Curic sun. SketchUp extension warehouse. Retrieved from https://extensions.sketchup.com/extension/49b56362-ada6-4bde-8213-8c68eb7763d1/curic-sun[21].

DIAL. (n.d.). DIALux: The software for your professional lighting design. Retrieved from https://www.dialux.com/en-GB/dialux.

Diehl, J. A., Sweeney, E., Wong, B., Sia, C. S., Yao, H., & Prabhudesai, M. (2020). Feeding cities: Singapore's approach to land use planning for urban agriculture. *Global Food Security*, 26, 100377. https://doi.org/10.1016/j.gfs.2020.100377.

Eaton, M., Harbick, K., Shelford, T., & Mattson, N. (2021). Modeling natural light availability in skyscraper farms. *Agronomy*. 11(9), 1684. https://doi.org/10.3390/agronomy11091684.

EnergyPlus. (n.d.). EnergyPlus. Retrieved April 16, 2023, from https://energyplus.net/.

Fu, F. (2018). Chapter six – Design and analysis of complex structures. In F. Fu (Ed.), *Design and Analysis of Tall and Complex Structures* (pp. 177–211). Butterworth-Heinemann. https://doi.org/10.1016/B978-0-08-101018-1.00006-X.

Gaur, A., & Lacasse, M. (2022). Climate data to support the adaptation of buildings to climate change in Canada. *Data*. 7(4), 42. https://doi.org/10.3390/data7040042.

Herrera, M., Natarajan, S., Coley, D. A., et al. (2017). A review of current and future weather data for building simulation. *Building Services Engineering Research and Technology*. 38(5), 602–627. https://doi.org/10.1177/0143624417705937.

IESVE. (n.d.). IESVE. [Website]. Retrieved April 16, 2023, from https://www.iesve.com/.

Jasak, H. (2009). OpenFOAM: Open source CFD in research and industry. *International Journal of Naval Architecture and Ocean Engineering*, 1(2), 89–94. https://doi.org/10.2478/IJNAOE-2013-0011.

Jylhä, K., Ruosteenoja, K., Jokisalo, J., Pilli-Sihvola, K., Kalamees, T., Mäkelä, H., Hyvönen, R., & Drebs, A. (2015). Hourly test reference weather data in the changing climate of Finland for building energy simulations. *Data in Brief*, 4. https://doi.org/10.1016/j.dib.2015.04.026.

Kok, M. A. (2020). Singapore's emerging agritech ecosystem. UNDP global Centre for Technology, Innovation and Sustainable Development. https://www.undp.org/policy-centre/singapore/blog/singapores-emerging-agritech-ecosystem.

Konjoian, P. (2014). The evolution of agriculture: Urban expansion is a new frontier. Greenhouse Product News. Retrieved from https://gpnmag.com/article/evolution-agriculture-urban-expansion-new-frontier/.

Kozai, T., Niu, G., & Takagaki, M. (Eds.). (2019). *Plant factory: An Indoor Vertical Farming System for Efficient Quality Food Production* (2nd ed.). Academic Press Inc. (Imprint of Elsevier Science Publishing Co Inc).

Krühne, U., Bodla, V. K., Møllenbach, J., Laursen, S., Theilgaard, N., Christensen, L. H., & Gernaey, K. V. (2012). Computational fluid dynamics at work – Design and optimization of microfluidic applications. In I. A. Karimi & R. Srinivasan (Eds.), *Computer Aided Chemical Engineering* (Vol. 31, pp. 835–839). Elsevier. https://doi.org/10.1016/B978-0-444-59507-2.50159-1.

Kubba, S. (2012). Building information modeling. In S. Kubba (Ed.), *Handbook of Green Building Design and Construction* (pp. 201–226). Butterworth-Heinemann. https://doi.org/10.1016/B978-0-12-385128-4.00005-6.

Ledesma, G., Nikolic, J., & Pons-Valladares, O. (2022). Co-simulation for thermodynamic coupling of crops in buildings. Case study of free-running schools in Quito, Ecuador. *Building and Environment*, 207(Part A), 108407. https://doi.org/10.1016/j.buildenv.2021.108407.

Lighting Analysts, Inc. (n.d.a). AGi32: Lighting analysis software. https://lightinganalysts.com/software-products/agi32/overview/.

Lighting Analysts, Inc. (n.d.b). ElumTools: Lighting design software for Autodesk Revit. https://lightinganalysts.com/software-products/elumtools/overview/.

LightUp. (n.d.). LightUp. Retrieved from https://www.light-up.co.uk/.

Mahan, J. R., McMichael, B. L., & Wanjura, D. F. (1995). Methods for reducing the adverse effects of temperature stress on plants: A review. *Environmental and Experimental Botany*, 35(3), 251–258. https://doi.org/10.1016/0098-8472(95)00011-6.

Meteorological Service Singapore (MSS). (n.d.). Retrieved from https://www.weather.gov.sg/climate-climate-of-singapore/.

Naranjani, B., Najafianashrafi, Z., Pascual, C., Agulto, I., & Chuang, P. A. (2022). Computational analysis of the environment in an indoor vertical farming system. *International Journal of Heat and Mass Transfer*, 186, 122460. https://doi.org/10.1016/j.ijheatmasstransfer.2021.122460.

NOAA (n.d.). Wind roses - Charts and tabular data. https://www.climate.gov/maps-data/dataset/wind-roses-charts-and-tabular-data.

Nurmalisa, M., Tokairin, T., Kumazaki, T., Takayama, K., & Inoue, T. (2022). CO2 distribution under CO2 enrichment using computational fluid dynamics considering photosynthesis in a tomato greenhouse. *Applied Sciences*, 12(15), 7756. https://doi.org/10.3390/app12157756.

Orsini, F., Kahane, R., Nono Womdim, R., & Gianquinto, G. (2013). Urban agriculture in the developing world: A review. *Agronomy for Sustainable Development*, 33, 695–720. https://doi.org/10.1007/s13593-013-0143-z.

Pacheco Diéguez, A., Gentile, N., von Wachenfelt, H., & Duboisba, M.-C. (2016). Daylight utilization with light pipe in farm animal production: A simulation approach. *Journal of Daylighting*, 3, 1–11. https://doi.org/10.15627/jd.2016.1.

Pfister, F., Bader, H.-P. & Scheidegger, R. & Baccini, P. (2005). Dynamic modelling of resource management for farming systems. *Agricultural Systems*, 86, 1–28. https://doi.org/10.1016/j.agsy.2004.08.001.

Raza, A., Razzaq, A., Mehmood, S. S., Zou, X., Zhang, X., Lv, Y., & Xu, J. (2019). Impact of climate change on crops adaptation and strategies to tackle its outcome: A review. *Plants*. 8(2), 34. https://doi.org/10.3390/plants8020034.

Rendel, J., Mackay, A., Manderson, A., & Neill, K. (2013). Optimising farm resource allocation to maximise profit using a new generation integrated whole farm planning model. *Proceedings of the New Zealand Grasslands Association*, vol. 75. Tauranga, New Zealand.

Rodrigues, E., Fernandes, M., & Carvalho, D. (2023). Future weather generator for building performance research: An open-source morphing tool and an application. *Building and Environment*, 233, 110104. https://doi.org/10.1016/j.buildenv.2023.110104.

SimScale. (n.d.). Computational fluid dynamics (CFD) software. Retrieved from https://www.simscale.com/product/cfd/.

Singapore Food Agency (SFA). (n.d.). Strengthening our food security: 30 by 30. Our Food Future. Retrieved from https://www.ourfoodfuture.gov.sg/30by30.

SketchUp. (n.d.). Retrieved from https://www.sketchup.com/.

Solemma LLC. (n.d.). ClimateStudio. Retrieved from https://www.solemma.com/climatestudio.

Teitel, M. (2020). An overview of methods for the reduction of heat load in protected crops under warm climatic conditions. *Acta Hortic*. 1296, 691–706. https://doi.org/10.17660/ActaHortic.2020.1296.88.

Trimble Inc. (n.d.). Sefaira. *SketchUp*. Retrieved from https://www.sketchup.com/products/sefaira/launch.

TRNSYS. (n.d.). TRNSYS – Transient system simulation tool. [Website]. https://www.trnsys.com/.

Urban Redevelopment Authority (URA). (n.d.). Privately-owned public spaces (POPS). Retrieved from https://www.ura.gov.sg/Corporate/Guidelines/Development-Control/gross-floor-area/GFA/Privately-OwnedPublicSpacesPOPS[51].

Vizeu da Silva, A. F. C., Godinho, A. O., Agreira, C. I. F., & Travassos Valdez, M. M. (2016). An educational approach to a lighting design simulation using DIALux evo software. *2016 51st International Universities Power Engineering Conference (UPEC)*, Coimbra, Portugal, pp. 1–6. https://doi.org/10.1109/UPEC.2016.8114127.

Wachter, S. B., Agutter, J., Syroid, N., Drews, F., Weinger, M. B., & Westenskow, D. (2003). The employment of an iterative design process to develop a pulmonary graphical display. *Journal of the American Medical Informatics Association*, 10(4), 363–372. https://doi.org/10.1197/jamia.M1207.

Wang, T., Wang, G., Innes, J. L., Seely, B., Chen, B. (2017). ClimateAP: an application for dynamic local downscaling of historical and future climate data in Asia Pacific. *Frontiers of Agricultural Science and Engineering*, 4(4), 448–458. https://doi.org/10.15302/J-FASE-2017172.

Wang, X., Cao, M., Hu, F., Yi, Q., Amon, T., Janke, D., Xie, T., Zhang, G., & Wang, K. (2022). Effect of fans' placement on the indoor thermal environment of typical tunnel-ventilated multi-floor pig buildings using numerical simulation. *Agriculture*, 12(6), 891. https://doi.org/10.3390/agriculture12060891.

Watkins, R., Levermore, G. J., & Parkinson, J. B. (2011). Constructing a future weather file for use in building simulation using UKCP09 projections. *Building Services Engineering Research and Technology*, 32(3), 293–299.

William, Y.E., An, H., Chien, S.-C., Soh, C.B., Ang, B.T.W., Ishida, T., Kobayashi, H., Tan, D., & Tay, R.H.S. (2022). Urban-metabolic farming modules on rooftops for eco-resilient farmscape. *Sustainability*. 14(24), 16885. https://doi.org/10.3390/su142416885.

Wong, C. E., Teo, Z. W. N., Shen, L., & Yu, H. (2020). Seeing the lights for leafy greens in indoor vertical farming. *Trends in Food Science & Technology*, 106, 48–63. https://doi.org/10.1016/j.tifs.2020.09.031.

World Economic Forum (WEF). (2022). How technology can help address challenges in agriculture. Retrieved from https://www.weforum.org/agenda/2022/03/how-technology-can-help-address-challenges-in-agriculture/.

Wu, X., Chien, S.-C., Sapa, M. B., & Zhang, Z. (2016) Assessment of natural ventilation performance of high-rise residential buildings in tropical climate, *CLIMA 2016*, Aalborg, Denmark.

Yilmaz, E., & Hu, J. (2018). CFD study of quadcopter aerodynamics at static thrust conditions. In *ASEE Northeast 2018 Annual Conference*. West Hartford, CT: University of Hartford.

7 Internet of Agricultural Things

Integration of Current Technological Tools for Indoor and Vertical Farming

Akmal Nizam Mohammed and Mohd Faizal Mohideen Batcha

INTRODUCTION

IoT, which stands for the Internet of things, refers to a network of interconnected physical devices that are embedded with sensors, software, and other technologies to collect and exchange data over the Internet. These devices can communicate with each other, as well as with humans, enabling data sharing, automation, and intelligent decision-making (Figure 7.1).

The concept behind IoT is to enable everyday objects to become "smart" by connecting them to the Internet, which opens up a wide range of possibilities and applications. Through IoT, objects can gather data from their surroundings, monitor conditions, and perform various tasks autonomously or based on remote instructions. This connectivity and data exchange enables enhanced functionality, efficiency, and convenience across multiple sectors and industries. IoT begins with devices that are equipped with sensors or actuators. These sensors can detect and measure various environmental or physical conditions such as temperature, humidity, light, motion, or pressure. They can also include actuators that can perform actions based on received instructions (Navarro et al., 2020).

The sensors collect data from the surrounding environment or the device itself. This data can be in the form of temperature readings, location coordinates, images, or any other relevant information. The sensors often have built-in processing capabilities to perform basic data processing tasks. The collected data is transmitted over a network, which can be a local network (e.g., Wi-Fi, Bluetooth) or a wide-area network (e.g., cellular networks, satellite). The data can be sent in real time or at scheduled intervals, depending on the specific application. Once the data reaches a cloud-based platform

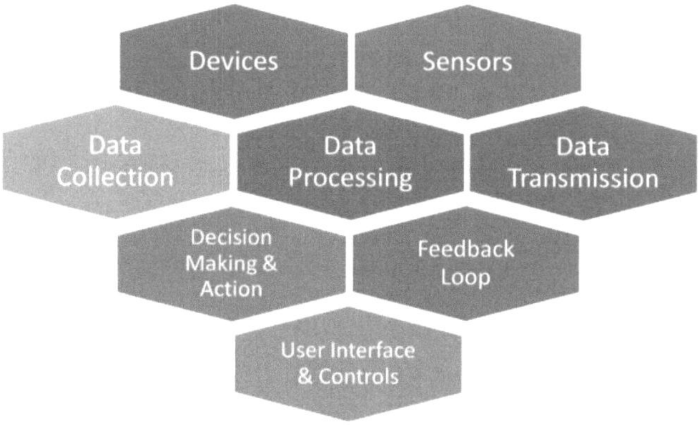

FIGURE 7.1 Major building blocks for an IoT system.

DOI: 10.1201/b23309-7

or a central server, it undergoes further processing and analysis. Advanced analytics algorithms and machine learning techniques can be applied to extract meaningful insights from the collected data. This analysis helps derive valuable information, identify patterns, detect anomalies, or make predictions.

Based on the insights and analysis from the collected data, decisions can be made and actions can be taken. This can involve triggering specific responses or commands to the connected devices or sending notifications to users or operators. Automated actions can also be performed based on predefined rules or algorithms. Users can access and interact with IoT systems through various user interfaces such as mobile apps, web-based dashboards, or voice assistants. These interfaces allow users to monitor and control the connected devices, receive alerts, view data visualizations, and configure system settings (Hamrita & Hoffacker, 2005; Jha et al., 2020).

The IoT system can continuously gather feedback by monitoring the performance of devices, evaluating user interactions, and analyzing data. This feedback loop helps improve the system's functionality, optimize processes, and enhance overall performance. Overall, IoT enables the seamless connection and communication between physical devices and digital systems, allowing for data-driven decision-making, automation, and enhanced capabilities in various domains such as agriculture, healthcare, transportation, smart homes, and industrial applications.

IOT IN AGRICULTURE – INDOOR FARMING

The Internet of agricultural things (IoAT) refers to the application of smart technologies and data analytics in crop farming, seed cultivation, animal husbandry, and other endeavors related to agriculture. It involves the use of interconnected devices, sensors, and software that collects and shares data about agricultural processes and conditions. Data such as weather predictions, soil conditions, and even environmental uncertainties can be used to increase output productivity, improve decision-making in farm management, and enhance the overall sustainability of the farming supply chain. In today's world of resource scarcity and the dire need for food security, the use of IoT in agriculture is progressively changing from being a luxury into an absolute necessity.

In the context of indoor farming, IoAT can be particularly useful in optimizing plant growth conditions and reducing waste. Indoor farming involves growing crops in controlled environments such as greenhouses, vertical farms, or other indoor facilities. By integrating IoT technologies into indoor farming, farmers can monitor and control various aspects of the growing environment, such as temperature, humidity, lighting, and irrigation.

For example, smart sensors can collect data on the temperature and humidity levels in the indoor farm, and send this information to a central computer system. The system can then adjust the environmental conditions to ensure that the plants are growing in optimal conditions. Similarly, automated irrigation systems can be used to deliver water to plants based on their individual needs, reducing water waste and ensuring that each plant receives the appropriate amount of water. IoAT can also be used to monitor crop health, detect pests and diseases, and provide real-time data on crop growth and yield. This data can be used to optimize crop management strategies and improve overall production. The integration of IoAT technologies in indoor farming has the potential to increase efficiency, reduce waste, and improve crop yields, making it a promising approach for sustainable agriculture.

The use of IoT in indoor farming has evolved over time, driven by advancements in technology and the growing need for efficient and sustainable agricultural practices. Here is a brief history of IoT for indoor farming:

Early Indoor Environmental Control Systems (1980s–1990s): The concept of controlling environmental parameters in indoor farming began to emerge in the 1980s. Basic systems were developed to monitor and control temperature, humidity, and lighting in greenhouses and controlled-environment agriculture facilities. These systems relied on wired connections and rudimentary sensors.

Introduction of Wireless Sensor Networks (2000s): In the early 2000s, wireless sensor networks started to gain traction in indoor farming. Wireless sensors enabled remote monitoring of environmental conditions, eliminating the need for extensive wiring. These sensors provided real-time data on temperature, humidity, light intensity, and other parameters, enhancing the ability to maintain optimal growing conditions.

Integration of IoT technologies (2010s): The advent of the IoT in the 2010s brought about significant advancements in indoor farming. IoT technologies allowed for seamless connectivity, data exchange, and control of various devices and systems. Sensors, actuators, and smart devices were integrated into a network, enabling the collection and analysis of large volumes of data.

Expansion of Sensor Applications (2010s): With IoT, the range of sensors used in indoor farming expanded. Besides basic environmental sensors, new types of sensors were introduced, including pH sensors, nutrient sensors, CO_2 sensors, and water quality sensors. These sensors provided precise measurements and facilitated real-time monitoring of key parameters essential for plant growth.

Emergence of Cloud Computing and Big Data Analytics (2010s): Cloud computing and big data analytics played a significant role in the evolution of IoT for indoor farming. Cloud platforms enabled the storage and processing of vast amounts of data collected from IoT devices. Big data analytics techniques allowed for the extraction of valuable insights and optimization of farming operations based on historical and real-time data.

AI and Machine Learning Integration (2010s): Artificial intelligence (AI) and machine learning (ML) became integral to IoT systems in indoor farming. AI algorithms and ML models were employed to analyze sensor data, identify patterns, make predictions, and provide actionable recommendations. This integration enhanced the automation and decision-making capabilities of indoor farming systems.

Advancements in Robotics and Automation (2010s–Present): Robotics and automation technologies have also made significant strides in indoor farming. IoT-enabled robots and automated systems are now used for tasks such as planting, harvesting, pruning, and crop monitoring. These advancements have improved efficiency, reduced labor requirements, and enabled precision farming practices.

Integration of IoT with Vertical Farming and Hydroponics (2010s–Present): Vertical farming and hydroponic systems have gained popularity in recent years, and IoT has played a crucial role in their development. IoT technologies have been integrated into vertical farming systems, providing precise control over lighting, nutrient delivery, and environmental conditions in multi-tiered growing setups.

Market Growth and Adoption (2010s–Present): The market for IoT in indoor farming has experienced significant growth in recent years. Numerous companies and startups have emerged, offering IoT solutions tailored to the needs of indoor farming. These solutions encompass hardware, software platforms, data analytics, and end-to-end IoT systems, catering to a wide range of indoor farming applications.

As technology continues to advance, the application of IoT in indoor farming is expected to expand further, enabling more efficient resource management, increased yields, and sustainable agricultural practices.

The size of the IoT industry for indoor farming worldwide is difficult to estimate precisely, as the industry is still relatively new and rapidly evolving. However, several market research reports suggest that the market for IoT in agriculture, including indoor farming, is growing rapidly and is expected to continue to do so in the coming years.

According to a report by MarketsandMarkets, the global IoT in agriculture market size was valued at USD 9.2 billion in 2020 and is expected to reach USD 20.9 billion by 2025, growing at a CAGR of 17.9% during the forecast period (Agriculture IoT Market Share, Scope & Industry Growth, Analysis, 2030, 2024). Another report by Grand View Research estimates that the global smart agriculture market, which includes IoT technologies for indoor farming, was valued at USD 9.2 billion in 2020 and is expected to reach USD 23.8 billion by 2028, growing at a CAGR of 12.9% during the forecast period (Indoor Farming Market Size, Share & Trends Analysis Report by Facility Type (Greenhouses, Vertical Farms, Others), by Component (Hardware, Software, Services), by Growing Mechanism, by Crop Category, by Region, and Segment Forecasts, 2023–2030, 2022).

These reports indicate that the IoT industry for indoor farming is a growing and significant market, with a significant potential for future growth as more indoor farms adopt these technologies to optimize plant growth, reduce costs, and improve overall efficiency.

PLANT GROWTH OPTIMIZATION

A wide range of crops can be grown in indoor farms, thanks to controlled environments that provide optimal conditions for plant growth. The choice of crops depends on factors such as market demand, crop yield, growth cycle, and the suitability of indoor farming methods. Here are some types of crops that are commonly planted in indoor farms (Robin, 2023; Lst, 2021; What Crops Can Be Grown in a Vertical Farm? | IGS, n.d.)

1. Leafy greens such as lettuce, kale, spinach, and arugula are popular choices for indoor farming. They have short growth cycles and high yields and are well-suited to hydroponic or vertical farming systems.
2. Various herbs like basil, mint, cilantro, parsley, and thyme are commonly grown on indoor farms. They are in high demand and can be grown year-round, providing a fresh and consistent supply to consumers.
3. Microgreens are young, tender greens that are harvested at an early stage of growth. They include varieties like micro basil, micro radish, micro arugula, and micro cilantro. Microgreens are nutrient-dense and are often used as garnishes or in salads.
4. Certain tomato varieties, particularly compact or dwarf varieties, can be successfully grown on indoor farms. They require adequate lighting and proper support systems, such as trellises or vertical setups.
5. English cucumbers or compact cucumber varieties can thrive in indoor farming environments. They require trellising or vertical growing systems for support and benefit from controlled temperatures and humidity.
6. Some indoor farms cultivate strawberries using hydroponic or aeroponic systems. These systems provide the necessary water and nutrient delivery to the plants, resulting in consistent yields throughout the year.
7. Peppers, including bell peppers, chili peppers, and sweet peppers, can be grown in controlled environments. They require specific temperature and lighting conditions for optimal growth.
8. Sprouts, such as alfalfa sprouts, bean sprouts, or broccoli sprouts, are commonly grown indoors. They have a short growth cycle and are typically cultivated hydroponically or using sprouting trays.
9. Indoor farms are well-suited for growing various mushroom varieties, including oyster mushrooms, shiitake mushrooms, and specialty mushrooms. Mushrooms require specific environmental conditions and substrate materials for cultivation.
10. Some indoor farms focus on cultivating specialty or exotic crops like micro-herbs, edible flowers, rare lettuce varieties, or niche vegetables. These crops cater to specific culinary or market demands and may require specialized growing techniques.

It is important to note that the suitability of crops for indoor farming can vary depending on the specific indoor farming techniques employed, such as hydroponics, aeroponics, or vertical farming. The choice of crops should be based on market demand, feasibility studies, and the resources available for indoor farm operations. Optimizing plant growth involves creating the ideal growing conditions for a specific crop to maximize its yield and quality. This can involve several factors, including the amount of light, water, nutrients, and temperature that the plants receive. With the help of IoAT technologies, indoor farmers can monitor and control these factors to create optimal growing conditions for their plants.

For example, they can use sensors to measure the amount of light that the plants are receiving and adjust the lighting system to ensure that they are getting the right amount of light for their growth stage. Similarly, they can monitor the temperature and humidity levels in the growing area and adjust the heating, ventilation, and air conditioning (HVAC) system to maintain the ideal conditions. This is important because different plants have different temperature and humidity requirements, and maintaining these conditions can help ensure that they grow at the desired rate.

IoAT technologies can also help indoor farmers optimize nutrient delivery to their plants. By monitoring the nutrient levels in the soil or water, they can adjust the amount and type of nutrients that the plants receive to ensure that they are getting the right balance of macronutrients (such as nitrogen, phosphorus, and potassium) and micronutrients (such as iron, zinc, and copper). Overall, by optimizing plant growth with the help of IoAT technologies, indoor farmers can improve their crop yields and quality, reduce resource waste, and ensure a more sustainable and efficient agricultural system.

Smart sensors are a key component of the IoAT in indoor farming. These sensors are designed to measure and monitor various environmental factors such as temperature, humidity, light, soil moisture, and nutrient levels in real time. Smart sensors are typically wireless and can be placed throughout the indoor farming environment to provide continuous monitoring of the growing conditions. They can also be connected to a central computer system or a cloud-based platform, which enables farmers to receive real-time data and make informed decisions about crop management.

For example, a smart sensor can measure the soil moisture levels in the growing medium and send the data to the computer system. Based on this information, the computer system can automatically adjust the amount of water delivered to the plants through the irrigation system to ensure that they receive the appropriate amount of water. Similarly, smart sensors can monitor the temperature and humidity levels in the growing environment, which can help farmers adjust the HVAC system to maintain the ideal conditions for plant growth. Smart sensors can also be used to detect changes in light intensity, which can indicate the need for adjustments to the lighting system.

Overall, smart sensors play a critical role in IoAT-enabled indoor farming by providing real-time data on environmental conditions and enabling farmers to optimize crop growth and yield. Smart sensors used in IoAT for indoor farming can capture various parameters related to the growing environment. These parameters may include:

1. Temperature: Sensors can measure the temperature of the air and/or the growing medium (e.g., soil, hydroponic solution). The ideal range for most indoor crops is between 65°F and 75°F (18–24°C) during the day and between 60°F and 68°F (15–21°C) at night. However, some crops may have specific temperature requirements outside of this range (Webb, 2023).
2. Humidity: Sensors can measure the relative humidity of the air, which is the amount of moisture in the air relative to its maximum capacity at a given temperature. The ideal range is generally between 50% and 70%, which helps to promote healthy plant growth and minimize the risk of mold and disease (Optimal Temperature & Humidity for Sprouts & Grown Plants, 2025). However, the ideal humidity range can also vary depending on the crop being grown and the stage of growth. For example, some crops may require higher humidity levels during the germination stage, while others may require lower humidity levels during flowering and fruiting.

3. Light Intensity: Sensors can measure the amount of light that the plants are receiving, which is important for controlling plant growth and optimizing crop yield. Typically, the red light used in indoor farming has a wavelength of around 600–700 nm. The peak absorption for red light by chlorophyll occurs at around 600 nm, while wavelengths between 600 and 700 nm are important for regulating plant growth and flowering. On the contrary, blue light used in indoor farming has a wavelength of around 400–500 nm. The peak absorption for blue light by chlorophyll occurs at around 440 nm, while wavelengths between 400 and 500 nm are important for promoting vegetative growth and regulating plant morphology (TCPi, 2022).
4. Carbon Dioxide (CO_2) Levels: Sensors can measure the concentration of CO_2 in the air, which is an important factor for photosynthesis and plant growth. The typical range of CO_2 values in an indoor farm is around 800–1000 ppm, although the optimal CO_2 level can vary depending on the specific crop being grown and the stage of growth. CO_2 is an essential component for plant growth, as it is used in the process of photosynthesis to produce energy and growth. Indoor farmers often supplement the CO_2 levels in their grow rooms to promote healthy plant growth and maximize crop yield (Greenhouse Carbon Dioxide Supplementation – Oklahoma State University, 2023).
5. Nutrient Levels: Sensors can measure the levels of nutrients such as nitrogen, phosphorus, and potassium in the soil or hydroponic solution. The typical range of nutrient concentration values used in indoor farming can vary depending on the specific crop being grown and the stage of growth. However, in general, most indoor farmers aim for nutrient concentration levels that fall within the following ranges (University of Massachusetts Amherst, 2016; Communications, 2018; Department of Jobs, Precincts and Regions, 2023).
   - Nitrogen (N): 150–250 ppm.
   - Phosphorus (P): 25–100 ppm.
   - Potassium (K): 101–300 ppm.
6. pH: Sensors can measure the acidity or alkalinity of the soil or hydroponic solution, which is important for plant nutrient uptake. The typical range of pH values used in indoor farming can vary depending on the specific crop being grown, but in general, most crops prefer a slightly acidic pH range between 5.5 and 6.5. Maintaining the correct pH level is important for plant growth because it affects nutrient uptake and overall plant health. If the pH is too high or too low, certain nutrients may become unavailable to the plant, which can lead to nutrient deficiencies and stunted growth (Meselmani, 2023).

Information on appropriate ranges for plant growth optimization is summarized in the following Table 7.1:

TABLE 7.1
Main Parameters Monitored for Indoor Farms and the Ideal Range of Plant Growth

Parameter	Range	Units
Temperature	15–24	Degree Celsius (°C)
Humidity	50–70	Percent (%)
Light wavelength: red light	660–730	Nanometers (nm)
Light wavelength: blue light	400–500	Nanometers (nm)
Carbon dioxide (CO_2) levels	800–1,200	Parts per million (ppm)
Nutrient levels: nitrogen	100–250	Parts per million (ppm)
Nutrient levels: phosphorus	25–100	Parts per million (ppm)
Nutrient levels: potassium	150–300	Parts per million (ppm)
pH (acidity/alkalinity)	5.5–6.5	

COMMERCIALLY AVAILABLE EQUIPMENT FOR IOT

By collecting and analyzing data from these parameters, indoor farmers can gain insights into the growing environment and make informed decisions about crop management. This can lead to increased crop yields, reduced resource waste, and more sustainable and efficient agriculture. There are several sensor brands that are commonly used in indoor farming, and the selection of a particular brand depends on factors such as the specific needs and goals of the farm, the budget, and the availability of support and resources. Some popular sensor brands used in indoor farming include:

1. Bosch: Bosch is a leading manufacturer of smart sensors for indoor farming, offering a range of sensors for measuring temperature, humidity, light, and air quality (SensorTec Environmental Sensors, n.d.).
2. Sensaphone: Sensaphone is a popular brand of environmental monitoring systems, including sensors for temperature, humidity, power, and water detection (http://www.aycmedia.com, n.d.).
3. Libelium: Libelium offers a range of sensor nodes and IoT platforms for indoor farming, including sensors for temperature, humidity, CO_2, and soil moisture (Libelium IoT, 2023).
4. Onset: Onset offers a range of data logging and environmental monitoring solutions for indoor farming, including sensors for temperature, humidity, light, and moisture (LKH Precicon, 2023).
5. TrolMaster: TrolMaster is a leading brand of environmental controllers and sensors for indoor farming, offering a range of sensors for temperature, humidity, CO_2, and light intensity (Growtent based on IdoSell – the best online selling solutions for your e-store (www.idosell.com/shop), n.d.).
6. Bluelab: Bluelab offers a range of sensors and controllers for indoor farming, including pH and nutrient sensors, as well as monitors for temperature and humidity (Hgshydro, n.d.).

Overall, there are many sensor brands available on the market, and it is important for indoor farmers to research and evaluate the features and capabilities of different sensors to determine which ones will best meet their specific needs and goals.

FEEDBACK MECHANISMS FOR THE IOT

The mechanism feedback loop for IoT in indoor farming involves a continuous cycle of data collection, analysis, and action (Figure 7.2).

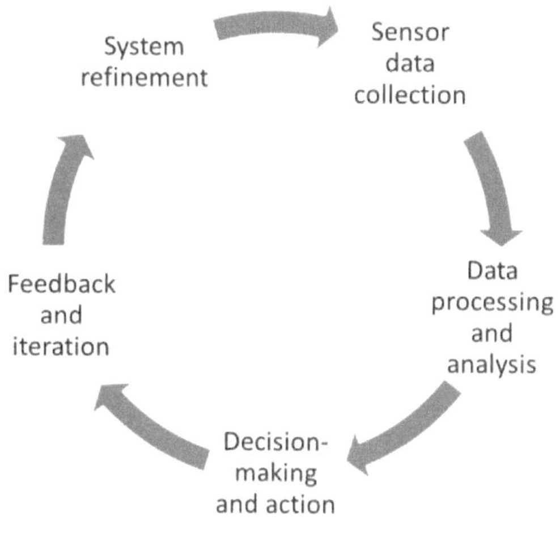

FIGURE 7.2 Feedback loop for Internet of things.

The feedback loop for IoT requires the following steps:

1. Sensor Data Collection: Smart sensors are deployed throughout the indoor farm to collect data on various environmental parameters, such as temperature, humidity, light, CO_2 levels, and nutrient concentration.
2. Data Processing and Analysis: The sensor data is processed and analyzed using IoT platforms, data analytics tools, and ML algorithms. This helps to identify patterns and trends in the data and to generate insights about the growing environment.
3. Decision-Making and Action: Based on the insights generated from the data analysis, farmers can make informed decisions about how to adjust the growing environment to optimize plant growth and maximize crop yield. This may involve adjusting lighting schedules, modifying nutrient solutions, or fine-tuning irrigation.
4. Feedback and Iteration: The actions taken based on the data analysis are monitored and evaluated to determine their effectiveness. If necessary, the feedback loop is reinitiated with new sensor data to further optimize the growing environment.
5. System Refinement: A review of system performance and data quality is periodically performed to look at possible areas for further improvement.

By continuously monitoring and optimizing the growing environment using smart sensors and IoT technology, indoor farmers can improve crop yields, reduce resource waste, and operate more sustainably and efficiently. The feedback loop is a critical mechanism for enabling this continuous improvement process. There are several providers of IoT systems for indoor farming, each with its own unique offerings and capabilities. Here are a few of the main providers:

1. CropX: CropX provides a range of IoT systems for precision irrigation and fertilization in indoor and outdoor farming. Their systems include soil moisture sensors, weather stations, and analytics software (CropX, 2024).
2. Motorleaf: Motorleaf offers a range of IoT solutions for indoor farming, including sensors for temperature, humidity, light, and CO_2, as well as predictive analytics and automation tools (AgFunder, 2019).
3. Agrilyst: Agrilyst provides a data analytics platform for indoor farmers, which includes sensors for tracking environmental conditions and plant growth, as well as analytics tools for optimizing crop yield and resource use.
4. Smart Farm Systems: Smart Farm Systems offers a range of IoT systems for indoor and outdoor farming, including sensors for soil moisture, weather conditions, and irrigation. They also offer automation tools for optimizing resource use and improving crop yield (Smart Farm, 2018).
5. Phytech: Phytech offers a plant monitoring system for indoor and outdoor farming, which includes sensors for measuring plant growth and stress, as well as analytics software for optimizing crop yield and resource use (Phytech | Connecting Growers to Their Plants, n.d.).
6. LumiGrow: LumiGrow offers a smart LED lighting system for indoor farming, which includes sensors for measuring light intensity and spectrum, as well as analytics software for optimizing lighting schedules and improving crop yield (LumiGrow Pilot Customers Have the Sun Manage Their Lights With Grow Light Sensor, n.d.).

These are just a few examples of the many providers of IoT systems for indoor farming. The selection of a particular provider depends on factors such as the specific needs and goals of the farm, the budget, and the availability of support and resources. The components for IoT in indoor farming can vary depending on the specific system and application, but generally, an IoT-enabled indoor farming system will include some or all of the following components:

1. Smart Sensors: These are devices that capture data on various environmental parameters such as temperature, humidity, light intensity, pH, and nutrient levels.
2. Data Storage and Analytics: Data collected by the sensors is stored in a centralized database or cloud-based platform for analysis and decision-making.
3. Automation and Control Systems: These are software and hardware systems that use the data collected by the sensors to automatically adjust environmental conditions such as lighting, temperature, and nutrient levels to optimize plant growth and health.
4. Communication Infrastructure: This includes the network of devices and systems used to transmit data between the various components of the IoT system, such as Wi-Fi, Bluetooth, or other wireless communication protocols.
5. User Interface: This is the software or application used by the user to access and interact with the system, such as a web-based dashboard or mobile app.

Together, these components enable indoor farmers to collect and analyze real-time data on environmental conditions, adjust and automate systems to optimize plant growth and health, and monitor and control the entire system remotely, leading to improved yields, reduced resource usage, and increased efficiency.

IOT SYSTEMS DEVELOPMENT USING RASPBERRY PI – AN EXAMPLE

Developing an IoT system using a Raspberry Pi is a popular and accessible approach. The Raspberry Pi is a small, affordable, and versatile single-board computer that can serve as the central processing unit for an IoT system. To start, the first step is to determine the purpose and objectives of the IoT system. The specific data to collect, the devices to be connected, and the actions to be performed based on that data need to be defined from the outset. The Raspberry Pi board is set up by installing the operating system (e.g., Raspbian), installing necessary software tools, and connecting to the Internet, either through Wi-Fi or Ethernet (Robocraze, 2023).

Next, the sensors or devices needed to collect data are identified and verified for compatibility. These sensors/devices are connected to the Raspberry Pi's general purpose input/output pins or other communication interfaces (e.g., USB). Concurrently, the software on the Raspberry Pi is configured to read data from the connected sensors/devices. This may involve writing code in a programming language such as Python or using libraries and frameworks specific to the sensors/devices being used. Proper configuration between the sensors and software would allow maximization of Raspberry Pi's processing capabilities to analyze the collected data. Algorithms from existing libraries can be used to process the data, detect patterns, or perform calculations (Instructables, 2017).

Once the data output has been verified as per expectations, a communication mechanism should then be established to send the processed data from the Raspberry Pi to a cloud platform or other endpoints. This can involve using protocols like MQTT or HTTP to transmit data over the Internet. A cloud-based platform or service to receive and store the data transmitted from the Raspberry Pi can be subscribed. This platform can provide storage, analytics, and visualization capabilities for the collected data.

To allow for easy access, a user interface such as a web application or mobile app needs to be developed to interact with the IoT system. This interface can display real-time data, allow users to control devices remotely, and provide insights from the analyzed data. Setting up a graphical user interface (GUI) for an IoT system involves creating a visual interface that allows users to interact with the system, monitor data, and control devices. From the beginning of development, the objectives and functionalities of the GUI need to be identified in terms of the key data to display, the user actions to support, and the overall look and feel of the interface. Popular options for GUI development include web-based frameworks like HTML/CSS/JavaScript (for responsive web interfaces) or frameworks like PyQt or Tkinter (for desktop applications) (Bennett, 2018).

Subsequently, the GUI will need to be connected to the backend of the IoT system. This involves establishing a connection to the IoT platform or the Raspberry Pi, typically through APIs or network protocols such as MQTT or HTTP. The GUI should be able to receive data updates and send control commands to the IoT devices. To this end, mechanisms to fetch and display real-time data from the IoT system could be implemented. Visual elements like charts, graphs, or gauges can be added to present the collected data in a user-friendly format displayed dynamically as new data is received.

To enable user interaction, interactive elements such as buttons, sliders, dropdown menus, or input fields can be added to allow users to control devices, adjust settings, or trigger specific actions in the IoT system. However, this may also require the implementation of error-handling mechanisms to notify users of any errors or issues. The system should be able to provide appropriate feedback, such as success messages, error messages, or progress indicators, to ensure users have a clear understanding of the system's state and actions (Shiverware, n.d.).

The next step is to test the GUI together with the IoT system thoroughly to ensure it functions as intended. Its usability, responsiveness, and compatibility need to be validated across different devices and screen sizes if necessary. User feedback is then gathered and iterative improvements are made based on usability testing and user experience. Once the GUI is tested and refined, it can then be deployed to the desired environment. This can involve hosting the web-based GUI on a server, packaging the desktop application for distribution, or deploying it to specific devices or platforms.

Another aspect of the IoT system is the database. Requirements for an IoT system depend on the specific needs and characteristics of the system. IoT systems generate a significant amount of data, especially when dealing with large-scale deployments or high-frequency data collection. Therefore, the database should be able to handle large data volumes efficiently and provide scalability options to accommodate future growth. Other considerations include (Education, 2021):

1. IoT data can be structured (e.g., sensor readings, device status) or unstructured (e.g., images, sensor logs); thus, the database needs to handle the data format and structure effectively. For structured data, relational databases like MySQL or PostgreSQL are commonly used, while NoSQL databases like MongoDB or Apache Cassandra are suitable for handling unstructured data.
2. IoT systems often deal with real-time data streams that require fast ingestion and processing. Databases should be able to handle high write throughput and provide efficient querying capabilities to analyze and extract insights from the data in real time.
3. IoT systems may collect data from various sources and formats, including sensor, location, multimedia, or social media data. The database needs to handle diverse data types and provide support for storing and querying different data formats efficiently.
4. As IoT systems rely on accurate and reliable data, a database should ensure data integrity and offer mechanisms for data consistency, durability, and fault tolerance. Features like replication, backup, and recovery options can enhance its reliability.
5. In some cases, IoT systems eventually would require advanced analytics and hence databases need to provide integration with analytics frameworks or have built-in analytical capabilities. This can include support for data aggregation, complex queries, ML algorithms, or integration with external analytics tools.
6. IoT systems often handle sensitive data, so database security is crucial. Databases must offer robust security features such as authentication, encryption, access control mechanisms, and compliance with relevant data protection regulations.
7. In determining the cost implications of the chosen database, factors such as licensing fees, maintenance costs, scalability options, and cloud-based offerings are considered. Some databases offer open-source versions or community editions, which can be cost-effective for certain IoT applications.
8. Operational requirements of the database, such as ease of administration, monitoring capabilities, backup and recovery processes, and compatibility with existing infrastructure, are evaluated before decisions and choices regarding it are made.

Continuing with the system itself, the functionality to automate actions or control devices based on predefined rules or user inputs is then developed. This can involve sending commands from the user interface to the Raspberry Pi, which then triggers actions on connected devices. In addition to database security, as mentioned earlier, the security of the IoT system is ensured by implementing authentication, encryption, and secure communication protocols. The privacy of collected data is protected by adhering to privacy regulations and best practices.

After completion of system development, the performance of the IoT system is monitored, feedback is gathered, and improvements are made gradually over time. Software updates, algorithm optimization, and new feature introductions are done based on user requirements and evolving needs.

IOT: COSTS AND POTENTIAL SAVINGS

The costs involved in setting up an IoT system for indoor farming can vary widely depending on the specific needs and requirements of the system. Some of the main factors that can impact the costs include (Inside Grower: The Cost of Indoor Farming, n.d.):

1. Size and Scale of the Farm: The larger the indoor farm, the more sensors, automation systems, and communication infrastructure will be required to monitor and control the system.
2. Type of Crops: Different crops may require different types of sensors and environmental controls, which can impact the costs.
3. Level of Automation: The more automation and control systems are used, the higher the cost of the system.
4. Data Storage and Analytics: The cost of storing and analyzing data collected by the sensors and other devices can vary depending on the amount of data and the complexity of the analysis required.
5. Communication Infrastructure: The cost of setting up and maintaining the communication infrastructure required for the system, such as Wi-Fi, cellular, or satellite connectivity.
6. User Interface: The cost of developing and maintaining the user interface used to access and interact with the system.

In general, the cost of setting up an IoT system for indoor farming can range from a few thousand dollars to tens of thousands of dollars or more, depending on the specific needs and requirements of the system. However, the potential benefits of increased yields, reduced resource usage, and improved efficiency can make the investment worthwhile for many indoor farmers. Indoor farms can use IoT to reduce energy usage in several ways, including (Gravitt, 2021; McDonald, 2022; Venkatesan et al., 2022):

1. Energy-Efficient Lighting: By using smart sensors and automation systems to adjust the lighting intensity and spectrum based on the plant's needs, indoor farms can reduce energy consumption and improve crop yields.
2. Efficient HVAC Systems: By using smart sensors and automation systems to control the HVAC systems, indoor farms can reduce energy usage while maintaining optimal growing conditions.
3. Real-Time Monitoring and Alerts: IoT-enabled systems can detect and alert farmers of any anomalies or issues with the growing environment, allowing for quick corrective action and avoiding prolonged energy waste.
4. Energy Monitoring and Management: IoT-enabled systems can monitor and manage energy usage by providing real-time data on energy consumption, identifying energy-intensive processes, and enabling the implementation of energy-saving strategies.
5. Renewable Energy Sources: Indoor farms can integrate renewable energy sources such as solar or wind power to reduce reliance on traditional energy sources and reduce their carbon footprint.

Calculating the return on investment (ROI) for an IoT system in an indoor farm requires consideration of various factors, including upfront costs, ongoing expenses, operational efficiencies, and potential revenue or cost savings. To calculate the ROI for an IoT system in an indoor farm (Guntaka & Saraswat, 2022; Shalimov, 2023; Vipond, 2023):

Determine Costs:

a. IoT Infrastructure: Calculate the costs of setting up the IoT infrastructure, including sensors, actuators, gateways, communication modules, and network infrastructure.
b. Software and Development: Consider the expenses related to developing or acquiring IoT software, including data analytics platforms, cloud services, and application development.
c. Installation and Integration: Account for the costs associated with installing and integrating IoT components into the indoor farm, such as labor, equipment, and any necessary modifications to the facility.
d. Maintenance and Upgrades: Estimate ongoing costs for maintaining and upgrading the IoT system, including sensor calibration, software updates, and hardware replacements.

Quantify Benefits:

a. Operational Efficiency: Identify how the IoT system improves operational efficiency in the indoor farm. For example, consider time savings, reduced labor costs, optimized resource utilization, and improved yield or quality of crops.
b. Cost Savings: Evaluate potential cost savings resulting from the IoT system, such as reduced energy consumption, optimized water usage, decreased waste, and improved inventory management.
c. Revenue Increase: If the IoT system enables higher crop yields, better quality produce, or access to premium markets, estimate the potential increase in revenue.

Calculate ROI:

a. Net Benefit: Calculate the net benefit of the IoT system by subtracting the total costs (from step 1) from the total benefits (from step 2).
b. ROI Calculation: Divide the net benefit by the total investment costs and multiply by 100 to get the ROI percentage. The formula is: ROI = (net benefit/total investment costs) × 100.

It is important to note that the ROI calculation can be complex and highly specific to the individual indoor farm's circumstances, market conditions, and business goals. The accuracy of the ROI calculation depends on the accuracy of the cost and benefit estimations, which may require historical data, industry benchmarks, or expert opinions.

Additionally, ROI is just one aspect of evaluating the value of an IoT system. Other considerations, such as improved decision-making capabilities, enhanced data insights, and long-term strategic advantages, should also be taken into account when assessing the overall impact of an IoT investment on the indoor farming operation.

IOT: INCORPORATION WITH AI

AI can greatly enhance IoT systems by leveraging advanced algorithms and ML techniques to analyze and derive meaningful insights from the vast amounts of data generated by IoT devices. Here are several ways AI can enhance IoT systems (Dhaduk, 2024; Chugh, 2024; Kashyapa, 2022; IoT Business News, 2023):

Data Analytics and Predictive Insights: AI algorithms can analyze IoT data to uncover patterns, trends, and correlations that humans may not easily identify. By applying techniques such as ML, deep learning, and predictive analytics, AI can extract valuable insights from

IoT data, enabling proactive decision-making and predictive maintenance. For example, AI can analyze sensor data to predict equipment failures, optimize resource allocation, or forecast crop yields in indoor farming.

Real-Time Monitoring and Anomaly Detection: AI algorithms can continuously monitor IoT data streams and detect anomalies or deviations from normal patterns. By training AI models on historical data, they can learn what constitutes "normal" behavior and raise alerts when unusual or potentially problematic events occur. This capability is particularly useful for detecting security breaches, equipment malfunctions, or environmental deviations in real time.

Autonomous Decision-making and Control: AI can enable autonomous decision-making and control in IoT systems. By integrating AI algorithms into the control mechanisms of IoT devices, systems can make intelligent decisions and take appropriate actions based on real-time data analysis. For example, in smart agriculture, AI can autonomously adjust irrigation levels based on soil moisture readings, weather forecasts, and crop requirements.

Personalization and Adaptive Systems: AI can help IoT systems personalize user experiences and adapt to individual preferences. By learning from user behavior and historical data, AI algorithms can tailor IoT system responses, recommendations, and automation to meet specific user needs. This personalization can enhance user satisfaction, optimize resource usage, and improve overall system performance.

Energy Optimization: AI algorithms can optimize energy consumption in IoT systems by analyzing data from energy sensors and making intelligent decisions. For example, AI can automatically adjust lighting, heating, and cooling systems in indoor farms based on occupancy, external weather conditions, and energy pricing to minimize energy waste and reduce costs.

Natural Language Processing (NLP) and Voice Control: AI-powered NLP techniques can enable natural language interaction with IoT systems. This allows users to control IoT devices or query system information using voice commands or text-based input. Voice assistants like Amazon Alexa or Google Assistant utilize AI to understand user commands and interact with IoT devices in a seamless manner.

Intelligent Data Filtering and Compression: AI can help reduce the amount of data transmitted and stored in IoT systems by intelligently filtering and compressing data. This optimization improves network bandwidth utilization, reduces storage requirements, and minimizes data transfer costs. AI algorithms can analyze sensor data in real time, identify relevant information, and transmit or store only the essential data. The integration of AI with IoT enables smarter, more efficient, and more autonomous systems, unlocking the full potential of the vast amount of data collected by IoT devices.

REFERENCES

AgFunder. (2019, August 16). How motorleaf is helping automate indoor farming. *AgFunderNews*. https://agfundernews.com/motorleaf-helping-automate-indoor-farming.

Agriculture IoT Market Share, scope & Industry Growth, Analysis, 2030. (2024, January 16). *MarketsandMarkets*. https://www.marketsandmarkets.com/Market-Reports/iot-in-agriculture-market-199564903.html.

Bennett, J. (2018, July 10). The IoT promises to bring about a revolution in the way we interact with devices around us. While many IoT devices will be hidden away, from sensors that measure manufacturing tolerances in a factory to hubs that control lighting around the home, there are a class of devices that need to provide some sort of [. . .]. *Ubuntu*. https://ubuntu.com/blog/graphical-environments-in-the-world-of-iot.

Chugh, S. (2024, January 31). Top 5 ways in which AI in IoT is transforming industries. Emeritus Online Courses. https://emeritus.org/blog/ai-in-iot/.

Syngenta (2018, December 15). Interpreting phosphorus and potassium levels. Syngenta – Know more, grow more. https://knowmoregrowmore.com/interpreting-phosphorus-and-potassium-levels/#:~:text=Healthy%20levels%20of%20P%20in,8%2D1%2F4%20lb.

CropX. (2024, February 7). CROPX agronomic farm management system. https://cropx.com/.

Department of Jobs, Precincts and Regions. (2023, September 20). Understanding soil tests for pastures. *Agriculture Victoria*. https://agriculture.vic.gov.au/farm-management/soil/understanding-soil-tests-for-pastures.

Dhaduk, H. (2024, January 10). AI for IoT: Paving the way to a connected and intelligent future. *Simform – Product Engineering Company*. https://www.simform.com/blog/ai-for-iot/.

Gravitt, D. (2021, September 1). Energy costs are eating up indoor farming budgets. Here's how to control them. *Schneider Electric Blog*. https://blog.se.com/sustainability/2021/09/01/energy-costs-are-eating-up-indoor-farming-budgets-heres-how-to-control-them/.

Greenhouse Carbon Dioxide Supplementation – Oklahoma State University. (2023, September 1). https://extension.okstate.edu/fact-sheets/greenhouse-carbon-dioxide-supplementation.html.

Growtent based on IdoSell – the best online selling solutions for your e-store (www.idosell.com/shop). (n.d.). Trolmaster Hydro-X Plus HCS-3- climate controller. *Growtent*. https://www.growtent.eu/product-eng-6308-TROLMASTER-HYDRO-X-PLUS-HCS-3-climate-controller.html.

Guntaka, M. L., & Saraswat, D. (2022). Overview of IoT applications in indoor farming. In *Automation, Collaboration, and E-Services* (pp. 621–637). https://doi.org/10.1007/978-3-031-10788-7_35.

Hamrita, T. K., & Hoffacker, E. C. (2005). Development of a smart wireless soil monitoring sensor prototype using rfid technology. *Applied Engineering in Agriculture*, 21(1), 139–143. https://doi.org/10.13031/2013.17904.

Hgshydro. (n.d.). Hydroponics store and indoor grow supplies. *HGS Hydro*. https://hgshydro.com/brand/bluelab.

https://www.aycmedia.com. (n.d.). Data center environmental monitoring system. *Sensaphone*. https://www.sensaphone.com/industries/data-center.

IBM Cloud Education. (2021, June 29). Structured vs. Unstructured data: What's the difference? *IBM Blog*. https://www.ibm.com/blog/structured-vs-unstructured-data/.

Indoor Farming Market Size, Share & Trends Analysis Report by facility type (Greenhouses, vertical farms, others), by component (Hardware, software, services), by growing mechanism, by crop category, by region, and segment Forecasts, 2023-2030. (2022, April 4). https://www.grandviewresearch.com/industry-analysis/indoor-farming-market.

Inside Grower: The Cost of Indoor Farming. (n.d.). https://www.inside-grower.com/Article/?articleid=26248.

Instructables. (2017, September 23). Data collection with Raspberry Pi. *Instructables*. https://www.instructables.com/Data-Collection-With-Raspberry-Pi/.

IoT Business News. (2023, December 21). 2024 IoT evolution: Cybersecurity, AI, and emerging technologies transforming the industry. https://iotbusinessnews.com/2023/12/21/63546-2024-iot-evolution-cybersecurity-ai-and-emerging-technologies-transforming-the-industry/.

Jha, D. N., Michalák, P., Wen, Z., Ranjan, R., & Watson, P. (2020). Multiobjective deployment of data analysis operations in heterogeneous IoT infrastructure. *IEEE Transactions on Industrial Informatics*, 16(11), 7014–7024. https://doi.org/10.1109/tii.2019.2961676.

Kashyapa, R. (2022, January 24). How integrated IoT with machine vision will revolutionize the industrial world? *Qualitas Technologies*. https://qualitastech.com/quality-control-insights/artificial-intelligence-machine-learning-in-iot/.

Libelium IoT. (2023, July 13). Smart agriculture and smart farming – Agricultural monitoring system. *Libelium*. https://www.libelium.com/iot-solutions/smart-agriculture/.

LKH Precicon. (2023, May 4). Onset data loggers. *LKH Precicon*. https://www.precicon.com.sg/partners/onset/.

Lst, I. (2021, September 9). What can be grown in a vertical farm? *Light Science Technologies*. https://lightsciencetech.com/what-can-be-grown-in-a-vertical-farm/.

LumiGrow Pilot Customers Have the Sun Manage Their Lights with Grow Light Sensor. (n.d.). LED professional – LED lighting technology. *Application Magazine*. https://www.led-professional.com/products/led-driver-ics-modules/lumigrow-pilot-customers-have-the-sun-manage-their-lights-with-grow-light-sensor.

McDonald, J. (2022, April 22). Vertical farms have the vision, but do they have the energy? *Emergingtechbrew*. Retrieved January 12, 2024, from https://www.emergingtechbrew.com/stories/2022/04/21/vertical-farms-have-the-vision-but-do-they-have-the-energy.

Meselmani, M. A. (2023). *Nutrient Solution for Hydroponics*. IntechOpen eBooks. https://doi.org/10.5772/intechopen.101604.

Navarro, E., Costa, N., & Pereira, A. (2020). A systematic review of iot solutions for smart farming. *Sensors*, 20(15), 4231. https://doi.org/10.3390/s20154231.

Optimal Temperature & Humidity For Sprouts & Grown Plants. (2025, January 1). *Mygardyn*. Retrieved February 12, 2024, from https://mygardyn.zendesk.com/hc/en-us/articles/360052627692-Optimal-Temperature-Humidity-For-Sprouts-Grown-Plants#:~:text=Once%20your%20plants%20are%20well,Humidity%3A%2050%2D70%25.

Phytech | Connecting Growers to their Plants. (n.d.). Plant-based farming. https://www.phytech.com/.

Robin. (2023, March 6). Vertical farm plants: What can you grow? *Avisomo*. https://avisomo.com/vertical-farm-plants-what-can-you-grow/.

Robocraze. (2023, November 7). Raspberry Pi in IOT. *Robocraze*. https://robocraze.com/blogs/post/raspberry-pi-in-iot.

SensorTec Environmental Sensors. (n.d.). Bosch | Mouser. https://my.mouser.com/new/bosch/bosch-environmental-sensors/.

Shalimov, A. (2023, March 1). IoT in agriculture: 9 technology use cases for smart farming (and challenges to consider). Eastern Peak - Technology Consulting & Development Company. https://easternpeak.com/blog/iot-in-agriculture-technology-use-cases-for-smart-farming-and-challenges-to-consider/.

Shiverware. (n.d.). 3 Types of interaction in IoT systems. https://shiverware.com/iot/3-interactions-iot.html.

Smart Farm. (2018, June 27). Product Overview – Smart farm. https://www.smartfarm.ag/products/product-overview/.

TCPi. (2022, November 16). Far red light is good for plant growth – TCP horticulture lighting. *TCP Lighting*. https://www.tcpi.com/far-red-light-its-benefits-on-plant-growth/#:~:text=Ranging%20from%20600%2D700%20nm,blue%20light%20for%20best%20results.

University of Massachusetts Amherst. (2016, October 26). Fertilizer calculations for greenhouse crops. *Center for Agriculture, Food, and the Environment*. https://ag.umass.edu/greenhouse-floriculture/fact-sheets/fertilizer-calculations-for-greenhouse-crops.

Venkatesan, S., Lim, J., Ko, H., & Cho, Y. (2022). A machine learning based model for energy usage peak prediction in smart farms. *Electronics*, 11(2), 218. https://doi.org/10.3390/electronics11020218.

Vipond, T. (2023, October 26). ROI formula (return on investment). *Corporate Finance Institute*. https://corporatefinanceinstitute.com/resources/accounting/return-on-investment-roi-formula/.

Webb, S. (2023, May 3). How cold is too cold for houseplants: A temperature guide. *Garden's Whisper*. https://gardenswhisper.com/how-cold-is-too-cold-for-houseplants/.

What Crops Can be Grown in a Vertical Farm? | IGS. (n.d.). https://www.intelligentgrowthsolutions.com/blog/what-crops-can-be-grown-in-a-vertical-farm.

8 A Review on Technological Advances and Challenges in Aquaponics Systems

Andrew Keong Ng and R. Mahkeswaran

INTRODUCTION

United Nations Committee on World Food Security defined food security as when "all people, at all times, have physical, social, and economic access to sufficient, safe and nutritious food that meets their dietary needs and food preferences for an active and healthy life" (Committee on World Food Security, 2017; Ng & Mahkeswaran, 2021a). The World Food Programme and World Health Organization highlighted that "the world is not on track to achieve zero hunger by 2030. If recent trends continue, the number of people affected by hunger will surpass 840 million by 2030" (Khim, 2020). Food production is further pressured by other calamities such as war, floods, droughts, storms, and pests. Some of these will intensify the effects of climate change and strain the food security of people from vulnerable populations (Khim, 2020), which will set the world back from achieving zero hunger in the future. Conventional farming practices are unsustainable. Huge natural habitats must be cleared for farmlands, which causes a loss of natural fauna and flora and can potentially threaten the ecosystem (Pradipta et al., 2022). Irrigated agriculture contributes to 40% of food production globally. However, it requires about 90% of global water consumption, of which only 55% is absorbed and used by plants, and the rest is lost as runoff (Pradipta et al., 2022). Runoff carries away fertile soil as well as excess fertilizers and pesticides downland, which spurs pollution worries and complaints. Overuse of fertilizers and pesticides to increase crop productivity can also kill beneficial insects and damage the ecosystem. Traditional aquaculture methods also put marine life in competition for space with the growing global population (Verma, 2017; Aishwarya et al., 2018). The shortcomings of conventional farming and the depletion of arable land reinforce the need to exploit other sustainable farming approaches (Aishwarya et al., 2018). Recent studies have demonstrated the feasibility of farming in urban residential and commercial areas, indoors and outdoors, as a business for revenue or an alternative food source (Podder et al., 2021; Rohr et al., 2019; Khan, 2018; Vernandhes et al., 2017).

Urban farming techniques include hydroponics, aeroponics, vertical farming, and aquaponics (Ng & Mahkeswaran, 2021a). Crops grown in hydroponics have their roots submerged in water dosed with nutrients, whereas aeroponics employs mists to spray nutrient-rich water directly onto roots (Jai et al., 2018; Rahman et al., 2018). Crops in vertical farming are grown in vertical layers on building surfaces and rooftops, in greenhouses and outdoors (Benke & Tomkins, 2017). Conversely, aquaponics integrates aquaculture and hydroponics (Kyaw & Ng, 2017). Fish, plants (vegetables and herbs), and nitrifying bacteria live in a symbiotic relationship within the aquaponics ecosystem. Fish consumes feed and excretes waste. Ammonia-rich waste is converted into nitrite and then nitrate by the nitrifying bacteria. Plants remove the nitrate by absorbing it as fertilizer, and the purified water returns to the fish tanks (Goddek et al., 2019; Mahkeswaran & Ng, 2020). The working principle of aquaponics allows fish and plants to be organically produced, which are healthy and chemical-free for consumption. Classic aquaponics operates in a single unidirectional closed loop where cultures within the ecosystem share water of the same parameters (Gibbons, 2020; Monsees et al., 2016). A more practical design of an aquaponics system is to allow the hydroponics and aquaculture

DOI: 10.1201/b23309-8

subsystems to have their separate water flow, which enables better control of water parameters in each loop for ease of maintenance (Gibbons, 2020; Monsees et al., 2016; Mahkeswaran et al., 2023). Such a design is known as a multi-loop or decoupled aquaponics system (Figure 8.1).

Maintaining ideal water parameter ranges is paramount in an aquaponics system. Any malfunction of a subcomponent or drastic change in water parameters can adversely affect the entire system. Hence, continuous system monitoring is indispensable; however, it can be taxing if done manually (Barosa et al., 2019). The recent COVID-19 global pandemic showed the vulnerability of workforce-dependent food systems. The workforce was limited to improve social distance and reduce COVID-19 transmission, which consequently lowered overall food production (Azra et al., 2021). Integrating technology into aquaponics systems can alleviate manual labor requirements and achieve precise system monitoring and control (Ke & Zhou, 2021). To further enhance and establish the potential of aquaponics systems, new technologies and innovations are being implemented. This chapter, therefore, aims to review the technological advancements in aquaponics and explore the challenges faced in their implementations. The literature review is limited to journal and conference papers published from 2017 to 2022 in popular scholarly databases like IEEE, SpringerLink, ScienceDirect, and Scopus. The number of literature papers is categorized (Figure 8.2) by emerging and disruptive technologies: Internet of things (IoT), automation, alternative resources, artificial intelligence (AI), green technology, extended reality, and advanced materials.

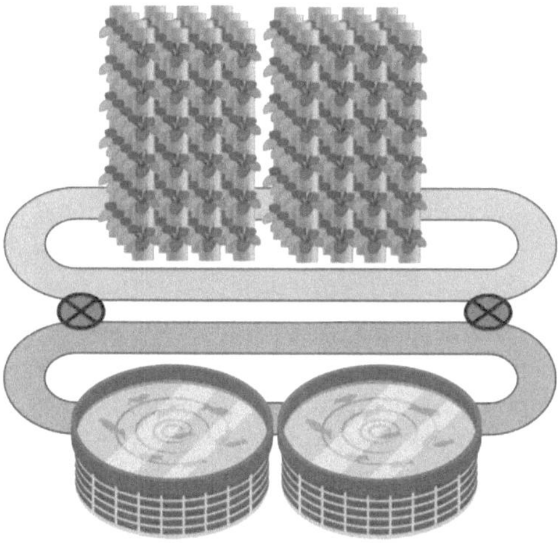

FIGURE 8.1 Diagram of a multi-looped aquaponics system.

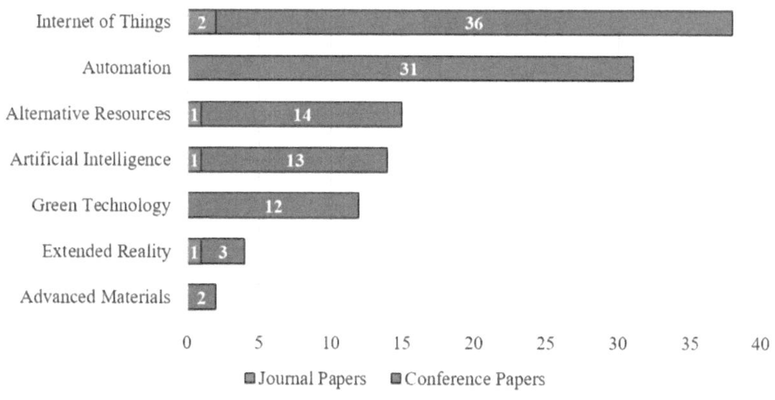

FIGURE 8.2 Breakdown of literature papers.

INTERNET OF THINGS

IoT is the most implemented enabling technology contributing to the industry 4.0 movement (Ng & Mahkeswaran, 2021a). As its name implies, IoT builds around a collection of embedded devices or Internet-enabled objects transferring data and maintaining communication over a network using Internet protocols (Manju et al., 2017). Anything can be part of an IoT network, provided it has been assigned a unique identifier or an Internet protocol address, resulting in numerous industrial applications in healthcare, transportation, manufacturing, and agriculture (Banjao et al., 2020; Kjellby et al., 2019; Manju et al., 2017), to name a few. One key advantage of IoT is its ability for devices to communicate with each other without the need for intervention (Butt et al., 2019), which is especially beneficial for aquaponics systems. Fish is the primary producer of nutrients in the aquaponics systems (Taufiqurrahman et al., 2020). The health and metabolism of the fish are directly related to the stability and conditions of the water parameters (Ke & Zhou, 2021). IoT can help monitor water quality and maintain water parameters, leading to a healthier growth of fish and plants. Failing to maintain water parameters can lead to crop loss, consequently reducing revenue (Banjao et al., 2020). Hence, there is great importance to keep constant track of the water quality in the system (Egargue et al., 2020; Ke & Zhou, 2021). A naturally balanced system must be simulated to ensure the ideal living conditions for the fish, plants, and beneficial bacteria. Real-time monitoring of system parameters can give an accurate insight into a farm's performance (Nichani et al., 2018). With the rapid development of technology, sensors and devices are improving their scalability, capacity, functionality, and affordability (Banjao et al., 2020; Kodali & Sabu, 2022; Nichani et al., 2018). As a result, data on aquaponics systems have been acquired by a wide range of sensors and microcontrollers in recent years (Ke & Zhou, 2021; Murdan & Joyram, 2021). Supported by the literature review, where 38 out of 50 papers adopted IoT, it is apparent that IoT is highly advantageous to aquaponics systems.

Abiotic conditions, for example, water pH, water dissolved oxygen (DO), water temperature, water electrical conductivity (EC), water pH, air humidity, and light intensity, are crucial for a healthy aquaponics system (Banjao et al., 2020; Butt et al., 2019). Water pH measures how acidic or alkaline a liquid is and affects critical processes like nitrification in aquaponics. Water DO measures the saturation level of oxygen in the water. Oxygen in the water is essential for healthy fish growth and plant root development. Nitrification requires about 4–8 mg/L of DO to be efficient (Ke & Zhou, 2021). Water temperature directly affects the nitrification rate; an increase in water temperature by 10°C can double the nitrification rate. The rate of metabolism in fish increases with water temperature as the fish require more energy to survive at higher temperatures (Ke & Zhou, 2021). Traditionally, these water parameters are evaluated weekly using chemical test kits and strips (Banjao et al., 2020). Conducting the tests manually can be laborious and time-consuming (Zaini et al., 2018). Sensors are a core aspect of IoT networks. The sensors obtain real-time data about their environment and share them over the Internet network (Vernandhes et al., 2017). Data obtained by IoT sensors can be timely analyzed to increase crop yields, reduce risks, and eliminate routine manual tasks (Butt et al., 2019; Nichani et al., 2018), enabling farm operators to perform countermeasures promptly if water parameters deviate from ideal conditions (Banjao et al., 2020). For example, chlorine can harm fish, especially sensitive organs such as gills. Chlorine sensors can help farmers verify that water is safe and chlorine-free before refilling tanks (Menon, 2020).

There are a wide range of sensors employed in an IoT-enabled aquaponics system. The typical sensors are water and air temperature sensors. Examples of water temperature sensors are DS18B20 (Ambrosio et al., 2019; Mahkeswaran & Ng, 2020; Mohd Ali et al., 2021; Murdan & Joyram, 2021; Rozie et al., 2020; Vernandhes et al., 2017) and LM35 (Elsokah & Sakah, 2019; Manju et al., 2017), whereas air quality sensors are DHT22 and DHT11 (Ntulo et al., 2021) used for measuring air temperature and humidity (Ambrosio et al., 2019; Arvind et al., 2020; Mahkeswaran & Ng, 2020; Murdan & Joyram, 2021).

Examples of water pH sensors used are B&C Electronics SZ 1093 (Murdan & Joyram, 2021), Winsense ISFET (Tolentino et al., 2019), Atlas Lab Grade pH Probe (Cooper et al., 2020) and PH4502C (Mohd Ali et al., 2021; Rozie et al., 2020). Some systems also utilize water level sensors to identify the water level of the fish tanks. Low water levels detected by sensors can trigger pumps for water top-ups. These include Omron K8AK-LS1 water level controller, double float ball switch sensors (Murdan & Joyram, 2021), P43 water level float switch sensor (Sriram & Sai Shibu, 2021), and HC-SR04 ultrasonic sensors (Arvind et al., 2020; Mahkeswaran & Ng, 2020; Rozie et al., 2020; Tolentino et al., 2019). Other water sensors include the YF-S201 flow rate sensor to monitor water flow (Nichani et al., 2018) and FC-28 moisture sensors to determine the wetness level of grow beds (Vernandhes et al., 2017). Moreover, some systems include DFROBOT total dissolved solid sensors (Cooper et al., 2020), turbidity sensors, MQ135 ammonia sensors (Rozie et al., 2020), and Atlas scientific K1 EC probes (Egargue et al., 2020). Ambient sensors include wind speed sensors (Barosa et al., 2019), light sensors like BH1750 (Arvind et al., 2020; Mohd Ali et al., 2021; Nichani et al., 2018), and other light-dependent resistors measuring light intensities on plants (Murdan & Joyram, 2021). DS3231 real-time clock module (Sriram & Sai Shibu, 2021) and DS1302 real-time clock accurately provide (Mohd Ali et al., 2021). Moreover, systems can incorporate a global positioning module to indicate the sensor location to the farmer, allowing better visualization of the overall farm (Barosa et al., 2019). In addition, passive infrared motion sensors can be applied to remotely detect any intruders entering the farm without the need for security personnel (Elsokah & Sakah, 2019). Since sensors only obtain data, forwarding the acquired data further up the IoT architecture requires an edge device connecting to the Internet (Menon, 2020). Edge devices can either process this data or send it to other IoT services, for instance, cloud storage or mobile applications (Arvind et al., 2020; Ke & Zhou, 2021).

Microcontrollers and microprocessors comprising efficient and low-powered central processing units are often used in aquaponics systems (Ke & Zhou, 2021). They are compact computing units equipped with input and output ports that allow communication and control with various compatible sensors and actuators. The onboard memory also enables it to be programmed to execute tasks (Murdan & Joyram, 2021). Microcontrollers such as Arduino Mega are commonly adopted; they consist of digital ports (Mohd Ali et al., 2021) for interfacing sensors and actuators. Add-on components can be mounted on extension boards such as Grove-Mega shield to help reduce and simplify wire connections (Mahkeswaran & Ng, 2020; Mohd Ali et al., 2021), boasting a wide range of sensor compatibility and potential scalability (Murdan & Joyram, 2021). Sensors can be directly connected to the pins of the microcontroller. However, the device has a maximum output of 5 V. Thus, relays are required to control actuators of higher voltages (Ambrosio et al., 2019). Smaller farm systems with only a few sensors can utilize cheap and compact Arduino Pro Mini (Cooper et al., 2020). Arduino Mega does not have built-in Wi-Fi capabilities and needs Wi-Fi modules, for example, NRF24L01 (Zaini et al., 2018) and ESP8266 (Riansyah et al., 2020). NodeMCU, a microcontroller built around the ESP8266 Wi-Fi module, uses MQ telemetry transport protocols to establish a connection to servers (Sriram & Sai Shibu, 2021; Taufiqurrahman et al., 2020). These boards are programmed using an Arduino integrated development environment that supports C and C++ languages and is equipped with a diverse collection of third-party libraries providing examples of how to operate them, thereby making it ideal to be adopted regardless of the programming background of users (Murdan & Joyram, 2021). Using the Arduino integrated development environment, Arduino Mega and NodeMCU can be connected via USB for easy debugging. Another component frequently used in IoT-enabled aquaponics systems as a CPU is the Raspberry Pi It is a compact and inexpensive microprocessor with built-in Wi-Fi and Bluetooth modules. Furthermore, it allows serial connections with compatible cameras, enabling live video monitoring and recording of aquaponics systems (Ong et al., 2019).

Besides that, Raspberry Pi is powerful enough to host local servers for data collection. Arduino boards and Raspberry Pi can be interconnected. The Arduino board receives and sends sensor values to the Raspberry Pi, which then uses Wi-Fi to upload data to real-time databases (Barosa et al., 2019). NodeMCU and Arduino Mega boards can be connected using a serial peripheral interface (Mahkeswaran & Ng, 2020). The Arduino Mega with many input pins can collect data from multiple sensors and transfer them to the NodeMCU, which then publish data to online cloud databases. In the absence of Wi-Fi connectivity, LoRaWAN transmission can be employed to send data from a remote location to a LoRaWAN gateway (Aishwarya et al., 2018; Murdan & Joyram, 2021).

Implementing multiple microprocessors in an IoT system extends the usage of source and sink nodes. Source nodes act as the CPU connects to several sensors. Multiple source nodes are fixed at crucial areas of the aquaponics systems. All the source nodes sense, collect, and transmit data to the sink node. The sink node is built using another microcontroller with more processing capabilities. The sink node queries the source nodes to send all sensor values and then publishes the data to the database server (Murdan & Joyram, 2021). An architecture of source-sink nodes can improve the system's scalability. If the farm expands, more source nodes can be introduced to enable coverage of new areas. Other microcontrollers adopted in aquaponics systems are WeMos D1 R1 (Riansyah et al., 2020), ESP32 (Banjao et al., 2020), and Intel Edison (Valiente et al., 2019). Lumisense IoT board utilizes a SIM900 GPRS modem to enable Internet connections (Manju et al., 2017). Microprocessors can double as local servers. Raspberry Pi can host databases such as MongoDB. Data can then be forwarded to services such as TelegramBot to notify farmers of abnormal sensor values (Rozie et al., 2020). Muhammad Fasih Uddin Butt et al. (2019) proposed an aquaponics system with an IoT framework that used multiple sensors to measure environmental parameters. A DHT temperature and humidity sensor was employed to obtain air parameter values. A DS18B20 water temperature sensor and a Sku: Sen0161 pH meter were used to obtain water parameter values. These values were sent to a NodeMCU microcontroller. The controller forwarded the data to a Firebase cloud server, which subsequently transferred the data to a mobile application and stored it in a Google spreadsheet. Using the mobile application, farmers triggered actuators with the help of relays and took corrective actions when necessary.

IoT systems reap the benefits of introducing cloud-based technologies. Cloud databases are required to process and make data available to other mobile devices (Zaini et al., 2018). Analyzing cloud data can give comprehensive insights into water parameters for the best growth of culture within the aquaponics system (Butt et al., 2019). Besides that, troubleshooting issues with poor crop yield can be performed (Vadivel et al., 2019) and future predictions of crop yield can also be forecasted (Kodali & Sabu, 2022). Data in cloud storage can be accessible by any authorized smart device from anywhere in the world (Kodali & Sabu, 2022; Ong et al., 2019). As an IoT system expands, it can employ cloud services, cloud computing, and big data analytics to improve its scalability and performance (Kodali & Sabu, 2022). Many cloud services are available, for instance, Microsoft Azure, IBM Watson, Oracle, Google Cloud, Cisco, ThinkSpeak, Salesforce, Bosch, GE Predix, and Amazon (Romli et al., 2017). Amazon provides AWS Cloud, AWS Lambda, and DynamoDB, allowing water parameters to be remotely monitored. AWS IoT Analytics and QuickSight are Amazon tools for data visualization and decision-making. If This Then That (IFTTT) services execute customized tasks using Amazon Alexa commands, for example, activating water pumps and fish feeders. Ravi et al. (Kodali & Sabu, 2022) used various environmental sensors to capture temperature, humidity, water level, LED status, and fish feeder status. The data was consolidated by an ESP-32 microcontroller, which also doubled as a local server. Using MQTT protocols, two-way data transfer was established to an AWS IoT core service. DynamoDB was utilized as data storage. Lambda functions were applied to carry out tasks and notify farmers when water levels were low. LED lights and the fish feeder were activated through the IFTTT application and Amazon Alexa commands. Amazon QuickSight was adopted to let farmers monitor the environmental parameters and actuator statuses.

A virtual private server is a virtual computer with its operating system hosted by a cloud service. Online databases and dashboards can be hosted on a virtual private server, allowing real-time data to be stored and visualized (Nichani et al., 2018). Services such as Firebase can enhance cloud storage capabilities. Being a backend service, Firebase empowers cloud services with various features. Real-time database management allows immediate reactions to data value changes such as cloud notifications to devices when data value exceeds a set threshold.

It also facilitates communication between microcontrollers and mobile applications. Moreover, Firebase can bridge mobile applications to cloud storage using application programming interface protocols (Barosa et al., 2019; Murdan & Joyram, 2021).

Mobile applications can be used to display data for easy visualization of the environment parameters of aquaponics systems (Zaini et al., 2018). Specially, IoT systems can adopt intuitive graphical user interfaces, human–machine interfaces, or interactive dashboards to display important information such as water parameters, actuator values, and routine tasks (Vernandhes et al., 2017). Live data of system parameters can be observed using mobile applications instead of manual readings (Tolentino et al., 2019) that can be troublesome when the farm is located at a remote location. Novice farmers may struggle with complicated chemical test kits (Banjao et al., 2020) while user-friendly dashboards (Sriram & Sai Shibu, 2021) can be used to monitor farm health from the comfort of a personal computer or mobile devices (Valiente et al., 2019). Live video feeds can also be viewed from mobile devices by incorporating cameras into aquaponic systems (Barosa et al., 2019). These applications can also provide farmers with recommendations using intelligent algorithms and alert functions (Menon, 2020). Alert functions can be in the form of emails, short text messages, and visual or audible push notifications (Ong et al., 2019). Remotely performing tasks, for example, feeding fish by activating fish feeders and displaying completed tasks, can consequently reduce labor time, cost, and human error, while increasing efficiency (Barosa et al., 2019; Kodali & Sabu, 2022; Manju et al., 2017; Nichani et al., 2018).

There are a wide range of tools to aid farmers in visualizing aquaponics parameters. Django framework, written in Python programming language, can be adopted to develop a web application to display farm parameters (Cooper et al., 2020). GUI applications obtain sensor values from a server and show them to farm operators. ThingSpeak, a software as a service platform (Sriram & Sai Shibu, 2021), can push data to mobile applications such as Virtuino to visualize data from anywhere and TelegramBot for triggering direct messages to farm operators for early warning and condition monitoring (Banjao et al., 2020). Besides that, ThinkSpeak has its online dashboard and cloud service (Banjao et al., 2020; Mohd Ali et al., 2021; Ntulo et al., 2021) that utilizes widgets and other user-friendly gauges or graphs to display sensor values from microcontrollers such Arduino Mega (Banjao et al., 2020; Barosa et al., 2019; Valiente et al., 2019). ThinkSpeak gauges allow operators to understand the current farm conditions, while ThinkSpeak graphs show the historical and real-time farm conditions, which enables troubleshooting if any issue occurs (Mohd Ali et al., 2021; Ntulo et al., 2021). An Android-based MIT App Inventor enables farmers to create interfaces using logic blocks. Pick and place different blocks together can create unique functions to suit the needed application (Vernandhes et al., 2017). Amazon Quicksight has various visualization tools like bar charts, pie charts, and pivot tables (Kodali & Sabu, 2022). Node-Red framework, a browser-based software, also has dashboard capabilities for remote monitoring of aquaponics water parameters (Taufiqurrahman et al., 2020). Furthermore, web-based GUIs can be created using HyperText Markup Language, Cascading Style Sheets, JavaScript, and PHP. Virtual measurement devices can display sensor values along with their optimal limits, allowing farmers to view quickly if gauge needles are positioned outside the optimal ranges (Ong et al., 2019). Ikonik is another android mobile application for visualizing an aquaponics system. Historical and real-time data can be displayed on its interface (Zaini et al., 2018). Aquaponics systems can be remotely controlled and monitored using a mobile application, Blynk (Riansyah et al., 2020; Sriram & Sai Shibu, 2021). Blynk allows farmers to customize the interface according to hardware freely (Mahkeswaran & Ng, 2020; Riansyah et al., 2020) and sends notifications to inform farmers of system conditions exceeding normal operating limits (Sriram & Sai Shibu, 2021). With the help of mobile, farm operators can remotely control relays with a touch of a screen, which in turn activates actuators, for instance, LED grow lights, fish feeders, circulation fans, water pumps, and misting systems (Murdan & Joyram,

2021; Vernandhes et al., 2017), thereby allowing farmers to remotely stop a process that may cause system malfunction. An example is the detection of chlorine during water top-ups to tanks, where the farm operator can remotely turn off controllable taps (Menon, 2020). Moreover, farmers can be notified if water parameters exceed ideal conditions so that they can take corrective measures (Butt et al., 2019). Plant heights can also be displayed on dashboards and notify farm operators if crops are ready for harvest (Mahkeswaran & Ng, 2020; Ong et al., 2019).

An IoT system carries valuable and sensitive data shared among many mobile devices. Security can increase the IoT system's reliability by improving secure communications between devices. For example, authentication ensures that only authorized personnel have access to farm data and dashboards. Furthermore, session timeouts prevent unauthorized personnel steal and use an existing user session (Rudd & Cunningham, 2021). Login identifications and passwords can be secured with SHA-256 encryption (Arvind et al., 2020; Ong et al., 2019; Riansyah et al., 2020); information transfer can be secured with WebSocket API; and hacking attempts can be tracked with Honeypot (Vadivel et al., 2019). However, implementing security infrastructure into an IoT system rises system setup and maintenance costs as well as energy consumption (Romli et al., 2017; Ng et al., 2023).

Despite the myriad benefits of IoT-enhanced aquaponics systems, there are challenges to overcome. Microcontrollers have limited processing capabilities (Rudd & Cunningham, 2021). Besides the need for constant Wi-Fi connection, there is an increased energy consumption when adopting IoT systems (Mohd Ali et al., 2021). Energy consumption increases with the number of devices, transmission range, and bandwidth of communication devices (Kjellby et al., 2019). For remote farms that cannot access the power line grid (Albright-Borden et al., 2019), IoT devices are operated on batteries, which will eventually run out and require routine replacement or recharging (Kjellby et al., 2019). A possible solution to this challenge is the implementation of energy harvesting capabilities, such as solar panels, to increase the self-sustainment of devices (Mahkeswaran & Ng, 2020). Furthermore, sensors are expensive and require frequent calibration, considerably increasing setup and maintenance costs for larger farms (Manju et al., 2017; Pastor et al., 2019). Besides that, some sensors are meant for laboratory use and cannot be exposed outdoors for prolonged periods. Using these sensors, contrary to their intended usage, places them at high risk of rapid deterioration and inaccurate readings (Butt et al., 2019). IoT devices must also be waterproofed or housed in a protective casing when deployed in aquaponics environments (Cooper et al., 2020). Due to manufacturing, many sensors have systematic errors that require a compensation value that changes with its lifespan (Arvind et al., 2020; Rozie et al., 2020; Zaini et al., 2018). Moreover, it is vital to ensure that only relevant data are uploaded to cloud systems. Failure to do so can result in accumulating junk data in cloud storage. Data movement from the sensor to the cloud typically takes several seconds, depending on the data travel length and traffic (Vernandhes et al., 2017). Some IoT systems that utilized ThingSpeak experienced delays of about 20 seconds, proving a challenge to monitor real-time data (Butt et al., 2019). Such a delay may compromise the opportunity to act when data exceeds allowable limits, which is especially dangerous in events like chlorine contamination, where quick actions must be taken (Menon, 2020). The introduction of edge computing or fog servers within the sensor-to-cloud chain can fulfill intermediate decision-making based on real-time data. Moreover, fog servers can decide which data to upload to cloud storage, doubling as a data filter (Romli et al., 2017)

Traditional farmers may not be supportive of implementing IoT systems to facilitate their aquaponics systems, which may be attributed to the lack of awareness and understanding of the IoT benefits (Elsokah & Sakah, 2019; Vadivel et al., 2019) or be daunted by the technicality of an IoT system (Barosa et al., 2019). IoT can be enhanced using AI or machine learning to enable systems to make decisions (Kodali & Sabu, 2022). Computer vision and image processing can be performed on video feeds obtained by smart cameras, aside from being sent to dashboards (Barosa et al., 2019; Mahkeswaran & Ng, 2020; Ong et al., 2019). IoT systems can also be improved by automation. Actuators can be automatically triggered based on threshold values (Kjellby et al., 2019), eliminating farmers' need to activate them manually onsite or remotely via dashboards.

AUTOMATION

The farming industry has seen increased accessibility and connectivity with automation, revolutionizing food production (Manju et al., 2017; Murdan & Joyram, 2021). Farm systems can become self-maintaining and independent with decisions based on data (Pasha et al., 2018), thereby reducing manual labor while maintaining control and precision (Yanes et al., 2020). There are different degrees to which farms can be automated. Semi-automated farms are remotely controlled by control inputs from the farmers. Fully automated farms take corrective actions independently, even when farm operators are unavailable or overlook the warning notifications (Butt et al., 2019). Fully automated farms operate based on threshold values. Actuators control external physical devices based on sensor readings compared with threshold values (Barosa et al., 2019; Mahkeswaran & Ng, 2020). Detection controllers can compare input values with set threshold values to run repetitive tasks (Ng & Mahkeswaran, 2021a). The workforce of the agricultural sector is aging. Being labor-intensive, agriculture is moving toward a labor shortage (Kodali & Sabu, 2022). Crucial tasks include feeding fish, maintaining water parameters, and ensuring unobstructed water flow throughout the system (Valiente et al., 2019). These tasks will get harder to manage as farms expand, reaching a point where manual monitoring is impracticable (Kodali & Sabu, 2022). As such, there is a need to automate and remotely manage processes as daily management of aquaponics systems is time and labor-intensive.

Fish, plants, and microorganisms within an aquaponics system have a range of environmental parameters for optimal growth. One of the critical environmental parameters is water quality which requires frequent testing (Manju et al., 2017; Pasha et al., 2018). Aquaponics systems coupled with automation capabilities help maintain the stability of water parameters (Pasha et al., 2018). With an understanding of environmental parameters, sensors can be used to obtain real-time data on the water parameters of aquaponics systems. Automation and control can be built around maintaining ideal conditions required for optimal growth by comparing obtained sensor values with the set thresholds (Friuli et al., 2021). These threshold markers allow automation systems to take reference when performing corrective actions (Manju et al., 2017) using microcontrollers like Arduino (Murdan & Joyram, 2021; Pasha et al., 2018). Flordeliza et al. (Valiente et al., 2019) integrated an IoT system with an aquaponics system equipped with automated LED grow lights, fish feeder, and water refilling. The IoT system was built with Intel Edison and allowed control of pH level and water temperature. Streaming video of fish behaviors, displaying real-time system parameters, and controlling actuators were realized through a web application and a mobile application. A growth comparison was made between tilapia grown in automated and traditional aquaponics systems. Results showed that greater growth tilapia was observed in the automated aquaponics system.

Actuators such as pumps, lights, and fans operate at voltages much higher than microcontrollers' output voltages. Hence, microcontrollers send signals to relays to aid the control of actuators (Ambrosio et al., 2019). Aside from relays, L298N H-Bridge motor drivers can control the speed and direction of motors by reverse voltages and pulse-width modulation (Murdan & Joyram, 2021). The pulse-width modulation can also control high-powered LED grow lights using MOSFETs (Ambrosio et al., 2019). pH values of aquaponics systems are maintained at around 6–8 pH levels (Tolentino et al., 2019). Peristaltic motors can control water pH levels to spray either acidic or alkaline solutions (Zaini et al., 2018). Peltier elements are thermoelectric plates that cool on one side and heat up on another when applied with electricity. These can be used to heat or cool water depending on whether the water is too cold or hot (Zaini et al., 2018). Servo motors are often used as automatic fish feeders as they allow precise control motor rotations. Servos dispense a consistent amount of feed each time it is activated (Aishwarya et al., 2018; Mahkeswaran & Ng, 2020; Nagayo et al., 2017; Ntulo et al., 2021; Pasha et al., 2018). Feeding can be done at regular daily intervals by preprogramming a microcontroller assisted with a real-time clock module (Nagayo et al., 2017). The feed can be precisely supplied to the fish, which is usually calculated based on the fish's weight. Traditional feeding methods rely on estimation, which can lead to underfeeding or overfeeding. Improper feeding amounts can cause fish malnutrition or water pollution (Nagayo et al., 2017). Some aquaponics systems work on pumping water into plant growth beds based on sensor moisture readings. When beds start to dry, water is automatically pumped to grow beds with the help of pumps or servos (Barosa et al., 2019). Bell siphons are often utilized to flood and drain plant grow beds. A flood-drain cycle every 15–20 minutes is best suited for optimal filtration and plant root growth, depending on the inlet flow rate to the plant grow bed. Too high inlet flow rate due to a stronger pump causes flood-drain cycles to be too short. Ball valves can be used to control the water flow by taking reference from water flow sensors (Romli et al., 2017). Water can be pumped back into tanks by water pumps controlled by relays (Aishwarya et al., 2018). Controlled pumping of clean water can help regulate water parameters, for instance, total dissolved solids, EC, and salinity. Low DO readings can be rectified using a spare air pump that can be activated automatically (Nagayo et al., 2017). Air temperature is required for plants to photosynthesize efficiently (Vernandhes et al., 2017).

Exhaust fans can be activated to direct hot air away from plants, cooling air around them. Conversely, fans can pull air from evaporative coolers to drop temperatures if the air around the plants is warmer than ideal. Exhaust fans or mist makers can be used to regulate air humidity (Nagayo et al., 2017; Vernandhes et al., 2017). Humidity is essential for plants. It determines how well transpiration and nutrient absorption processes occur (Vernandhes et al., 2017).

It is important to know actuation statuses; software like LabVIEW can be adopted to monitor sensor values and actuator statuses. Increased accessibility of aquaponics systems can be accomplished by implementing web or mobile applications (Vernandhes et al., 2017). Farmers can remotely monitor and control the system using a smartphone (Aishwarya et al., 2018) with applications like Blynk to showcase an intuitive dashboard with farm data and hardware control through a smartphone (Mahkeswaran & Ng, 2020). Statuses of these actuators can also be reflected on mobile applications, viewed from smartphones by farmers (Barosa et al., 2019; Zaini et al., 2018). Mobile applications can also determine how long actuators stay on and off (Barosa et al., 2019). Therefore, farmers can remotely control the environmental parameters of the farm (Barosa et al., 2019). Notifications can be sent to farmers' smartphones to alert them if an actuator is activated or switched off (Zaini et al., 2018). Microcontrollers can also utilize GSM modules to send mobile notification messages of actuation statuses (Nagayo et al., 2017). Web-based applications with automation capabilities can be created with Raspberry Pi equipped with MySQL database, WebSocket protocols, and Apache Web server. OpenWrt can be employed to develop an interactive application that provides valuable information on aquaponics parameters. Actuators, such as pumps, lights, and fans, can also be controlled using this web application (Pasha et al., 2018) or set to work in a timer-based mode. Other advancements include using Amazon Alexa or Google Assistant. Hands-free remote control of actuators can be done through voice commands with the help of Amazon API gateways (Kodali & Sabu, 2022).

In the current state of automation, there are some technical limitations. The complexity of automation systems increases with more moving parts and electronic devices (Mohd Ali et al., 2021).

Higher maintenance may be incurred without proper condition monitoring, limiting its application in large commercial-scale farms (Yanes et al., 2020). The workforce in agriculture is an aging population. Thus, complex automation systems may pose a steep learning curve for farm operators. Sensors must be adequately calibrated at regular intervals to avoid measuring inaccuracies (Zaini et al., 2018). Energy usage and cost increase with the number of actuators introduced (Mohd Ali et al., 2021; Nagayo et al., 2017). Larger numbers of devices also introduce some delay in the response of actuators when triggered remotely or automatically. Delays also vary depending on network traffic (Vernandhes et al., 2017). As the computing power of controllers increases, more advanced control methods can be implemented. Extending into AI, computer vision systems can be employed using a camera to detect diseases on plants and automatically inform farmers (Barosa et al., 2019). A methodology like decision support systems has the potential to automate farms completely (Friuli et al., 2021). Adaptive control is another aspect to be explored (Ke & Zhou, 2021). A closed-loop control system can be employed where obtained sensor values are repeatedly compared with desired values; actuation is then performed in accordance with the difference in the two values (Ambrosio et al., 2019). While simple threshold automation bears good results, introducing more dynamic control through AI-driven fuzzy logic can precisely determine the degree of control rather than just switching on and off actuators (Pasha et al., 2018).

ALTERNATIVE RESOURCES

With urbanization, the constant energy demand has been increasing (Egargue et al., 2020). Integration of technologies into aquaponics systems has proved to have many benefits; however, the system's energy consumption will correspondingly increase (Mohd Ali et al., 2021). One of the concerns of technology integration into the aquaponics system is energy and resource usage. Using alternative resources not only increases the viability of the aquaponics system but also limits agricultural effects on the environment (Mohd Ali et al., 2021). One solution is the incorporation of solar or photovoltaic (PV) panels into aquaponics systems (Albright-Borden et al., 2019; Cooper et al., 2020; Kjellby et al., 2019). PV panels consist of an array of solar cells. An electrical current is generated when sunlight hits these cells. It charges batteries with the help of a charge controller by converting solar energy into electrical energy. Charge controllers protect the battery by limiting current intake, preventing overcharging, and improving lifespan. Small-scale aquaponics systems have been able to be wholly off-grid by PV panels (Murdan & Joyram, 2021). This could be revolutionary if commercial-scale aquaponics systems could entirely rely on just PV panels and open the possibility of farming in areas where power grids are unavailable (Pastor et al., 2019). Tropical countries, such as the Philippines and Malaysia (Mohd Ali et al., 2021), have rich agricultural history and access to a large extensive array of alternative energies. Most notably, solar energy is abundant, boasting up to 5.5 kWh per square meter daily (Egargue et al., 2020). Implementing alternative energy resources for the systems can also avert failure in a power shortage. A power shortage could be catastrophic for an aquaponics system dependent on the pump's functionality. IoT nodes powered solely by batteries need routine replacing or recharging. Over the years, there has been an increase in the usage of energy harvesting nodes allowing nodes to operate for years without human interaction. Typically, IoT networks must bear trade-offs for higher transmission ranges with higher energy consumption which can be lessened using alternative energy (Kjellby et al., 2019).

PV panels integrated into aquaponics systems have shown promising results (Albright-Borden et al., 2019). PV panel systems can be integrated by selecting appropriate panel sizes, batteries, solar charge controllers, and inverters based on the system energy consumption (Mohd Ali et al., 2021; Murdan & Joyram, 2021; Nagayo et al., 2017). The typical usage of PV panels usually involves harvesting energy from the sun and charging batteries. Effective selection of battery type and capacity can power aquaponics systems throughout the night (Albright-Borden et al., 2019). Safety chips like BQ25560 IC from Texas instruments can optimize PV installations. It comprises under and over-voltage protection and maximum power point tracking to monitor energy obtained from PV panels (Kjellby et al., 2019). Zener diodes can be a voltage limiter to protect circuits (Deepthi et al., 2021). During the day, solar panels can provide energy to devices and charge batteries. At night, the battery supplies energy to devices and sensors, allowing constant operation (Deepthi et al., 2021). Furthermore, aquaponics systems can be designed with a smart control system using signal processing to switch power supply between grid or PV panels. Effective implementation can make aquaponics systems less dependent on the grid and more on PV panels, increasing cost-effective operations. Real-time monitoring systems can extend from observing water parameters to system conditions, for example, power consumption and condition monitoring of power supplies and PV panels. Jerome Christian C. Egargue et al. (2020) developed an automatic aquaponics system equipped with PV panels and integrated with intelligent control for switching the power supply. The system could switch between being grid or PV-dependent for energy. The system was initialized to prioritize operation on the battery while being charged by solar energy. Once the battery was detected to be in a low state, the control system utilized a transistor to trigger a relay to switch the power supply to the grid. The PV panels then continued charging the batteries. Once the batteries were adequately charged, the control system switched back the power to operate on the battery. The control system program identified the conditions for the 'low' and 'high' state of the battery, which minimized dump energy (Egargue et al., 2020).

Aside from energy harvesting, solar power can be used for other applications like bio-solar-purification. Wastewater can be treated using heat or ultraviolet light from solar radiation. Solar energy can also be used to regulate air or water temperatures in aquaponics systems. Radiative sky cooling systems exploit colder temperatures of the night. Functioning like a heat sink that dissipates heat using vanes can help passively cool aquaponics systems without electricity, which facilitates the incorporation of aquaponics systems into buildings, giving rise to the concept of building integrated agriculture (Samuel et al., 2020). Electrical water heaters are commonly used to heat the water temperature to ideal conditions for fish. Usually consisting of a heating coil, electrical water heaters can consume considerable energy, especially in colder climates where the heater will be on more often. It may be inefficient to employ PV panels to power water heaters, which heat the water. Instead, solar heaters with controllers can be employed to increase the temperature of the water. Solar water heaters are much more efficient by directly using solar radiation to heat water. Appropriate control must be implemented to regulate the level of heating. If uncontrolled, solar heaters can reach very high temperatures, potentially harming the fish. Moreover, innovative protective shields can be used to inhibit solar exposure of the water heaters when target temperatures are met, preventing overheating. Alternatively, controllable valves can be used to mix hot water from the solar heater with cold water until the target water temperature is met before supplying it to the fish tank (Dbouk & Khalife, 2020). Fareed Ismail and Jasson Gryzagoridis (2018) developed a solar-powered aquaponics system with a PV/thermal hybrid system. The system utilized recirculating water from the aquaculture subsystem to cool PV panels. The heated water then returned to the fish tank, maintaining ideal water temperatures for fish growth and cooled PV panels to maintain their performance, as efficiencies are inversely proportional to their temperature. A comparison was made between the cooled and uncooled PV panels of the system. Results indicated a loss of only 4% in efficiency in 85 minutes for the cooled panels but a loss of 20% in 15 minutes for the uncooled ones, proving cooling panels with water were more effective than air.

Freshwater is a vital yet scarce resource. Water efficiencies of aquaponics systems can be improved by innovation. Rooftops can be lined with several mat layers of filtration material topped off with carpeting vegetation facilitating the collection of rainwater to be filtered and funneled to a collection point, further increasing water efficiencies of aquaponics systems. Only 2.5% of the earth's water is freshwater. The remaining 96.5% is salt water found in oceans and seas. Most aquaponics systems use fresh water. The abundance of salt and brackish water available gives rise to the viability of saline aquaponics. Also known as maraponics, the practice involves rearing euryhaline aquatic species with halophytes. Euryhaline are organisms that tolerate a range of saline environments. Grouper, mussels, shrimp, and seabass are some euryhaline cultures that can be reared in maraponics systems. Halophytes are salt-tolerant plants that can grow in water with salt concentrations of up to 1M NaCl. Some examples of maraponics halophytes are glasswort, saltgrass, and pepperweed. An increase in the implementation of maraponics can positively impact seafood industries. However, more research must be conducted to determine the effectiveness of these systems (Spradlin & Saha, 2022). Lastly, using wind turbines to harness wind energy, micro-hydro systems to transform the energy of flowing water into electricity, and hydrogen fuels (Pastor et al., 2019) can be other approaches to power aquaponics systems.

The efficiencies of PV panels are the most significant limiting factor. Panel efficiencies drop with rising temperature. Roughly every 1°C increase from the panel rating temperature results in a drop of efficiency by 0.5%, especially in hotter climates where panel temperatures can reach up to 70°C (Ismail & Gryzagoridis, 2018). Proper fault management can therefore be practiced, for instance, battery monitoring for tracking panel performance (Kjellby et al., 2019). It is also essential to consider the size and placement of the PV panels. PV panels can be attached to nearby roofs to eliminate competition with plants for light (Albright-Borden et al., 2019). Besides that, incorrect battery selection can lead to inefficiencies like excessive dump energy. To lessen dump energy, it is crucial to understand how much energy the aquaponics system consumes using data loggers or power monitoring systems before determining battery sizes (Mohd Ali et al., 2021). System energy consumption can change throughout the year. Systems with automatic climate control may have actuators like fans to cool air temperature and keep plants cool. These actuators will be operated much longer during the summer, increasing energy usage compared to winter (Nagayo et al., 2017). Accumulation of dust and dirt on panels can hinder panel performance. Regular cleaning of panels can alleviate this (Egargue et al., 2020). A climate with cloudy skies can also reduce the effectiveness of PV panels (Pastor et al., 2019). Solar trackers can help PV panels maintain ideal incident angles about the sun's location throughout the day by using sensors and actuators to tilt PV panels automatically (Tharamuttam & Ng, 2017; Vadivel et al., 2019). Installation of PV panels can increase start-up costs. Farmers should know the long-term benefits of investing in PV panels (Albright-Borden et al., 2019; Egargue et al., 2020).

ARTIFICIAL INTELLIGENCE

As the world is moving toward industrial 4.0, AI is indispensable and can substantially benefit agriculture (Aquino et al., 2020). Neural networks, evolutionary computations, and fuzzy structures are cornerstones of computational or AI (Aquino et al., 2020). Much research has suggested adopting AI and machine learning to address some agricultural issues, adding value to an existing automated aquaponics system (Cooper et al., 2020). Google supported farmers with AI systems to help increase crop production (Aquino et al., 2020). The inclusion of drones on farms is predicted to amount to $82 billion in economic activity in the US in the next decade. IoT systems with simple automation are inadequate to manage aquaponics systems productively. With the development of AI technology, more complex adaptive control can be implemented (Ke & Zhou, 2021). AI and machine learning can be trained using information from IoT cloud databases, for instance, AWS, Microsoft Azure, Google Cloud, and IBM Cloud (Arvind et al., 2020). Machine learning methods, for example, linear regression, support vector regression, and CART decision trees, can be explored to extrapolate the growth rate of fish and plants (Ghandar et al., 2021). Vadivel R et al. (2019) proposed a system where sensors within the aquaponics system obtained environmental readings and forwarded them to cloud storage. The data was made available for machine learning algorithms. Linear regression was adopted to predict the health of the plants based on the water quality.

Monitoring and analyzing water quality is vital in an aquaponics system because water quality is associated with the health of fish and plants. However, many sensors are required for a holistic awareness of water parameters. Furthermore, the sensors are costly and often not viable. As such, some studies proposed the use of aquaphotomics and spectroscopy to measure nitrate, phosphate, and potassium concentrations and sensors to measure pH, EC, and temperature values. With the help of algorithms, such as adaptive neuro-fuzzy inference systems, minimum redundancy maximum relevance algorithm, and univariate feature ranking for regression using F-tests, models were built to correlate pH, EC, and temperature values with nitrate, phosphate, and potassium levels in the water. With the trained models, future readings of nitrate, phosphate, and potassium levels can be easily obtained without purchasing actual sensors, significantly reducing setup costs (Lauguico et al., 2020). Fish are susceptible to sudden changes in water temperature. Some weather or climates can induce abrupt temperature changes that conventional control systems are not responsive enough. An adaptive boosting (AdaBoost) regression algorithm can be used to optimize water temperature control. Adaboost operates by using results generated by other machine learning algorithms that are labeled as 'weak learners'. Adaboost algorithms can be evaluated by calculating the mean square error and R-squared values from the predicted temperature results (Taufiqurrahman et al., 2020). In addition, an intelligent algorithm can be developed to automatically recommend farmers with various suggestions and options, providing a means of recommending a course of action to farmers (Menon, 2020) and aiding inexperienced farmers to pick up the trade while being assisted by software promptly.

Computer vision, a subgenre of AI, has been a significant contributor to agricultural technological advancements. Computer vision can provide a new nondestructive and nonintrusive way of monitoring and detecting crop issues (Aquino et al., 2020). Images of plants can be captured using a camera and preprocessed via image enhancement and segmentation in Raspberry Pis (Tolentino et al., 2019). Preprocessing of images includes gray scaling, thresholding, image enhancement, and color space conversion. Leaf images of plants are then contoured and segmented to help the system identify leaf features and consequently locate the unhealthy plant. Feature-based K-means algorithm can be used for conducting segmentation on the images. Texture feature extraction can be done via gray level co-occurrence matrix and then classified by a support vector machine (Barosa et al., 2019). Disease identification works on the principle of detecting spots and irregularities in the leaves. Brown and black spots on the leaves are symptoms of plant disease. RapidMiner – a machine learning and data analytics platform – can be utilized to identify the disease. A report of the diseases identified can be automatically generated and sent to the farmers' mobile application. Hence, manual labor can be reduced, as well as early disease detection can be accomplished, which can potentially save crops (Barosa et al., 2019). The leaf canopy area is one of the defining traits of crop growth. Crop growth models can be created using neural network discriminant function analysis and quadratic support vector networks. Online tools like Infragram can be used to assess photosynthesis activities through near-infrared plant images (Sriram & Sai Shibu, 2021). Detecting the intensity of green pigmentation on leaves can indicate if there is any nitrogen deficiency. Furthermore, computer visions can be adopted to automate fish feeding; the fish feeder can operate based on the behavior of the fish to limit overfeeding fish. Fish counting can also be realized using computer vision. Using hybrid models hidden in an artificial neural network can consistently identify fish using underwater imageries.

AI can be implemented for effective control of an aquaponics system. Fuzzy logic methods can be implemented to control the water flows of the system. The duty cycle of alternating current motors of water pumps can be controlled more precisely for more dynamic control (Sriram & Sai Shibu, 2021). Regulating the speed of the pump using fuzzy logic can also provide an energy-efficient means to control the temperature and pH of an aquaponics system (Sriram & Sai Shibu, 2021). Supplementation of the nutritional needs of plants can also be done using a motor pump controlled by a fuzzy logic system. AB mix nutrition can be precisely dosed into the water based on EC readings obtained by sensors (Yuhasari et al., 2021).

There is an inherent difficulty in applying AI to aquaponics systems because they comprise multiple living organisms. A possible solution would be the construction of a simulation model. Simulation models can provide system responses across a period. Data can be fed for routine recalibration of the simulation model, increasing its accuracy and effectiveness (Ghandar et al., 2021). However, predictions can only provide a certain degree of accuracy. Excessive training to increase model accuracy can lead to overfitting, where the model closely matches the training data, creating problems for future predictions. However, ensembled learning that uses multiple learning models can be adopted to lessen the effects of overfitting (Taufiqurrahman et al., 2020). False positives should be kept at a minimum because they reduce the usefulness of machine learning models. Improper calibration of sensors can lead to an increase in false positives (Arvind et al., 2020). Computer vision implemented for underwater imagery can be challenging. Besides lower visibility underwater, high false positive rates can occur as ripples in the water may be mistaken as fish. Fish constantly move in different directions and bend their bodies differently, making it hard to train a robust model with a standard set of fish images (Arvind et al., 2020).

GREEN TECHNOLOGY

In contrast with alternative energy that focuses on harvesting energy from other renewable sources to meet the energy demands of aquaponics systems, green technology involves applying energy-efficient strategies to improve the utility of aquaponics system energy (Tang et al., 2021). These energy-conscious systems have a smaller carbon footprint and longer lifespans (Rudd & Cunningham, 2021). The application of green technology extends to choosing the appropriate energy-efficient components and construction materials, which can lead to energy savings. In any PV system, the selection of a suitable battery is essential. Lithium-ion batteries of the same capacity ratings as lead-acid batteries can charge and discharge more energy. Lithium-ion batteries have a higher depth of discharge (Mohd Ali et al., 2021). Some green strategies involve repurposing waste energy. The heat generated by PV panels can be utilized to increase the water temperature of aquaponics systems, which can be done by routing water across the panel heat sinks using customized backing (Ismail & Gryzagoridis, 2018). IoT devices adopting sleep-wake strategies operate on the concept that when sensor nodes are not being used, they can be switched to a low-power sleep mode to be more energy efficient. Microcontrollers can operate in a sleep cycle mode, waking up once during preset intervals to obtain sensor values before going into a low-power sleep, further optimizing energy usage (Pastor et al., 2019). Rolf A. Kjellby et al. (2019) developed a wireless sensor network of self-powered IoT devices that utilized a custom ultra-low power multi-hop protocol. Nodes were built using the ultra-low power nRF52840 microcontroller. Adopting the multi-hop protocol allowed energy-efficient data transfer as low as 267 μW between devices and gateways within the network. Data were consolidated in a cloud server where analysis was conducted to maximize crop growth in an aquaponics system.

Microcontrollers send information to other parts of the IoT infrastructure using communication protocols. Introducing alternative low-energy protocols like Bluetooth low energy (Pastor et al., 2019) and long-range wireless area networks as opposed to Wi-Fi can lower energy consumption (Rudd & Cunningham, 2021). Security of Internet-enabled microcontrollers and other IoT devices is also important because they carry valuable system information that unauthorized devices may intercept. However, establishing a robust cybersecurity infrastructure can considerably increase the system's energy consumption. Common security methods like transport layer security operate on a handshake between the data sender and receiver. Whenever data are sent, it is authenticated upon receipt, thereby incurring additional energy usage and posing a setback to aquaponics systems using battery-operated microcontrollers.

Actuators like pumps, lights, and fans consume high energy (Pastor et al., 2019). Energy savings can be achieved by controlling these actuators to only function when necessary. For example, running water pumps continuously for water circulation in an aquaponics system can consume high energy. Thus, water pumps can be controlled to operate only for a set duration of time every hour based on the size and energy capacity of the system (Ambrosio et al., 2019; Cooper et al., 2020; Murdan & Joyram, 2021). Hence, energy consumption can be lowered while maintaining adequate water circulation throughout the system. More fine-tuned control of water pumps in an aquaponics system can be made by controlling the pump's duty cycle rather than just turning it off and on in intervals. Fachrul Rozie et al. (2020) proposed an intelligent aquaponics system where the pump's duty cycle was controlled by a controller using the fuzzy logic method. Important parameters, such as water temperature and ammonia levels, were inputs to the fuzzy logic controller. Employing an alternating current dimmer and a zero-crossing switch, the pump's duty cycle was regulated accordingly, which then controlled the speed of the motor in the water pump. System energy efficiency consequently increased due to decreased pump's energy consumption since the pump was not running continuously at maximum speed. Using grow lights in an outdoor aquaponics system can improve plant growth rates while reducing energy consumption. Indoor aquaponics systems must utilize grow lights to supply 14–16 hours of light for vegetables. An outdoor aquaponics system only activates the grow light when the ambient light intensity drops below the minimum lux requirement of the vegetables. Therefore, these systems can see up to 42.9% energy savings compared to indoor aquaponics systems (Ong et al., 2019). Aside from the benefits of green technologies, they must be coupled with some energy harvesting capabilities to be truly self-sustaining (Kjellby et al., 2019).

EXTENDED REALITY

Extended reality is an umbrella term for technologies incorporating the digital world into the physical world (Wang, 2022). These technologies can be broadly classified into virtual reality, augmented reality, and mixed reality. Virtual reality allows immersion into a computer-generated 3D virtual world created using software like AutoCAD (Farshid et al., 2018). Augmented reality superimposes virtual data and objects over the actual world to allow a better understanding of reality by providing more information about it (Farshid et al., 2018). Mixed reality combines part of the actual world and the digital world to create a hybrid reality (Farshid et al., 2018). Creating these virtual worlds can stimulate a greater sensory perception of reality. Auxiliary information is provided to the users where it may not be apparent if observed with just the eyes, thereby augmenting engagement. Extended reality systems can impact education by using aquaponics systems to teach students (Ng & Mahkeswaran, 2021b). Juan Garzon et al. (2020) developed an augmented-reality-based aquaponics system to enhance environmental educational programs. An augmented reality instructional application for learning was developed using software packages, for example, Unity and Vuforia. The application provided games, videos, and graphics to improve learning. Based on ten student volunteers' feedback, the application motivated them to learn and foster sustainable agriculture. Thus, these educational programs can bring awareness to the general population about the need for sustainable farming and the importance of nature (Garzon et al., 2020). A digital twin is a virtual representation of an actual real-world physical product, system, or process. It is beneficial for predicting and maintaining operational performance in real-time. Although most digital twin applications are found in the manufacturing industry, they can also be applied to aquaponics. Data of environmental parameters obtained by sensors are sent to a Wi-Fi-enabled microcontroller. Next, the data are preprocessed and uploaded to cloud storage, where the digital twin captures them to update and change its model accordingly to its physical counterparts. In doing so, farmers can obtain valuable insights into the aquaponics system, such as predicting future events and outcomes like pump malfunction (Ghandar et al., 2021). Studies on applying digital twins to aquaponics systems are limited which may be due to the complexity of aquaponics encompassing numerous living organisms (Ghandar et al., 2021).

ADVANCED MATERIALS

Fabricated materials have made their way into enhancing aquaponics systems. With the use of three-dimensional printing, mounts for sensors can be customized and printed precisely to the requirements of the sensors. Printing material made from polylactic acid is also environmentally friendly (Mahkeswaran & Ng, 2020). Another advanced material adopted for growing bed media in aquaponics systems is superabsorbent hydrogel (SAH). This material can absorb large amounts of water and expand several times its size. The water is then slowly released along with any absorbed nutrients over time. Marco Friuli et al. (2021) utilized SAH made from natural polymers as grow media for an aquaponics system. The material could retain water and nutrients and deliver them to the plants. The usage of SAH allowed the gradual release of water and nutrients to the plant roots, reducing the irrigation frequency needed from the aquaculture subcomponent. A comparison was made between the water retention ability of traditionally used expanded clay and SAH. Results showed that plants grown using SAH thrived even when watered every 2 days intervals using a pump while the plants growing in the expanded clay wilted. With less usage of the irrigation pump, energy consumption was reduced.

The adoption of new materials can be advantageous to aquaponics systems. However, ensuring that these materials do not alter the water parameters or harm the culture is critical. Microwave telemetry can be employed to determine if the quality of water changes before and after introducing these materials as a means of verification.

CONCLUSION

This chapter presents and discusses technological advancements in aquaponics systems and the challenges faced in their implementation. The literature from 2017 to 2022 shows that the top three technologies implemented in aquaponics systems are IoT, automation, and alternative resources, which are likely attributable to their direct influence in addressing the shortcomings of traditional aquaponics systems. On the contrary, AI, green technology, extended reality, and advanced material are the least four implemented technologies because they are still in the early implementation stage in aquaponics systems. On top of these technologies, other new technologies are big data, blockchain, robotics, drones (Sadiku et al., 2020), 4D printing (Maraveas et al., 2022), AI of things (Chen et al., 2020), and sixth-generation wireless technology (Zhang et al., 2022). These technologies can potentially redefine the future of aquaponics systems, further accelerating crop productivity to meet the global demand for food.

REFERENCES

Aishwarya, K. S., Harish, M., Prathibhashree, S., & Panimozhi, K. (2018). Survey on automated aquponics based gardening approaches. *International Conference on Inventive Communication and Computational Technologies – Proceedings*, 1377–1381. https://doi.org/10.1109/ICICCT.2018.8473155

Albright-Borden, R., Nelken, P., Sparagana, S., Thompson, S., Wang, J., Doyle, L., Lee, H., Parker, M. T., & Wilson, S. K. (2019). Combating food insecurity with large scale Aquaponics: a case study in silicon valley. *2019 IEEE Global Humanitarian Technology Conference*, 1–5. https://doi.org/10.1109/GHTC46095.2019.9033047

Ambrosio, A. Z. M. H., Jacob, L. H. M., Rulloda, L. A. R., Jose, J. A. C., Bandala, A. A., Sy, A., Vicerra, R. R., & Dadios, E. P. (2019). Implementation of a closed loop control system for the automation of an aquaponic system for urban setting. *IEEE 11th International Conference on Humanoid*, Nanotechnology, Information Technology, Communication and Control, Environment, and Management, 1–5. https://doi.org/10.1109/HNICEM48295.2019.9072729

Aquino, H. L., Sybingco, E., Bandala, A. A., & Dadios, E. P. (2020). Trend forecasting of computer vision application in aquaponic cropping systems industry. *IEEE 12th International Conference on Humanoid*, Nanotechnology, Information Technology, Communication and Control, Environment, and Management.

Arvind, C. S., Jyothi, R., Kaushal, K., Girish, G., Saurav, R., & Chetankumar, G. (2020). Edge computing based smart aquaponics monitoring system using deep learning in IoT environment. IEEE Symposium Series on Computational Intelligence, 1485–1491. https://doi.org/10.1109/SSCI47803.2020.9308395.

Azra, M. N., Noor, M. I. M., Ikhwanuddin, M., & Ahmed, N. (2021). Global trends on Covid-19 and food security research: A scientometric study. In *Advances in Food Security and Sustainability* (Vol. *6*, pp. 1–33). Elsevier. https://doi.org/10.1016/bs.af2s.2021.07.005

Banjao, J. P. P., Villafuerte, K. S., & Villaverde, J. F. (2020, December 3). Development of cloud-based monitoring of abiotic factors in aquaponics using ESP32 and internet of things. *IEEE 12th International Conference on Humanoid*, Nanotechnology, Information Technology, Communication and Control, Environment, and Management. https://doi.org/10.1109/HNICEM51456.2020.9400083.

Barosa, R., Hassen, S. I. S., & Nagowah, L. (2019). Smart aquaponics with disease detection. *Conference on Next Generation Computing Applications*, 1–6. https://doi.org/10.1109/NEXTCOMP.2019.8883437

Benke, K., & Tomkins, B. (2017). Future food-production systems: Vertical farming and controlled-environment agriculture. *Sustainability: Science, Practice and Policy*, *13*(1), 13–26.

Butt, M. F. U., Yaqub, R., Hammad, M., Ahsen, M., Ansir, M., & Zamir, N. (2019). Implementation of aquaponics within IoT framework. 2019 SoutheastCon, 1–6. https://doi.org/10.1109/SoutheastCon42311.2019.9020390

Chen, C. J., Huang, Y. Y., Li, Y. S., Chang, C. Y., & Huang, Y. M. (2020). An AIoT based smart agricultural system for pests detection. *IEEE Access*, *8*, 180750–180761. https://doi.org/10.1109/ACCESS.2020.3024891.

Committee on World Food Security. (2017). Global Strategic Framework for Food Security & Nutrition. https://openknowledge.fao.org/server/api/core/bitstreams/7b5ef32b-5224-4d1b-9e85-3f567102a3cc/content

Cooper, J., MacOne, C., & Pham, C. (2020). Aquaculture: A cost-effective automated aquaponic gardening system. IEEE MIT Undergraduate Research Technology Conference. https://doi.org/10.1109/URTC51696.2020.9668889.

Dbouk, H. M., & Khalife, F. G. (2020). Autonomous solar-based aquaponics: Towards agricultural sustainability in Lebanon and the region. *5th International Conference on Renewable Energies for Developing Countries*, 1–4. https://doi.org/10.1109/REDEC49234.2020.9163832

Deepthi, A. S., Niranjanaa, A., Hari Haran, A., Austin Joel, A., Dhinakaran, S. B., Monisha Thangam, K., & Anitha, N. (2021). Integrated smart system for urban farming. *International Conference on Advancements in Electrical, Electronics, Communication, Computing and Automation*. https://doi.org/10.1109/ICAECA52838.2021.9675617.

Egargue, J. C. C., Pacaigue, F. A., Galicia, R. G. F., & Magwili, E. G. V. (2020). Development of an automated aquaponics system with hybrid smart switching power supply. *IEEE Region 10 Annual International Conference, Proceedings/TENCON, 2020-November*, 544–549. https://doi.org/10.1109/TENCON50793.2020.9293853.

Elsokah, M. M., & Sakah, M. (2019). Next generation of smart aquaponics with internet of things solutions. *19th International Conference on Sciences and Techniques of Automatic Control and Computer Engineering*, 106–111. https://doi.org/10.1109/STA.2019.8717280

Farshid, M., Paschen, J., Eriksson, T., & Kietzmann, J. (2018). Go boldly!: Explore augmented reality, virtual reality, and mixed reality for business. *Business Horizons*, *61*(5), 657–663.

Friuli, M., Masciullo, A., Blasi, F. S., Mita, M., Corbari, L., & Surano, I. (2021). A 4.0 sustainable aquaponic system based on the combined use of superabsorbing natural hydrogels and innovative sensing technologies for the optimization of water use. 6th International Forum on Research and Technology for Society and Industry - Proceedings, 429–434. https://doi.org/10.1109/RTSI50628.2021.9597250.

Garzon, J., Baldiris, S., Acevedo, J., & Pavon, J. (2020). Augmented reality-based application to foster sustainable agriculture in the context of aquaponics. *IEEE 20th International Conference on Advanced Learning Technologies*, 316–318. https://doi.org/10.1109/ICALT49669.2020.00101.

Ghandar, A., Ahmed, A., Zulfiqar, S., Hua, Z., Hanai, M., & Theodoropoulos, G. (2021). A decision support system for urban agriculture using digital twin: A case study with aquaponics. *IEEE Access*, *9*, 35691–35708. https://doi.org/10.1109/ACCESS.2021.3061722.

Gibbons, G. M. (2020). An Economic Comparison of Two Leading Aquaponic Technologies using Cost Benefit Analysis: The Coupled and Decoupled Systems, ProQuest LLC. https://www.proquest.com/openview/63af971870871d79247af7d50657604c/1?pq-origsite=gscholar&cbl=44156

Goddek, S., Joyce, A., Kotzen, B., & Burnell, G. M. (2019). *Aquaponics Food Production Systems: Combined Aquaculture and Hydroponic Production Technologies for the Future*. Springer Nature. https://doi.org/10.1007/978-3-030-15943-6

Ismail, F., & Gryzagoridis, J. (2018). Optimising photovoltaic system by direct cooling and transferring heat to aquaculture medium to boost aquaponics food production in needy communities. *In 2018 International Conference on the Industrial and Commercial Use of Energy (ICUE)*. https://www.researchgate.net/profile/Fareed-Ismail/publication/363158136_Optimising_photovoltaic_system_by_direct_cooling_and_transferring_heat_to_aquaculture_medium_to_boost_aquaponics_food_production_in_needy_communities/links/638dab45658cec2104b0015c/Optimising-photovoltaic-system-by-direct-cooling-and-transferring-heat-to-aquaculture-medium-to-boost-aquaponics-food-production-in-needy-communities.pdf

Jai, N., Dontha, B., Tripathy, A., & Mande, S. S. (2018). Near real time-sensing system for hydroponics based urban farming. *3rd International Conference for Convergence in Technology*, 1–5. https://doi.org/10.1109/I2CT.2018.8529380

Ke, Z., & Zhou, Q. (2021). Research progress of intelligent monitoring and control in aquaponics. *2021 International Conference on Information Science, Parallel and Distributed Systems - Proceedings*, 177–180. https://doi.org/10.1109/ISPDS54097.2021.00041.

Khan, F. A. (2018). A review on hydroponic greenhouse cultivation for sustainable agriculture. *International Journal of Agriculture Environment and Food Sciences*, *2*(2), 59–66.

Khim, W. (2020). Disaster risk reduction in times of COVID-19: What have we learned? Disaster Risk Reduct. Times COVID-19 What Have We Learn, 2018–2021. https://doi.org/10.4060/cb0748en

Kjellby, R. A., Cenkeramaddi, L. R., Frøytlog, A., Lozano, B. B., Soumya, J., & Bhange, M. (2019). Long-range & self-powered IoT devices for agriculture & aquaponics based on multi-hop topology. IEEE 5th World Forum on Internet *of Things*, 545–549. https://doi.org/10.1109/WF-IoT.2019.8767196

Kodali, R. K., & Sabu, A. C. (2022). Aqua monitoring system using AWS. *International Conference on Computer Communication and Informatics*. https://doi.org/10.1109/ICCCI54379.2022.9740798.

Kyaw, T. Y., & Ng, A. K. (2017). Smart aquaponics system for urban farming. *Energy Procedia*, *143*, 342–347.

Lauguico, S., Baldovino, R., Concepcion, R., Alejandrino, J., Tobias, R. R., & Dadios, E. (2020). Adaptive neuro-fuzzy inference system on aquaphotomics development for aquaponic water nutrient assessments and analyses. *Proceedings of the 12th International Conference on Information Technology and Electrical Engineering*, 317–322. https://doi.org/10.1109/ICITEE49829.2020.9271736.

Mahkeswaran, R., & Ng, A. K. (2020). Smart and sustainable home aquaponics system with feature-rich internet of things mobile Application. *6th International Conference on Control, Automation and Robotics*, 607–611. https://doi.org/10.1109/ICCAR49639.2020.9108041.

Mahkeswaran, R., Ng, A. K., Toh, C., & Toh, B. (2023). Multi-loop aquaponics systems: A review and proposed multi-loop agrogeological aquaponics system. *Journal of Advanced Agricultural Technologies*, (Vol. *11*, issue no. 1), 1–7.

Manju, M., Karthik, V., Hariharan, S., & Sreekar, B. (2017). *Real* time monitoring of the environmental parameters of an aquaponic system based on Internet of Things. Third International Conference on Science Technology Engineering & Management, 943–948. https://doi.org/10.1109/ICONSTEM.2017.8261342

Maraveas, C., Bayer, I. S., & Bartzanas, T. (2022). 4D printing: Perspectives for the production of sustainable plastics for agriculture. In *Biotechnology Advances* (Vol. *54*). Elsevier Inc. https://doi.org/10.1016/j.biotechadv.2021.107785.

Menon, P. C. (2020). IoT enabled aquaponics with wireless sensor smart monitoring. *4th International Conference on IoT in Social, Mobile, Analytics and Cloud - Proceedings*, 171–176. https://doi.org/10.1109/I-SMAC49090.2020.9243368.

Mohd Ali, M. F., Asrul Ibrahim, A., & Mohd Zaman, M. H. (2021). Optimal sizing of solar panel and battery storage for a smart aquaponic system. *19th IEEE Student Conference on Research and Development: Sustainable Engineering and Technology towards Industry Revolution*, 186–191. https://doi.org/10.1109/SCOReD53546.2021.9652782.

Monsees, H., Kloas, W., & Wuertz, S. (2016). Comparison of coupled and decoupled aquaponics-Implications for future system design. Abstract from Aquaculture Europe.

Murdan, A. P., & Joyram, A. (2021, July 1). An IoT based solar powered aquaponics system. *Proceedings of the 13th International Conference on Electronics, Computers and Artificial Intelligence*. https://doi.org/10.1109/ECAI52376.2021.9515023.

Nagayo, A. M., Mendoza, C., Vega, E., Al Izki, R. K. S., & Jamisola, R. S. (2017). An automated solar-powered aquaponics system towards agricultural sustainability in the Sultanate of Oman. *IEEE International Conference on Smart Grid and Smart Cities*, 42–49. https://doi.org/10.1109/ICSGSC.2017.8038547.

Ng, A. K., Chin, J. Z., & Mahkeswaran, R. (2023). Energy conservation strategies for increased sustainability and energy efficiency in urban farming. *Proceedings of the IEEE International Conference on Service Operations and Logistics, and Informatics*, 1–7. https://doi.org/10.1109/SOLI60636.2023.10425538

Ng, A. K., & Mahkeswaran, R. (2021a). Emerging and disruptive technologies for urban farming: A review and assessment. *Journal of Physics*: Conference Series, 2003(1), 012008.

Ng, A. K., & Mahkeswaran, R. (2021b). Fostering computational thinking and systems thinking through aquaponics capstone projects. *IEEE International Conference on Engineering, Technology and Education, Proceedings*, 1039–1044. https://doi.org/10.1109/TALE52509.2021.9678854.

Nichani, A., Saha, S., Upadhyay, T., Ramya, A., & Tolia, M. (2018). Data acquisition and actuation for aquaponics using IoT. *3rd IEEE International Conference on Recent Trends in Electronics, Information & Communication Technology*, 46–51. https://doi.org/10.1109/RTEICT42901.2018.9012260

Ntulo, M. P., Owolawi, P. A., Mapayi, T., Malele, V., Aiyetoro, G., & Ojo, J. S. (2021, October 7). IoT-Based smart aquaponics system using arduino uno. *International Conference on Electrical, Computer, Communications and Mechatronics Engineering*. https://doi.org/10.1109/ICECCME52200.2021.9590982.

Ong, Z. J., Ng, A. K., & Kyaw, T. Y. (2019). Intelligent outdoor aquaponics with automated grow lights and internet of things. *IEEE International Conference on Mechatronics and Automation*, 1778–1783. https://doi.org/10.1109/ICMA.2019.8816577

Pasha, A. K., Mulyana, E., Hidayat, C., Ramdhani, M. A., Kurahman, O. T., & Adhipradana, M. (2018). System design of controlling and monitoring on aquaponic based on Internet of Things. *4th International Conference on Wireless and Telematics*, 1–5. https://doi.org/10.1109/ICWT.2018.8527802

Pastor, Z., Akim, R., Do, J., Singh, M., Pastor, Z., Alghamdi, G., Nguyen, L., Tong, Q., Decuir, J., & Alnamlah, K. (2019). Aquaponics water monitoring and power system. *2019 IEEE Global Humanitarian Technology Conference*, 1–4. https://doi.org/10.1109/GHTC46095.2019.9033016

Podder, A. K., Al Bukhari, A., Islam, S., Mia, S., Mohammed, M. A., Kumar, N. M., Cengiz, K., & Abdulkareem, K. H. (2021). IoT based smart agrotech system for verification of Urban farming parameters. *Microprocessors and Microsystems*, *82*, 104025.

Pradipta, A., Soupios, P., Kourgialas, N., Doula, M., Dokou, Z., Makkawi, M., Alfarhan, M., Tawabini, B., Kirmizakis, P., & Yassin, M. (2022). Remote sensing, geophysics, and modeling to support precision agriculture - Part 2: Irrigation management. Water, 14(7), 1157.

Rahman, F., Ritun, I. J., Biplob, M. R. A., Farhin, N., & Uddin, J. (2018). Automated aeroponics system for indoor farming using Arduino. *Joint 7th International Conference on Informatics, Electronics & Vision and 2nd International Conference on Imaging, Vision & Pattern Recognition*, 137–141. https://doi.org/10.1109/ICIEV.2018.8641026

Riansyah, A., Mardiati, R., Effendi, M. R., & Ismail, N. (2020, September 3). Fish feeding automation and aquaponics monitoring system base on IoT. *6th International Conference on Wireless and Telematics - Proceedings*. https://doi.org/10.1109/ICWT50448.2020.9243620.

Rohr, J. R., Barrett, C. B., Civitello, D. J., Craft, M. E., Delius, B., DeLeo, G. A., Hudson, P. J., Jouanard, N., Nguyen, K. H., & Ostfeld, R. S. (2019). Emerging human infectious diseases and the links to global food production. *Nature Sustainability*, *2*(6), 445–456.

Romli, M. A., Daud, S., Kan, P. L. E., Ahmad, Z. A., & Mahmud, S. (2017). Aquaponic growbed siphon water flow status acquisition and control using fog server. *IEEE 13th Malaysia International Conference on Communications*, 224–228. https://doi.org/10.1109/MICC.2017.8311763

Rozie, F., Syarif, I., & Al Rasyid, M. U. H. (2020). Design and implementation of intelligent aquaponics monitoring system based on IoT. *International Electronics Symposium: The Role of Autonomous and Intelligent Systems for Human Life and Comfort*, 534–540. https://doi.org/10.1109/IES50839.2020.9231928.

Rudd, S., & Cunningham, H. (2021). Low-energy authentication with selective privacy for heterogeneous IoT devices in smart-farms. Conference of Open Innovation Association, FRUCT, 2021-October, 230-238. https://doi.org/10.23919/FRUCT53335.2021.9599987.

Sadiku, M. N. O., Ashaolu, T. J., & Musa, S. M. (2020). Emerging technologies in agriculture. *International Journal of Scientific Advances*, *1*(1). https://doi.org/10.51542/ijscia.v1i1.6.

Samuel, A. K., Mohanan, V. P., Sempey, A., Garcia, F., Lagiere, P., Bruneau, D., & Mahanta, N. R. (2020). A self-sustainable home: 'BAITYKOOL' developed for the extreme warm climate an experimentation of active & passive strategies. *Advances in Science and Engineering Technology International Conferences*, 1–7. https://doi.org/10.1109/ASET48392.2020.9118326

Spradlin, A., & Saha, S. (2022). Saline aquaponics: A review of challenges, opportunities, components, and system design. In *Aquaculture* (Vol. *555*). Elsevier B.V. https://doi.org/10.1016/j.aquaculture.2022.738173.

Sriram, G. Y., & Sai Shibu, N. B. (2021). Design and implementation of automated aquaponic system with real-time remote monitoring. Advanced Communication Technologies and Signal Processing. https://doi.org/10.1109/ACTS53447.2021.9708396.

Tang, C., Xu, Y., Hao, Y., Wu, H., & Xue, Y. (2021). What is the role of telecommunications infrastructure construction in green technology innovation? A firm-level analysis for China. Energy Economics, 103. https://doi.org/10.1016/j.eneco.2021.105576.

Taufiqurrahman, A., Putrada, A. G., & Dawani, F. (2020, December 14). Decision tree regression with AdaBoost ensemble learning for water temperature forecasting in aquaponic ecosystem. *6th International Conference on Interactive Digital Media*. https://doi.org/10.1109/ICIDM51048.2020.9339669.

Tharamuttam, J. K., & Ng, A. K. (2017). Design and development of an automatic solar tracker. *Energy Procedia*, *143*, 629–634.

Tolentino, L. K. S., Fernandez, E. O., Jorda, R. L., Amora, S. N. D., Bartolata, D. K. T., Sarucam, J. R. V, Sobrepeña, J. C. L., & Sombol, K. Y. P. (2019). Development of an IoT-based aquaponics monitoring and correction system with temperature-controlled greenhouse. International SoC Design Conference, 261–262. https://doi.org/10.1109/ISOCC47750.2019.9027722

Vadivel, R., Parthasarathi, R. V, Navaneethraj, A., Sridhar, P., Nafi, K. A. M., & Karan, S. (2019). Hypaponics-monitoring and controlling using Internet of Things and machine learning. *1st International Conference on Innovations in Information and Communication Technology*, 1–6. https://doi.org/10.1109/ICIICT1.2019.8741487

Valiente, F. L., Garcia, R. G., Domingo, E. J. A., Estante, S. M. T., Ochaves, E. J. L., Villanueva, J. C. C., & Balbin, J. R. (2019, March 12). Internet of Things (IoT)-based mobile application for monitoring of automated aquaponics system. *IEEE 10th International Conference on Humanoid*, Nanotechnology, Information Technology, Communication and Control, Environment and Management. https://doi.org/10.1109/HNICEM.2018.8666439.

Verma, A. K. (2017). Impacts of unsustainable farming on environment. SN Page No., 59.

Vernandhes, W., Salahuddin, N. S., Kowanda, A., & Sari, S. P. (2017). Smart aquaponic with monitoring and control system based on IoT. *Second International Conference on Informatics and Computing*, 1–6. https://doi.org/10.1109/IAC.2017.8280590

Wang, S. S. Y. (2022). Use of extended reality in medicine during the Covid-19 pandemic. In *Extended Reality Usage during COVID 19 Pandemic* (pp. 1–14). Springer. https://doi.org/10.1007/978-3-030-91394-6_1

Yanes, A. R., Martinez, P., & Ahmad, R. (2020). Towards automated aquaponics: A review on monitoring, IoT, and smart systems. *Journal of Cleaner Production*, *263*, 121571.

Yuhasari, R., Mardiati, R., Ismail, N., & Gumilar, S. (2021). Fuzzy logic-based electrical conductivity control system in aquaponic cultivation. *7th International Conference on Wireless and Telematics - Proceedings*. https://doi.org/10.1109/ICWT52862.2021.9678423.

Zaini, A., Kurniawan, A., & Herdhiyanto, A. D. (2018). Internet of Things for monitoring and controlling nutrient film technique (NFT) aquaponic. *International Conference on Computer Engineering, Network and Intelligent Multimedia*, 167–171. https://doi.org/10.1109/CENIM.2018.8711304

Zhang, F., Zhang, Y., Lu, W., Gao, Y., Gong, Y., & Cao, J. (2022). 6G-enabled smart agriculture: A review and prospect. *Electronics (Switzerland) 11*(18). https://doi.org/10.3390/electronics11182845.

9 Microalgae-Based Recirculating Aquaculture System

Shy Chyi Wuang

INTRODUCTION

Aquaculture production has steadily increased over the past decades to support the growing population and affluence. As the wild stocks deplete, aquaculture becomes essential to fish and other seafood. According to the Food and Agriculture Organization (FAO), aquaculture production contributed 49.2% of global aquatic production in 2020 (FAO, 2022). Asian countries accounted for 70% of the production, followed by the Americas, Europe, Africa, and Oceania. China remained the largest producer, accounting for 35% of the total production. The expansion of aquaculture has boosted the overall growth of inland aquaculture from 12% in the 1980s to 37% of the total aquaculture production in 2020. Global aquaculture production in 2020 was a high record of 122.6 million tons, including 87.5 million tons of aquatic animals and 35.1 million tons of algae.

In intensive aquaculture, high-density farming and excessive feeding add load to the water, and consistent water changes are necessary, especially in the lack of other water treatment methods. Good water quality is the prerequisite to healthy aquaculture, and, therefore, large quantities of clean, fresh water are used in intensive aquaculture. Surplus feed and fish excreta will decompose if not removed from the water. This will reduce the water's oxygen content and increase the nutrient levels in the receiving water, rendering treatment or replenishment of the water necessary (Gal et al., 2003). The continuous discharge of aquaculture wastewater rich in nutrients (e.g., nitrogen and phosphorus) into the surrounding environment will cause water eutrophication, such as algae blooms and red tides (Cao et al., 2007), and impact the downstream water purification processes. Sustainable water treatment technologies are needed to reduce nutrient and chemical discharge into receiving waters.

The main expenses in intensive aquaculture are feeds and water treatment. The typical recirculating aquaculture systems (RAS) use biological treatment whereby bacteria convert the dissolved wastes to cell mass and other stable end products. Bacteria-based nitrification will convert the nitrogenous wastes (ammonia and nitrite) to nitrates (the less harmful form). By allowing for partial recycling of water, RAS reduces water consumption. However, the bacteria biomass and related fouling issues pose other problems that add to the high operating costs and additional measures to tackle them. The setup cost for RAS is also high. Current RAS does not supply oxygen to the aquatic animals; extensive aeration is required to ensure sufficient dissolved oxygen.

ALGAE APPLICATIONS IN AQUACULTURE

Unlike bacteria-based nitrification, microalgae assimilate the nutrients and convert them into biomass. The microalgae-based RAS for aquaculture can increase dissolved oxygen for the aquaculture species. Nie and co-workers (2020) have described the progress of aquaculture wastewater treatment using microalgae and the efficacy of different microalgae species. Due to their high photosynthetic efficiency and productivity, microalgae have also been recognized as a feedstock

DOI: 10.1201/b23309-9

for biofuels, bioremediation agents, and carbon dioxide sequestrators. Specific species such as *Spirulina platensis* and *Chlorella vulgaris* have also been widely used in health supplements due to their rich bioactive constituents. The use of algae has expanded from the traditional ones in aquaculture and biofuels to a multitude of functional foods and feeds, as well as natural alternatives for bioactive compounds.

Many ingredients in aquafeeds are relatively common, including fishmeal, fish oil, soybean meal, and oilseed cakes (De Silva and Hasan, 2007). Fishmeal is widely used in aquafeeds due to its substantial content of high-quality proteins and essential amino acids. Most crop plant proteins have the disadvantage of being deficient in certain amino acids and, therefore, cannot be good substitutions for fishmeal. Algae, on the other hand, generally contain all the essential amino acids, although there are significant variations across species. For example, surveying 19 tropical seaweeds by Lourenço and co-workers (2002) and another study of 34 edible seaweed products (Dawczynski et al., 2007) found that these species contained all the essential amino acids. Others have also reported similar findings (Wong and Cheung, 2000; Ortiz et al., 2006). A comprehensive study of 40 microalgae species from 7 algal classes (Brown et al., 1997) found that all these species have similar amino acid compositions and are rich in essential amino acids. Algae have been demonstrated to be a possible supplementation or complete replacement for protein in aquafeeds (Habib et al., 2008) and can replace 50% of fishmeal with comparable feed conversion efficiency in silver seabream (El-Sayed, 1994). Besides amino acids, algae contain high oil levels. Algae oil is an essential source of omega-3 polyunsaturated fatty acids (PUFAs), the most important being docosahexaenoic acid (DHA). Although widely used as a source of omega-3 PUFAs, fish oils suffer from unpleasant odor, taste, and poor oxidative stability. Algae-sourced DHA is a natural alternative as fish obtain their high omega-3 PUFA content through the consumption of algae. Therefore, microalgae have been used as live feeds for finfish, shellfish, and invertebrate species and aquafeeds for zooplankton. Besides essential amino acids and fatty acids, algae are rich sources of vitamins (Takeuchi et al., 2002).

Many review papers have reported the use of microalgae in aquaculture applications (Nie et al., 2020; Han et al., 2019; Nagappan et al., 2021). Han and co-workers (2019) have described microalgae-assisted aquaculture and wastewater remediation. Nutritional profiles of microalgae were discussed alongside their potential for aquaculture feeds. Nie et al. (2020) discussed the major challenges of aquaculture waste nutrient removal using microalgae, including the diverse environmental variations and non-optimized growth rates. Nagappan and co-workers (2021) reported the other advantages of algae meal, including their ability to function as pre- and probiotics and immunostimulants, and some challenges, such as their low palatability. With the increasing focus on microalgae research, more widespread applications in aquaculture could be expected soon.

Despite having good potential as a bulk feedstuff for aquafeeds, the current substantial costs of algae biomass production have limited their widespread commercial use. Its commercial viability will depend on several factors, including its economical supply of sufficient quantities and the quality of essential amino acids and valuable oil content. With more concerted efforts in bridging the gaps preventing the widespread use of algae biomass, the demand for fishmeal and fish oil can be eased with algae biomass as a partial replacement or total substitution.

We have demonstrated earlier that algae remediation from aquaculture-used water is an effective way of water treatment, and the accumulated algae biomass could be used in biofuels and fertilizer applications (Wuang et al., 2016a,b). In the first report (Wuang et al., 2016a), *Chlorella* sp. and *Nannochloropsis* sp. achieved fast ammonia removal in fish water. Analysis of the lipid productivities of both species revealed the varying proportions of saturated fatty acids (SFAs) and PUFA. The calorific values of the obtained bio-oils were high and comparable to earlier reports. In the second study (Wuang et al., 2016b), the supplementation of *S. platensis* for leafy vegetables found enhanced plant growth in all tested vegetables compared to the controls. When compared to the performance of chemical fertilizer, the spirulina-based fertilizer performed comparably in most plant growth parameters and favorably for one tested species—Arugula. Seed germination (when measured by the seedling's dry weight) also improved for all tested vegetables except white crown. This work evidenced the usefulness of *S. platensis* in fish water treatment and its applicability as an agricultural fertilizer.

As fast-growing photosynthetic organisms, microalgae convert atmospheric carbon dioxide to carbohydrates and other products during photosynthesis. In the process, microalgae assimilate nutrients such as nitrogen and phosphorus to synthesize bioproducts. In rivers or waterways where excess fertilizers are washed into, algae growth is a common sight. In aquacultural applications, it is ideal for the algae biomass accumulated from the treatment of wastewater to be reused as feeds for the aquatic livestock to achieve circularity. In this work, an attempt was made to bridge the gaps in the use of algae feeds by evaluating the feasibility of an economical means of algae biomass production using used aquaculture water (i.e., microalgae-based RAS) and assessing the aquafeeds formulated with the accumulated algae biomass. The nutritional analyses of the feeds were performed to ascertain the possibility of a partial or complete replacement of fishmeal and fish oil with algae meal. The vital advantage of microalgae use is their ability to utilize all forms of nitrogen while maintaining acceptable pH and dissolved oxygen levels. In indoor vertical farming, the use of microalgae can extend to algae-based aquaponics, with symbiotic benefits in sustainable water treatment and the production of concomitant fish feed constituents. The following sections of this chapter will detail the establishment of the microalgae-based RAS and the assessment of the accumulated algae biomass for aquaculture feeds.

Suitability of Algae Strains in Bioremediation of Fish Water

Four algae species were used in early experiments to identify suitable algae species with high bioremediation potential in fish water. These include *Chlorella* sp., *Dunaliella salina*, *Nannochloropsis* sp., and *S. platensis*. Figure 9.1 shows the optical microscopy images of these four algae. Other than *S. platensis*, the other algae are small and round, around 2–5 μm in diameter.

While microalgae generally assimilate the nutrients in fish water, not all species can effectively assimilate ammonia, nitrite, and nitrate. Preliminary studies were done to ascertain the efficacy of nutrient assimilation by the different microalgae species in laboratory-scale experiments using fish water. It was found that *Chlorella* sp. and *Nanochloropsis* sp. were both effective and efficient in assimilating all the nitrogenous nutrients in fresh water and salt water, respectively. *S. platensis* and *D. salina* were found to be less effective in removing all the nitrogenous nutrients.

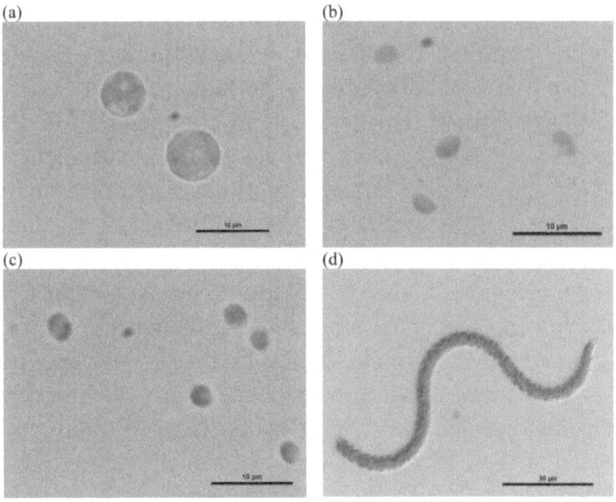

FIGURE 9.1 Optical microscopy images of algae: (a) *Chlorella* sp., (b) *Dunaliella salina*, (c) *Nannochloropsis* sp., and (d) *Spirulina platensis*.

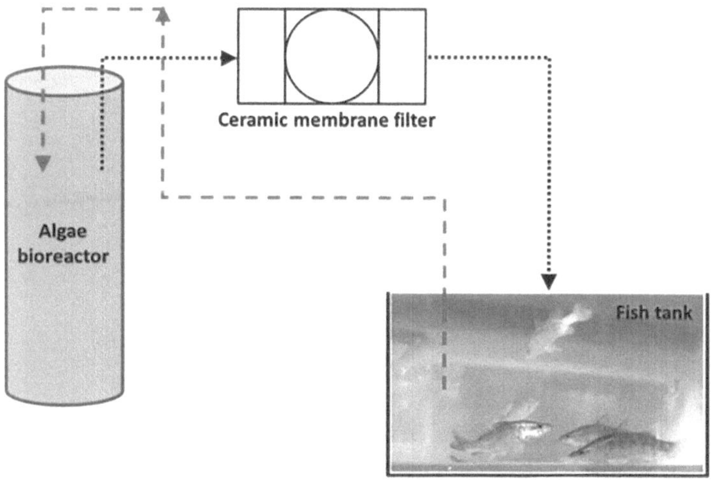

FIGURE 9.2 Schematic of microalgae-based recirculating aquaculture system (RAS).

SETUP OF MICROALGAE-BASED RAS

Figure 9.2 shows the schematic diagram of the microalgae-based RAS. Fish water was pumped into the algae bioreactor. After stipulated periods when the nutrient levels fell, the microalgae-containing water would be pumped through a ceramic membrane filter unit to filter out the algae before the treated water was recirculated into the fish tank. Excess algae biomass was harvested using a ceramic membrane filter, and a new batch of fish water would be routed to the algae bioreactor.

As microalgae assimilate nutrients during growth, the microalgae density determines the nutrient removal capacity. To identify the appropriate algae concentration for optimal fish water treatment, various batch trials were conducted before the operation of the microalgae-based RAS. The fish water was first pumped into the 80-L algae bioreactor. Stipulated amounts of microalgae were added to the reactor to achieve a final concentration of 1, 2, 3, and 4 million algae cells per mL, respectively. For algae quantification, stipulated volumes of algae suspension were withdrawn, and direct counts were performed using a Neubauer hemocytometer and an Olympus light microscope. The nutrient (ammonia, nitrite, and nitrate) concentrations and algae cell densities were monitored daily. Ammonia, nitrite, and nitrate concentrations were determined using standard methods with HACH reagents, including Nessler's reagent, NitriVer 3 Nitrite, and NitraVer 5 nitrate reagent powder. Concentration readings were obtained using the HACH DR3900 spectrophotometer. Fluorescent lamps provide 1,000–3,000 lux illumination, typical of indoor lighting and similar to the natural light intensity on an overcast day. Temperature has a diurnal range of 28–30°C. Batch studies confirmed that *Nannochloropsis* sp. could generally remove ammonia, nitrite, and nitrate from fish wastewater. As shown in Figure 9.3, the nutrient reductions depended on algae densities. From the testing, an algae concentration of 3×10^6 cells/mL worked best to treat fish water from sea bass rearing. This concentration was subsequently used in the operation of the microalgae-based RAS.

WATER TREATMENT IN MICROALGAE-BASED RAS

The fish tank's water passed through a cloth filter before it was routed to the algae bioreactor. After stipulated periods, the treated water from the bioreactor passed through a ceramic membrane to filter off the microalgae before it returned to the fish tank. The algae from the ceramic membrane

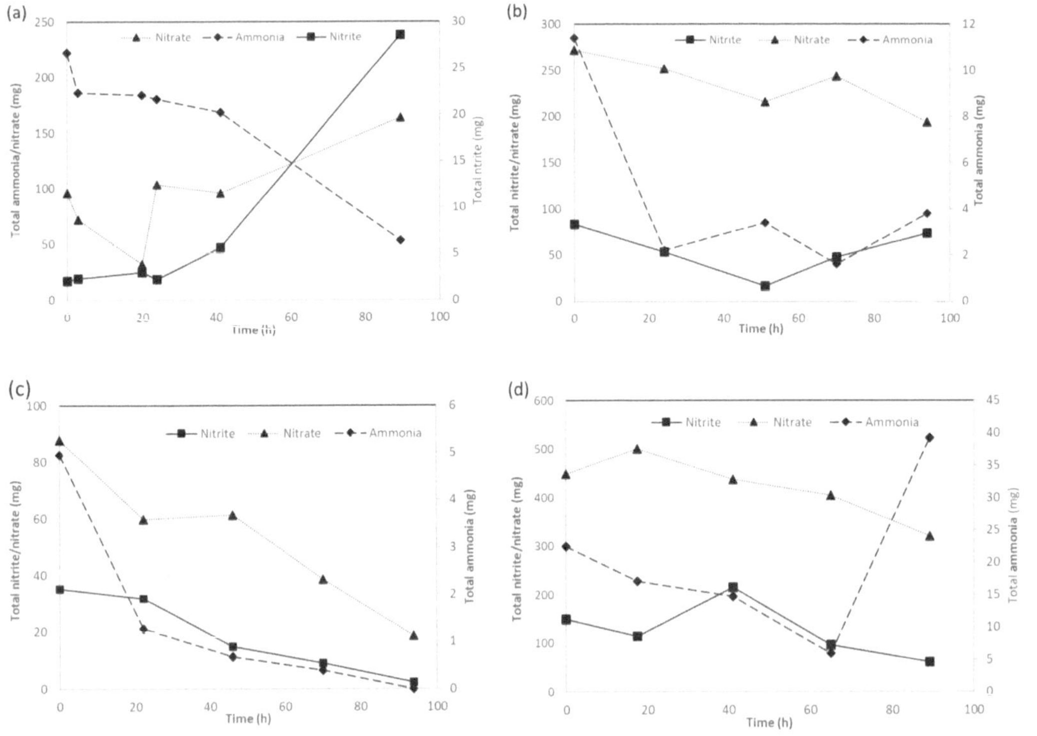

FIGURE 9.3 Batch studies with different starting *Nannochloropsis* sp. concentrations (cells/mL): (a) 1 million, (b) 2 million, (c) 3 million, and (d) 4 million.

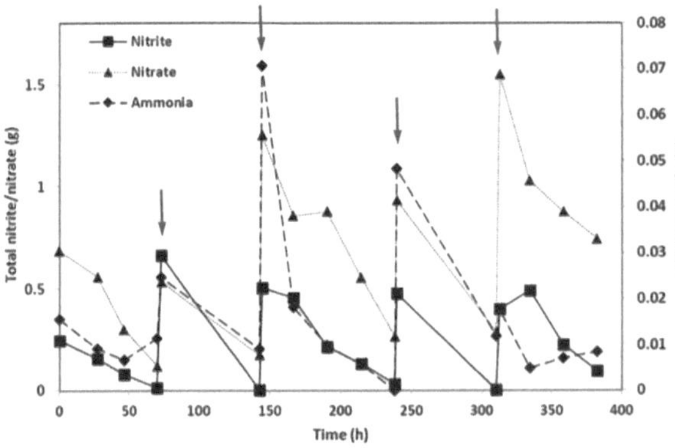

FIGURE 9.4 Total amount of ammonia, nitrite, and nitrate in the microalgae-based recirculating aquaculture system (RAS). Arrows indicate batch changes of fish water.

were retrieved via filter rinsing. Depending on the treatment time, fresh seawater might be required to meet the desired water change frequency. The harvested algae biomass was subsequently dried for use in fish feed formulation. The dried biomass was used for nutritional analysis and fish feed formulation.

Figure 9.4 shows the total amount of ammonia, nitrite, and nitrate in the microalgae-based RAS using *Nannochloropsis* sp. Consistent declining trends were observed for the ammonia, nitrite, and nitrate levels in fish water. The presence of ammonia directly interfered with the nitrate uptake by

TABLE 9.1
Yields of Algae Biomass

Run No	Algae Species	Duration (day)	Accumulated Algae Biomass (g)	Biomass Productivity (g/m³ h)
1	*Nannochloropsis sp.*	22	15.03	0.407
2	*Nannochloropsis sp.*	23	12.44	0.322
3	*Nannochloropsis sp.*	12	9.38	0.543

Nannochloropsis sp. (Hii et al., 2011), as ammonia was the preferred nitrogen source. The preferential uptake of ammonia and the suppression of nitrate (and possibly nitrite) uptake by ammonia availability have been reported elsewhere (Maguer et al., 2007) and observed in our earlier report (Wuang et al., 2016a). However, when ammonia levels become insufficient to meet growth requirements, the algae cells utilize other nitrogen sources. As cells assimilate nutrients during growth, more incredible algae growth could achieve higher nutrient reduction in the fish water. Using *Nannochloropsis* sp., bioremediation of the fish water was achieved, in addition to biomass accumulation for fish feed formulations.

There were other methods for harvesting, namely, sand filtration, modified starch flocculation, polyester felt filter bags, and adhesion of microalgae to solid surfaces. Sand filtration and modified starch flocculation methods posed additional technical difficulties and complications that rendered both methods impractical to incorporate into the system.

Table 9.1 shows the yields of algae biomass for three different runs. The average biomass productivity for *Nannochloropsis* sp. was around 0.413 $g/m^3 h$.

MICROALGAE BIOMASS FOR FISH FEEDS

There are no international standards for fish feed compositions. However, it has been recognized that the essential nutrients for fish should include protein, fat, carbohydrates, vitamins, and minerals. A complete diet would typically consist of the required protein (18–50%), lipid (10–25%), carbohydrate (15–20%), ash (< 8.5%), phosphorus (< 1.5%), water (< 10%), and trace amounts of vitamins and minerals (Craig and Helfrich, 2009). Proteins are required for growth and energy. They provide the amino acids required for many syntheses and functions. Amino acids are crucial for all metabolic processes, including optimal transport and storage of all nutrients. Protein deficiency will lead to slower growth and poorer health. On the contrary, carbohydrates and fats provide energy. The right amount of fat can also improve the taste and texture of the fish. Vitamins and minerals are essential in enhancing the natural resistance of fish and their feed conversion ratios.

Various fish feed formulations were made with the constituents' composition shown in Table 9.2. The proportion of algae was varied to identify a suitable quality of feed.

The carbohydrate, protein, and fat analyses of the algae (*Nannochloropsis* sp.) biomass were performed. Inorganic and metallic constituents were also determined. In addition, their amino acid and lipid profiles were investigated. Crude protein analysis was performed using the Kjedahl method (AOAC Official Method 2001.11). Total carbohydrate content was determined using a Total Carbohydrate Assay Kit (MAK104, Sigma-Aldrich) with prior sample hydrolysis. The inorganic and metallic constituents were performed using inductively coupled plasma spectrometry. Pacific Lab Singapore performed amino acid analysis of the algae biomasses. The extraction of lipids from the dried biomass was adapted from the Folch procedure (Folch et al., 1957) as follows. The extracted product was analyzed with gas chromatography to determine its fatty acid composition. Chromatographic analysis of fatty acid methyl ester (FAME) profiles in the extracted product was

TABLE 9.2
Compositions of (a) Fish Feed Formulations and (b) Mineral Mix

(a)

	Composition of Fish Feed (%)						
Feed Name	*Nannochloropsis* sp.	Other Algae	White Cornmeal	Wholemeal Flour	Oil	Mineral Mix	Soy Bean
10% N	10	—	17	62	10	1	—
5% N + 5% Alg SB1	5	5	40	20	4	1	25
5% N + 5% Alg SB2	5	5	25	20	4	1	40

(b)

Chemical Name	Proportion (%)
Calcium carbonate	1
Manganese(II) sulfate monohydrate	1
Zinc sulfate heptahydrate	2.5
Copper(II) sulfate pentahydrate	0.2
Iron(II) sulfate heptahydrate	1.8
Potassium iodide	0.007
Monosodium phosphate	70.7
Magnesium sulfate	3.5
Sodium chloride	15.8
Calcium phosphate	3.5

performed using a GC-2010 gas chromatograph (Shimadzu, Japan) with a flame ionization detector. Individual FAME was identified by comparing its retention time with the retention times of a 37-component FAME mix (Sigma-Aldrich, Bellefonte, USA) and quantified using internal standards. The composition of the extracted product is expressed in terms of its fatty acid proportion in percentages.

Table 9.3 details the nutritional analyses of the algae biomass, while Table 9.4 shows their inorganic and metallic constituents.

Nannochloropsis sp. contains all the essential amino acids and substantial amounts of inorganic/metallic constituents. It also contains a high proportion of PUFA and monounsaturated fatty acids (MUFA) (59.5%), including eicosapentaenoic acid (EPA) and DHA, compared to SFAs (40.5%). This high PUFA content will serve well in fish nutrition. Besides playing an essential part in healthy brain function and growth, EPA and DHA act as a source of energy, insulate against heat loss, and provide cushion to tissues and organs. Notably, the EPA content in *Nannochloropsis* sp. is high and forms the bulk of PUFAs. The comprehensive fatty acids profile of *Nannochloropsis* sp. suggests its potential to substitute the use of fish oil in aquafeeds.

Many studies have reported the feasibility and applicability of algae-based fish feeds in specific aquatic species (González-Félix et al., 2016; Macias-Sancho et al., 2014; Norambuena et al., 2015; Fonseca et al., 2023). Here, we have investigated the comprehensive nutritional constitutions of these feeds regarding both macro- and micronutrients. Macronutrient deficiency is rare in today's context. Compared to macronutrients, micronutrients have more subtle effects on fish well-being, and their interactions with other dietary components can affect the immune system in fish (Lall, 2000), which is vital in disease prevention. Zinc plays important roles in the growth,

TABLE 9.3
Nutritional Analysis of Algae Biomass

Component	*Nannochloropsis* sp.
Total carbohydrate (%)	20.76
Total lipid (%)	14.12
Lipid constituents (expressed as % of total lipid)	0.170
Caproic acid	0.155
Caprylic acid	0.180
Capric acid	0.319
Lauric acid	0.152
Tridecanoic acid	2.350
Myristic acid	0.183
Myristoleic acid	0.283
Pentadecanoic acid	0.134
Pentadecenoic acid	7.212
Palmitic acid	7.711
Palmitoleic acid	0.210
Heptadecanoic acid	0.318
Heptadecenoic acid	0.316
Stearic acid	0.162
Elaidic acid	0.945
Oleic acid	0.152
Linolelaidic acid	1.806
Linoleic acid	0.166
Arachidic acid	0.275
Linolenic acid	0.192
γ-Linolenic acid	0.160
Eicosadienoic acid	0.207
Behenic acid	0.172
Eicosatrienoic acid	1.666
Tricosanoic acid	0.246
Docosadienoic acid	0.164
Lignoceric acid	7.279
Eicosapentaenoic acid (EPA)	0.167
Docosahexaenoic acid (DHA)	
Crude protein (%)	35.62
Amino acids (expressed as % of total algae biomass)	
Aspartic acid	2.721
Threonine	1.382
Serine	1.257
Glutamic acid	4.085
Glycine	1.938
Alanine	2.337
Cystine	0.260
Valine	1.973
Methionine	0.755
Iso-leucine	1.390
Leucine	2.877
Tyrosine	1.193
Phenylalanine	1.738
Histidine	0.822
Lysine	1.956
Arginine	1.815
Tryptophan	0.581
Proline	1.608

TABLE 9.4
Inorganic and Metallic Constituents of Algae Biomass

Component	*Nannochloropsis* sp.
Calcium	0.92%
Phosphorus	0.81%
Magnesium	0.55%
Sodium	1.03%
Potassium	0.87%
Iron	3470.76 ppm
Manganese	231.29 ppm
Zinc	226.34 ppm
Selenium	19.94 ppm

reproduction, development, vision, and immune function of various fish species (Watanabe et al., 1997). Iron is an essential nutrient for fish, and its availability can affect the ability of pathogens to infect a host. Low free iron concentrations in mucus membranes and other tissues may help defend against bacterial infections. Iron deficiency can cause microcytic anemia in fish and make the host more susceptible to infectious agents (Lall, 2000). Copper is involved in the activity of enzymes such as cytochrome oxidase and superoxide dismutase. Selenium, in fish, is vital for optimal growth, feed conversion, glutathione peroxidase activity, and immune function. Manganese is widely distributed in fish tissue and is necessary for normal brain functions and proper lipid and carbohydrate metabolism (Watanabe et al., 1997). Chromium is necessary for normal carbohydrate and lipid metabolism. Supplemental dietary chromium has increased weight gain, energy deposition, and liver glycogen content in tilapia-fed glucose diets (Shiau and Lin, 1993). Other elements like sodium and potassium maintain fluid balance and homeostasis. Magnesium plays important roles as enzyme co-factors and structural components of cell membranes. Calcium and phosphorus are primarily present in the bones, teeth, and scales of fish. Even though feeds are usually high in these components, most fish can also obtain dissolved calcium from their aquatic environment through the gills.

The nutritional value of the various formulated fish feeds was compared with existing commercial feeds, as illustrated in Table 9.5. The formulated fish feeds generally had a lower proportion of proteins and a higher proportion of carbohydrates. In addition, a mineral mix was necessary to ensure suitable amounts of inorganic and metallic constituents. Increasing the proportion of algae meal in the aquafeeds would improve its quality in terms of nutritional profiles. However, for a sustainable supply of aquafeeds, the algae biomass productivities must be able to keep up with the usage of algae meal in feeds.

Considering the observed biomass productivities (0.38–0.41 g/m^3 h), the proportions of algae meal in the fish feeds should be kept to 5–10%. With the addition of mineral mix and oil, the micro-nutritional requirements of the formulated feeds became mostly adequate (see Table 9.5). However, the protein content remained much lower than commercial feeds. To improve the protein content of the aquafeeds, soybean was introduced (5% N + 5% Alg SB1 and SB2). As seen in Table 9.5, a high proportion of soybean (in 5% N+5% Alg SB2), in place of wholemeal flour, helped to increase the protein content of feeds slightly and decrease the carbohydrate content to comparable levels as the commercial aquafeeds. Two commercial feeds, sinking Feed A and floating Feed B, were used as references to gauge the nutritional adequacy of the algae-based aquafeeds.

TABLE 9.5
Constitution of Fish Feeds

Component	Formulated Feeds with *Nannochloropsis* sp. (N) and/or Other Algae Meal (Alg) and Oil and Mineral Mix: 10% N	5% N + 5% Alg SB1	5% N + 5% Alg SB2	Commercial Feed A (Sinking)	Commercial Feed B (Floating)
Ca (%)	2.656	0.218	0.206	1.634	0.712
Mg (%)	0.452	0.184	0.204	0.156	0.150
Na (%)	0.532	1.138	1.343	0.524	0.282
K (%)	0.589	0.740	0.995	0.680	0.630
Fe (%)	0.023	0.010	0.011	0.032	0.013
Mn (%)	0.018	0.003	0.004	0.003	0.007
Zn (%)	0.017	0.008	0.008	0.010	0.006
Se (%)	$<1 \times 10^{-3}$	$<1 \times 10^{-3}$	$<1 \times 10^{-3}$	$<1 \times 10^{-3}$	$<1 \times 10^{-3}$
Cu (%)	0.012	0.006	0.001	0.001	0.0015
Cr (%)	$<1 \times 10^{-3}$	$<1 \times 10^{-3}$	$<1 \times 10^{-3}$	$<1 \times 10^{-3}$	$<1 \times 10^{-3}$
P (%)	0.073	0.131	0.159	1.548	0.180
Carbohydrate, total (%)	56.47	57.54	39.77	11.59	39.61
Crude protein (%)	14.18	17.88	23.29	49.50	30.79
Lipid, total (%)	10.31	8.53	7.67	8.27	2.99

COST ANALYSIS

The average biomass productivity for *Nannochloropsis* sp. was about 0.413 g/m^3 h. From the studies, it was observed that four healthy-sized fish (700 g–1 kg) consumed about 30 g of fish feed per week during the experimental study. The following assumptions were made with study findings and literature references in assessing functional fish feed yields and its imparted reduced reliance on fishmeal.

1. Average production of algae biomass of 0.413 g/m^3 h
2. The proportion of algae in fish feeds is 15%
3. Fish density of 100 kg/m^3 or 100 $fish/m^3$
4. The average consumption of feed of 10 g/week/fish

In 1 week, the mass of algae harvested will be around 69.4 g/m^3. This amount can translate into 462.6 g of formulated fish feeds (derived from algae growth in 1 m^3 of fish water). The mass of fish feed required to meet consumption is 1,000 g/m^3. Therefore, the algae fish feeds can meet approximately 46% of the feed consumption, which could allow for savings from lower feed costs.

APPLICATIONS OF MICROALGAE-BASED RAS

A microalgae-based RAS was set up, consistently reducing the primary contaminants (nutrients) in fish water, including ammonia, nitrite, and nitrate. Sufficient residence time in the algae bioreactor was required to reduce the contaminants. Treated water was routed back to the fish tanks while excess microalgae biomass was harvested and used to prepare algae fish feeds. Nutritional analysis of the algae fish feeds found that their quality compared favorably with commercial fish feeds except for lower protein content.

Based on the biomass production rate and fish feed consumption, an economic analysis was performed to assess the commercial feasibility of the approach. Approximately 46% of the consumed fish feeds can be obtained, potentially resulting in close to 50% savings in fish feed costs. While the technical feasibility of the microalgae-based system has been established, the challenges to overcome include imparting robustness to the microalgae-based RAS and the need to customize the system for each aquatic species. Different water treatment durations will be required depending on the species and growth density. A sound monitoring system would be vital in optimizing the microalgae-based RAS application in field trials before robust operation can be achieved and controlled. Using microalgae as entities to treat fish wastewater in a complete RAS and produce fish feed constituents will greatly reduce fresh water demand in indoor vertical farming without excessive use of chemicals or energy.

ACKNOWLEDGMENTS

The work was supported financially by the National Research Foundation (Singapore) and Social and Innovation Research Fund, TOTE Board.

REFERENCES

Brown M.R., Jeffrey S.W., Volkman J.K., Dunstan G.A. Nutritional properties of microalgae for mariculture. *Aquaculture* (1997) 151, 315–331.

Cao L., Wang W., Yang Y., Yang C., Yua Z., Xiong S., Diana J. Environmental impact of aquaculture and countermeasures to aquaculture pollution in China. *Environ. Sci. Pollut. Res. Int.* (2007) 14, 452–462. https://doi.org/10.1065/espr2007.05.426.

Craig S., Helfrich L.A. *Understanding Fish Nutrition, Feeds, and Feeding*. Virginia Cooperation Extension. (2009) Publication. Virginia Tech, Virginia State University. 420–256.

Dawczynski, C., Schubert, R., Jahreis, G. Amino acids, fatty acids, and dietary fibre in edible seaweed products. *Food Chem.* (2007) 103, 891–899.

De Silva S.S., Hasan M.R. Feeds and fertilizers: The key to long-term sustainability of Asian aquaculture. In Hasan M. R., Hecht T., De Silva S. S. and Tacon A. G. J., (Eds), *Study and Analysis of Feeds and Fertilizers for Sustainable Aquaculture Development*. FAO Fisheries Technical Paper No. 497. Rome, FAO. 2007. 504p.

El-Sayed A.F.M. Evaluation of soybean meal, spirulina meal and chicken offal meal as protein sources for silver seabream (*Rhabdosargus sarba*) fingerlings. *Aquaculture* (1994) 127, 169–176.

Folch J., Lees M., Stanley G.H.S. A simple method for the isolation and purification of total lipids from animal tissues. *J. Biol. Chem.* (1957) 226, 497–509.

Fonseca F., Fuentes J., Vizcaíno A.J. et al. From invasion to fish fodder: Inclusion of the brown algae Rugulopteryx okamurae in aquafeeds for European sea bass *Dicentrarchus labrax* (L., 1758). *Aquaculture* (2023) In Press. https://doi.org/10.1016/j.aquaculture.2023.739318.

Food and Agriculture Organisation of the United Nations (FAO). *The State of World Fisheries and Aquaculture* (2022). Towards Blue Transformation, https://www.fao.org/3/cc0461en/cc0461en.pdf.

Gal D., Szabo P., Pekar F., Varadi L. Experiments on the nutrient removal and retention of a pond recirculation system. *Hydrobiology* (2003) 506-509, 767–772.

González-Félix M.L., Maldonado-Othón C.A., Perez-Velazquez M. Effect of dietary lipid level and replacement of fish oil by soybean oil in compound feeds for the shortfin corvina (*Cynoscion parvipinnis*). *Aquaculture* (2016) 454, 217–228.

Habib M.A.B., Parvin M., Huntington T. C., Hasan M. R. A review on culture, production and use of Spirulina as food for humans and feeds for domestic animals and fish. *FAO Fisheries and Aquaculture Circular.* No. 1034. Rome, FAO. 2008. 33p.

Han P., Lu Q., Fan L., Zhou W. A review on the use of microalgae for sustainable aquaculture. *Appl. Sci.* (2019) 9, 2377. https://doi.org/10.3390/app9112377.

Hii Y.S., Soo C.L., Chuah T.S., Mohd-Azmi, A., Abol-Munafi A.B. Interactive effect of ammonia and nitrate on the nitrogen uptake by *Nannochloropsis* sp. *J. Sustain. Sci. Manag.* (2011) 6, 60–68.

Lall, S.P. Nutrition and health of fish. In: Cruz -Suárez, L.E., Ricque-Marie, D., Tapia-Salazar, M., Olvera-Novoa, M.A., Civera-Cerecedo, R., (Eds.). *Avances en Nutrición Acuícola V. Memorias del V Simposium Internacional de Nutrición Acuícola.* 19–22, Noviembre, 2000. Mérida, Yucatán, Mexico. pp 13–23.

Lourenço, S.O., Barbarino, E., De-Paula, J.C., Pereira, L.O.S., Marquez, U.M.L. Amino acid composition, protein content and calculation of nitrogen-to-protein conversion factors for 19 tropical seaweeds. *Phycol. Res.* (2002) 50, 233–241.

Macias-Sancho J., Poersch L.H., Bauer W., Romano L.A., Wasielesky W., Tesser M.B. Fishmeal substitution with Arthrospira (*Spirulina platensis*) in a practical diet for Litopenaeus vannamei: Effects on growth and immunological parameters. *Aquaculture* (2014) 426-427, 120–125.

Maguer J.F., Helguen S., Madec C., Labry C., Corre P.L. Nitrogen uptake and assimilation kinetics in Alexandrium minutum (dynophyceae): Effect of n-limited growth rate on nitrate and ammonium interactions. *J. Phycol.* (2007) 43: 295–303.

Nagappan S., Das P., Quadir M.A., Thaher M., Khan S., Mahata C., Al-Jabri H., Vatland A.K., Kumar G. Potential of microalgae as a sustainable feed ingredient for aquaculture. *J. Biotechnol.* (2021) 341, 1–20.

Nie X., Mubashar M., Zhang S., Qin Y., Zhang X. Current progress, challenges and perspectives in microalgae-based nutrient removal for aquaculture waste: A comprehensive review. *J. Clean. Prod.* (2020) 277, 124209.

Norambuena F., Hermon K., Skrzypczyk V., Emery J.A., Sharon Y., Beard A., Turchini G.M. Algae in fish feed: Performances and fatty acid metabolism in Juvenile Atlantic Salmon. *PLoS One* (2015) 10(4), e0124042. https://doi.org/10.1371/journal.pone.0124042.

Ortiz J., Romero N., Robert P., López Hernández J., Bozzo C., Navarrete E., Araya J.C., Osorio A., Ríos A., Ortiz Viedma J. Dietary fiber, amino acid, fatty acid and tocopherol contents of the edible seaweeds *Ulva lactuca* and Durvillaea Antarctica. *Food Chem.* (2006) 99, 98–104.

Shiau S.Y., Lin S.-F. Effect of supplemental dietary chromium and vanadium on the utilization of different carbohydrates in tilapia, *Oreochromis niloticus* X 0 aureus. *Aquaculture* (1993) 110, 321–330.

Takeuchi T., Lu J., Yoshizaki G., Satoh S. Effect on the growth and body composition of juvenile tilapia *Oreochromis niloticus* fed raw Spirulina. *Fish. Sci.* (2002) 68, 34–40.

Watanabe T., Kiron V., Satoh S. Trace elements in fish nutrition. *Aquaculture* (1997) 151, 185–207.

Wong K.H., Cheung, P.C.K. Nutritional evaluation of some subtropical red and green seaweeds: Part I—proximate composition, amino acid profiles and some physico-chemical properties. *Food Chem.* (2000) 71, 475–482.

Wuang S.C., Khin M.C., Chua P.Q. D., Luo Y. D. Use of Spirulina biomass produced from treatment of aquaculture wastewater as agricultural fertilizers. *Algal Res.* (2016b) 15, 59–64.

Wuang S.C., Luo Y.D., Wang S., Chua P.Q.D., Tee P.S. Performance assessment of biofuel production in an algae-based remediation system. *J. Biotechnol.* (2016a) 221, 43–48.

10 Variety Improvement for Closed Controlled Environments

Customizing Genetics for Specific Growing Conditions

Kevin M. Folta

INTRODUCTION

Not all cultivated plant varieties are created equal. There is significant variation across the genetics of strawberries, tomatoes, or lettuce. Each variety is specifically bred, trialed, or adapted to various field environments, as the Central Valley of California is, in all ways, a distant environment from the horticulturally productive Niagara Peninsula and the winter production fields of Mexico.

The best way to start a chapter on breeding for controlled environments is to provide an example of how breeding transformed opportunities in an equal alien environment—Florida, USA. The temperate fall and spring interrupted by warm-wet winters provide an excellent canvas for high-value horticultural crop production if the crop can tolerate the oppressive summer heat, sandy soils, and unmatched assault from pests and pathogens. Blueberries (*Vaccinium corymbosum*) are an excellent case study.

In the 1930s, "highbush" blueberries gained traction as a delicious summer fruit, netting high prices for farmers in New Jersey and other points north. Not to miss a new opportunity in a profitable crop, breeders in North Carolina created specific lines to fit their region, and, soon, farmers could produce blueberries 3 weeks before the New Jersey crop, bringing huge profits to the farmers who grew them.

Ralph Sharpe was a faculty member in the University of Florida Fruit Crops Department in the 1950s. He wanted to know if Florida farmers could also profit from growing blueberries for an even earlier market, supported by a warm pseudo-Mediterranean climate. Sharpe trialed the North Carolina and New Jersey varieties in Florida, and they failed miserably, producing fruit weeks after the producers in other states. Why?

Florida's environment was utterly alien. Sandy soils, strange pests, relentless pathogens, and a remarkably few hours of flower-inducing winter chill meant low and slow yields. Time spent near freezing tells the plant that winter has concluded, and it is a favorable time to emerge from dormancy. From there, the plant can reproduce (for the plant) or make fruit (for its human overlords).

Sharpe examined the species resident in the swamps and hammocks of the Sunshine State and witnessed that blueberry's native relatives thrive. These small, compact *Vaccinium* species possessed the genetic software to perform in this challenging environment—but their fruits were small and unfit for a commercial market. There were genetic resources compatible with this unusual environment; it was simply a question of marrying Florida's native heat tolerance genetics to the improved high-production fruit varieties from the north, sequestering the best traits in one genetic package.

DOI: 10.1201/b23309-10

Smash cut to the 1980s, and the Florida Highbush Blueberry was born, a genetic amalgam of north meets south, the fruit traits of the north, the resilience traits of the south. Today, these efforts have resulted in the Southern Highbush Blueberry, a profitable crop grown throughout the southern USA, Mexico, South America, and parts of Europe.

The point is simple. To maximize production and profit, a plant's genetics must dovetail precisely with the environment. This is no surprise to those who study the interface of speciation and ecology. Genetic changes happen over time, as centuries and millennia shape genetics to adapt to new niche opportunities in a persistent backdrop of environmental change.

The glacial transition of genetics into a new environment was wholly ignored when seeds bred for the field were scattered under canopies of funny-colored LEDs in a plant factory. "It's just a plant, after all." In a controlled environment, farmers watched as varieties proven in the field germinated with enthusiasm, but ultimately their products fell short of expectations.

The first forays into indoor vertical cultivation taught us that it is not enough to just control the environment—it is necessary to optimize plant genetics to match artificial demands Mother Nature never conceived of. At the same time, there is an immense and virtually untapped opportunity in inventing plants that may thrive in the alien-controlled environment, potentially outproducing their field-bound relatives.

THE MOST COMPLICATED MACHINE ON THE FARM

The plant is the most complex of technologies supporting controlled-environment agriculture (CEA). The plant is home to a well-orchestrated set of chemical reactions that transform a minor atmospheric gas, water, light, and a pinch of minerals into the chemical backbones of molecules used in structure and metabolism. The stage for the chemical reaction is established by an integrated cluster of sensors that continually monitor the ambient environment and adjust growth, development, and metabolism to fit prevailing conditions. As sessile organisms, plants must possess the means to identify (or even anticipate) environmental change and then launch a gene expression program that continually adjusts physiology to optimize performance in each environment. Whereas significant public controversies are rooted in the addition of a gene or two to create "genetically modified organisms," a change in the environment affects how dozens or perhaps thousands of genes are expressed, making plants truly (and non-controversially) *environmentally modified organisms* (Carvalho and Folta, 2014a).

Discussions of light and plant biology typically start with photosynthesis. Of course, this is important, as an energy threshold is necessary to drive all plant metabolic processes. However, an excess of light energy beyond the cell's capacity is wasted as heat through non-photochemical quenching or the generation of free radicals that can damage the photosynthetic hardware, a process known as photoinhibition (Ruban, 2016). While anyone who dipped a cuvette of chlorophyll extract into a Spec20 in high school can tell you that chlorophyll best absorbs blue and red wavelengths, photosynthesis is driven primarily by wavebands between 400 and 700 nm, the light qualities known *as photosynthetically active radiation*, or PAR (McCree, 1972). While photosynthesis is always a factor in plant growth, it happens on a backdrop that is defined by the action of other light sensors that control the stage on which photosynthesis performs.

THE SENSOR COLLECTION

A network of sensors receives and integrates the many inputs that can affect plant growth and crop productivity. Gravity, temperature, partial pressure of O_2/CO_2, humidity, the position of neighboring plants, and specific wavebands of light are just a few of the cues that shape plant growth. Light itself is a complex milieu of information, as energies from discrete wavebands each deliver explicit signals from the environment to optimize growth to match conditions. Ultraviolet, blue, green, red, and far-red light wavebands each drive signaling cascades that ultimately affect gene expression and physiology. Therefore, these sensors constitute a series of togglable inputs that may be controlled to affect plant growth and development and, ultimately, crop productivity.

Starting from lower energy wavebands, red and far-red light are sensed by a biochemical switch that may be flipped between active and inactive states (Quail, 2002; Smith, 1994). The phytochromes are inactive in the dark and reside in the cytosol of the cell. Upon illumination with red light, the phytochromes transit to the nucleus where they interact with the machinery governing gene expression (Huq et al., 2003). Far-red light reverts phytochromes to an inactive state. These are general rules, as some phytochromes are tuned to transduce far-red light signals, controlling specific responses in far-red enriched environments, such as those experienced in the shade of leaves (Pierik and de Wit, 2014; Smith and Whitelam, 1997). A family of specialized phytochromes (phyA-phyE in the model *Arabidopsis thaliana*) control traits like chloroplast development, stem elongation, leaf expansion, pigment accumulation, proximity of neighbors, and many aspects of metabolism.

Blue light signals are sensed by a different family of receptors that initiate another set of physiological events. The cryptochromes respond to blue light signals (Lin, 2000). Cryptochrome 1 (Cry1) regulates stem elongation and various aspects of vegetative growth. Like the phytochromes, Cry1 has profound effects on gene expression and ultimately metabolism. The cryptochrome 2 (Cry2) receptor is most known for regulating the transition to flowering in response to blue light (Guo et al., 1998). The phototropins control plant processes linked to the optimization of photosynthesis. Phototropin 1 (phot1) regulates directional growth, stomatal opening, and leaf expansion and position (Briggs et al., 2007). Its partner phototropin 2 (phot2) regulates the position of chloroplasts within the cell and also plays a role in the stomatal opening (Christie, 2007). Phototropins also control aspects of photosynthesis-related gene expression (Folta and Kaufman, 2003) as well as compensatory reactions to low-light environments (Takemiya et al., 2005). Other blue light sensors share similar activation chemistry with the phototropins, yet they differ in the downstream processes they regulate. The proteins ZEITLUPE, FKF1, and LKP2 all contribute to different aspects of circadian clock regulation or flowering (Demarsy and Fankhauser, 2009).

The role of green light has been understated, but research shows it does have discrete roles in plant physiology (Folta and Maruhnich, 2007; Klein, 1992; Smith et al., 2017). Green-light signals seem to oppose the effects of processes controlled by red or blue light, particularly in low-light environments, such as within a canopy of leaves (Sellaro et al., 2011; Wang et al., 2013; Zhang et al., 2011). Many green-light responses persist in all photoreceptors mutant backgrounds studied, suggesting that the pathway remains to be elucidated. However, a suite of low-fluence-rate blue light responses are modulated through the cryptochrome photoreceptors, as the chromophore shifts to a green-light absorbing state (Banerjee et al., 2007). Green light can then reverse a blue light response. The presence of green light in a controlled environment makes the leaves appear green to the human eye, allowing a better opportunity to inspect for damage or other signs of stress (Kim et al., 2004).

The effects of ultraviolet (UV) light have also been understated, yet these wavebands have powerful control of plant growth and development. UV signals are sensed through a photosensor called UVR8 (Jenkins, 2014). In the environment, UV signals represent a significant stress, a signal of full sun that can damage many aspects of plant physiology (Li et al., 2013). As such, the plant responds to UV signals in two ways, both as a photomorphogenic signal and as a stress signal. The photomorphogenic effects include changes in plant stature that are consistent with what happens in response to blue or red wavebands—reduced hypocotyl elongation (Favory et al., 2009), increased photosynthetic efficiency (Davey et al., 2012), and control of stomatal aperture (Eisinger et al., 2000). UV exposure also changes lateral root development and release of axial buds. The most relevant traits affected are the accumulation of attractive dark-colored photoprotective pigments related to the production of flavonoids associated with health (Favory et al., 2009; Kliebenstein et al., 2002). UV wavebands may adversely affect some horticultural traits, as they inhibit leaf expansion (Dillenburg et al., 1995; Lindoo and Caldwell, 1978).

Controllable Factors

There are essentially five parameters of light that may be controlled inside the vertical farm. Each must be carefully optimized for any given crop and even any variety. Ultimately these are the variables that may be manipulated to scale production while keeping costs down and management minimal.

Light Quality (Wavelength): As mentioned above, different wavebands of light control particular aspects of plant growth, development, metabolism, and ultimately horticultural productivity, mediated by specific light sensors, as well as PAR waveband contribution to photosynthesis.

Light Quantity (Fluence Rate): In the parlance of light effects on plant biology it is discussed as a particle, quantified in micromoles of photons per square meter (fluence rate). The concept of fluence rate is contrasted against intensity, which is a subjective measurement based on the limits of human vision. Fluence rates allow light to be delivered at a specific dose per unit time. There are measurable thresholds for where responses are activated and saturated, and these need to be elucidated for each crop or variety.

Light Duration (Photoperiod): Plants evolved on a rotating planet and therefore have an ingrained sense of timing that informs the time of day. Plants in temperate latitudes use day length to judge seasonality. The number of daylight hours provides an important environmental cue that can activate specific developmental events, such as the transition to flowering. In the controlled environment the length of the day may be optimized to maximize growth against energy costs. Economically important developmental transitions, such as the change from the production of leaves to the production of flowers, may be limited or activated by day length, keeping long-day flowering herbs in leaf production, while adjusting the timing of flower production to meet a market opportunity.

Waveband Combinations: The biochemical circuits that control plant response to light may be co-activated, as they are in full sunlight. Activation of multiple photosensory systems can ignite synergistic responses, which may be positive or negative, depending on the trait.

Light Pulse Intervals: Light may be pulsed to save energy. At rapid pulse rates (milliseconds) the pluses simply translate to lower fluence rates with increased dark intervals. Some advantages to these strategies have been observed. Longer pulses (minutes-hours) have been delivered, showing that such treatments are highly disruptive to normal physiology (Garner and Allard, 1931; Song et al., 2019). Pulse rates on the order of seconds seem to have minimal effects on plant growth and development, and increasing the duration of the dark period between pulses, while keeping the light period static, may be a means to substantially cut energy costs (Song et al., 2019).

A comprehensive understanding of how plants interact with the light environment as well as the variable ways light inputs may be manipulated presents a daunting problem—how can we begin to match the seemingly endless optimization options to any given variety? But therein lies a substantial opportunity. Together, the electronic canopy of the agricultural controlled environment is not just a way to grow plants but, rather, to control how plants grow. Understanding how light from the environment specifically connects to important horticultural traits affords a virtually untapped opportunity to manipulate plant growth and development, the timing of crop production, and the quality of products for the consumer.

REFOCUSING PRIORITIES

There is a great opportunity in breeding for controlled environments. Field-grown crops must surrender to the ever-changing elements of the environment. Traditional breeders must prioritize resistance to disease, tolerance to weather extremes, and escaping the influence of insect pests. Without checking these basic boxes, tasty fruits and high yields are completely irrelevant.

This reality frames the opportunity of the controlled environment. Plants designed for CEA are divorced from the wild fluctuation of the field, and breeders can focus on traits relevant to the consumer rather than the producer. In the past, plants needed to meet the demands of the field and

TABLE 10.1
Breeding Priorities for the Field Compared to Controlled Environments

Trait	Field Priority	CEA Priority	Remarks
Disease tolerance	+++	+	CEA offers a potentially lower incidence of fewer potential diseases, making scouting/treatment easier
Pest resistance	+++	+	Pests are limited by physical/chemical barriers and sound operating practices. Problems easily rectified
Postharvest quality	+++	+	Shorter supply chains deprioritize the need for breeders to focus on shipping, storage, and shelf life
Flavors and aromas	+	+++	Sensory quality needs to be acceptable in the field but can be optimized by managing genetics/environment
Colors	+	+++	Accumulation of attractive and potentially healthful pigments can be prioritized and optimized in CEA
Nutraceuticals	+	+++	Breeding can focus on the accumulation of potentially healthful compounds, like flavonoids and glucosinolates

produce products acceptable to the consumer. That relationship has now flipped, as controlled environments offer the opportunity to grow crops acceptable to the producer and meet the demands of the consumer (Table 10.1).

For example, when breeding for the field, one of the priorities is acceptable fruit/vegetable products regardless of environmental variation. Produce must be marketable whether grown under drought, heat, floods, and other uncontrollable stressors, taking genetics × environment (GxE) interaction into high regard (Corlouer et al., 2019). Breeders were tasked with producing a product that was always acceptable, as product trait stability in response to changing environment was a key metric in selection. But when the environment can be controlled, removing the variability of E in G×E, it allows breeders to select for ideotypes that maintain an optimal trait presentation in a steady, static environment. There are many opportunities to create consumer-centered products when environmental flux is removed from the equation. Some examples follow.

Colors and Aromas

The controlled environment provides an opportunity to adjust the spectrum and directly affect the traits that appeal to human senses. The attractive pigments and aromas that entice consumers and ensure repeat purchases are also outputs of the interaction between genetics and environment. The purple pigments associated with healthy compounds are controlled by light, namely, by the blue and UV portions of the spectrum. Horticultural products like pak choi may be significantly improved with UV-A (315–400 nm) and blue light treatment (Huang et al., 2022). Addition of blue wavebands also induces deeper colors in purple leaf lettuces (Shao et al., 2020a) and turnips (Zhou et al., 2013). The addition of UV-A can significantly increase the presence of purple pigments in kale (Jiang et al., 2022). These examples are just a few that demonstrate the ability to skew a spectrum to create colors that are attractive to consumers and potentially create unique products that are not possible in the field.

While visual quality is what sells the product at market, flavors and aromas encourage repeat purchases (Baldwin, 2002). Flavors and aromas are generated by the interaction of sugars, acids, and volatile compounds with the gustatory and olfactory systems, forming the perception of a given fruit, vegetable, or herb (Tieman et al., 2012). Sugars, acids, and volatiles are products of metabolism, and their accumulation may be influenced by light (Campbell et al., 2020; Kim et al., 2023). Sugar abundance is a product of photosynthesis. Transport, assimilation, and retention in specific crop organs or tissues are also light dependent (Lemoine et al., 2013). Volatile compounds are low-vapor pressure chemicals that produce the complex perception of a given fruit, vegetable, or herb. Their production is also a product of secondary metabolism, gated through various pathways, some of which are light-regulated (Simkin et al., 2004; Xiang et al., 2022). For instance, the unique flavor profile of basil is constructed by the presence of various compounds that are expressed in a wavelength-dependent manner (Carvalho et al., 2016; Hammock et al., 2021; Litvin et al., 2020). Basil is just one example of how the sensory quality of a horticultural food product may be enhanced by the light environment. Discrete portions of the spectrum have been shown to control the sense of spicy flavors from glucosinolates (Carvalho and Folta, 2014c), as well as potentially suppress compounds that may negatively affect flavor.

Control of Flowering Time

While the juvenility-maturity transition creates a competence to flower, the actual transition from vegetative growth (adding more branches and leaves) to reproductive growth (flowering) is a carefully orchestrated process. It must be, as plants that flower at the wrong time may miss the services of pollinators or may set fruits and seeds during unfavorable conditions (Shivanna and Tandon, 2014). Plants requiring outcrossing use environmental cues to synchronize flowering so that sexually compatible individuals are presenting gametes at the same time. In a commercial context, timing or controlling the transition to flowering dictates the production of flowers for cut flower production or fruits and vegetables to be consumed. Therefore, the precise control of flower time is of great economic value.

In the field, flowering control is dictated by the confluence of at least four independent pathways that merge environmental and internal information (Jung et al., 2017). These pathways are of great interest in plant genetic improvement because farmers can potentially benefit from flowering to fill specific market windows (Jung and Müller, 2009). One of the pathways with significant influence is light (Cerdan and Chory, 2003). Different wavebands of light have a differential influence on flowering time. In the model plant, *A. thaliana*, red light represses flowering whereas blue and far red enhance it (Valverde et al., 2004). This same pattern of flowering based on waveband-specific induction or repression has been observed in other crops as well and could serve as an elegant means to control the timing of flowering.

For example, a blue light night break induces flowering on an otherwise vegetative chrysanthemum plant (Park and Jeong, 2020). The role of red and far-red light has long been described in poinsettia production, where bright red bracts are in demand specifically before the winter holidays (Zhang and Runkle, 2019). Other ornamental flowers may be induced by manipulating the red to far-red light ratio (Craig and Runkle, 2013). These are just several examples from ornamental crops where the timing of flowering can be controlled to match market windows. Similar opportunities exist in fruit and vegetable production, by inducing the production of flowers on fruiting plants or by repressing flowering on vegetables and herbs, where flowering marks the end of saleable vegetative structures and can even affect aspects of flavor, rendering them unmarketable (Ryder, 1996).

Nutraceutical Compounds

The consumption of plant products is associated with long-term health and decreased incidence of degenerative disease. Experiments examining plant responses to space travel illustrated that different light qualities could change the antioxidant levels in sprouts, countering the potential effects of ionizing radiation (De Micco et al., 2021). Increasing the amount of far-red light induced the accumulation of aliphatic glucosinolates and antioxidant levels in kale sprouts (Carvalho and Folta, 2014b). Variation in the frequency of light pulses affects the antioxidant capacity of lettuce, along with the accumulation of specific metabolites associated with health (Carotti et al., 2021)). The accumulation of ascorbic acid, carotenoids, phenolic compounds, soluble sugars, and anthocyanin may be modulated by specific light wavelengths (Bian et al., 2015). There are many additional examples of how variations in the light environment or other stresses can modulate the accumulation of healthful metabolites.

DE-EMPHASIZING FIELD PRIORITIES

Plant varieties developed for the field must meet a wide suite of standards to be commercially viable. Many of these become irrelevant in the controlled environment. For instance, a closed environment offers protection from many stresses that limit production in field-grown crops. This realization allows breeders to return to the creative integration of traits from wild species or old varieties that may not offer viable horticultural opportunities for the field yet may be valuable under controlled conditions.

Postharvest Quality

Another consumer complaint is the postharvest life of horticultural products. Field-grown produce must possess postharvest quality, meaning that the products harvested must be able to retain their marketable quality despite the challenges of a supply chain. Field-grown produce oftentimes transits a country or the globe, based on seasonality, labor, and environmental constraints. Again, the controlled environment limits the necessity to prioritize postharvest quality, as supply chains are typically short, and products arrive at retail no more than a few days after harvest.

Biotic Pressure

Plant breeders traditionally place tremendous priority on pest and pathogen resistance, as a plant that can't survive the field has little commercial value. The costs of pest/pathogen control, both from the costs of inputs and the labor/fuel to apply them, are high. There also are the potential environmental impacts of application and their public perception. Together this means that crops selected for the field must start with a durable genetic disease/pest resistance package. Recent efforts in genomic selection and understanding the genetic basis of disease/pest resistance have hastened breeding efforts, allowing breeders to slide their focus to consumer-center traits (Folta and Klee, 2016; Lado et al., 2019; Luby and Shaw, 2009). Despite an improvement, breeding for resistance is a challenge and an ongoing struggle, as most mechanisms are subject to evolutionary pressures that allow pets/pathogens to eventually evade genetic controls. The controlled environment presents a different landscape for combating pests and pathogens, starting with an opportunity for preemptive prophylactic control, followed by aggressive, even automated scouting for a limited range of potential problems. These realities again allow plant breeders to shift priorities toward profitable production and consumer traits.

DESIGNER GENES AND PROGRAMMABLE PLANTS

Growers have traditionally been beholden to plant performance in each environment. The seeds are planted, the plant grows, the products are produced, and the only management input is protection from pests and pathogens and some fertilizer to maximize growth. The controlled environment offers a unique situation to not just grow plants but, rather, to control how plants grow (Folta and Childers, 2008; Paradiso and Proietti, 2022). Because we know that specific light energies affect discrete processes in plant biology, light offers a means to manipulate plant growth and development. How can adjustment of the electronic canopy allow growers to affect their end products?

Whereas breeding for the field is a quest for phenotypic stability under any environmental condition, breeding for controlled environments offers the opportunity to do exactly the opposite—how can breeders maximize regulated plasticity? Can plants be developed that respond differently to different environmental inputs, producing products that are different visually or metabolically from one set of genes? Can tugging at the light spectrum create new products with variations in flavor, aroma, or nutrient content, based on a single set of genetic instructions? It is analogous to crews constructing two completely different homes from the same set of blueprints, only because one home was built in Toronto and one in Las Vegas. The environment would dictate how the blueprints were interpreted.

Genetic engineering plays a central role in creating next-generation controlled environment crops. Genetic engineering can be grouped into two broad categories. Transgenic and cisgenic technologies allow the introduction of known gene sequences that encode a given trait. Those may be under the control of their native regulatory elements (promoters, introns, terminators, etc.) or expressed from constructs that implement heterologous constitutive or regulated promoters with known regulation patterns. The use of RNA interference is a useful tool to negate the expression of a given gene. While a wide application of these technologies has been limited to a small suite of agronomic crops due to the enormous cost of deregulation, the landscape is slowly changing, as over two decades of safe implementation is shifting public sentiment (Evanega et al., 2022).

Gene editing using site-specific nucleases (e.g., CRISPR/Cas, TALEN) offers the opportunity to perform custom mutations in specific genes. Whereas plant breeding has traditionally been constrained by the genetic variability found within a population of sexually compatible plants, gene editing now allows the creation of lesions in key genes that affect horticulturally relevant traits. Techniques now exist that permit this process without the addition of a transgene, curtailing the need for costly and time-consuming deregulation. New varieties may also be carefully assessed for off-target gene edits, making the process even less risky than traditional breeding. The following traits are just some attractive targets that may be adjusted using genetic engineering, hastening the production of new varieties for controlled environments.

Bypassing Juvenility

Most multicellular organisms experience change through time that is encapsulated in specific life periods. Just like in animals, plants undergo a period of juvenility, defined primarily by reproductive incompetence. If high-value, smaller-stature horticultural crops (e.g. blueberry) are going to be grown in CEA, there will be a critical need to suppress juvenility, as there is limited value in growing a plant for years before it is competent to flower and produce fruit. Aside from the idea of importing mature plants, complete with the perils of bringing in disease and pests, it would be of value to hasten the time to reproductive maturity. Fortunately, the main driver of juvenility is highly conserved and well defined and could be manipulated through genetic engineering.

The transition from juvenility to reproductive competence in flowering plants is dictated by the reciprocal control of two small RNA species, microRNAs—miRNA156 and miRNA172 (Wu et al., 2009). The miRNA156 controls genes that maintain juvenile morphology and physiology. The miRNA172 controls genes that drive reproductive competence and floral induction. Suppression of miRNA156 and/or induction of miRNA172 has been shown to induce early flowering in species normally requiring longer juvenility periods. These two small RNAs serve as master regulators and are outstanding targets for creating plants that lack the repressor, overexpress the inducer, or both.

Creating New Colors and Aromas

As mentioned earlier, deep colors and complex sensory quality are important to the consumer, as they encourage higher consumption, but also are associated with health (Pojer et al., 2013; Wallace and Giusti, 2013). The attractive pigments and volatile compounds in produce may also be adjusted using genetic engineering. Plants may be engineered to increase flux through the relevant pathways to increase the output of desirable end products. For instance, several genes affecting the anthocyanin biosynthetic pathway have been overexpressed to produce more colorful outcomes (Tanaka and Ohmiya, 2008). Gene editing was used to create a pink-fleshed tomato by editing a deletion in *SlMYB12*, a transcription factor controlling the production of naringenin chalcone in the fruit epidermis (Yang et al., 2019). It has been demonstrated that it is possible to control the accumulation of pigments and aromas by driving the expression of rate-limiting genes using light-regulated promoters, producing plants that respond to specific wavelengths in the controlled environment, and producing specific flavors or colors on demand.

Design of Dwarf Varieties

The profitability of the controlled environment is highly tied to plant size. Large vegetative crops are unsuitable for vertical farms because they quickly absorb space, and they present challenges to efficient lighting and energy consumption. However, modern genetics may play a role in adjusting plant size. The control of plant stature is well understood with respect to environmental and genetic mechanisms. Blue and far-red wavebands greatly suppress stem elongation in early development. Blue light continues to suppress elongation, whereas far-red later contributes to internode elongation. The precise mixing of wavelengths can constrain plant size, investing energy only in the wavebands that keep plants short.

At the same time the genetic control of plant stature is also well understood and lies primarily in the control of elongation growth via the gibberellins. The gibberellins are a class of plant hormones that have diverse roles, from germination to activation of cell division, sucrose partitioning (source/sink), and many aspects of plant development. But, most of all, gibberellins are known to control elongation growth. The genes that control sensitivity to gibberellins have been well characterized, and several cereal crops have been dwarfed using site-directed mutagenesis (CRISPR/Cas, TALEN, etc.). The same technology may be applied to make any horticultural crop a dwarf variety, as already demonstrated in tomatoes (Tomlinson et al., 2019) and bananas (Shao et al., 2020b). As tissue culture and transformation techniques continue to improve in horticultural crops, the use of dwarfing strategies may increase the likelihood that high-value fruit crops may be grown in controlled environments.

Fruit without Fertilization

The production of many fruits and vegetables is dependent on pollination. The typical vectors of pollen transit, bees, moths, and wind are typically absent from the controlled environment, limiting the types of fruits that may be produced without significant intervention or hand pollination. However, there are ways that plants may produce fruit in the absence of fertilization.

Parthenocarpy is the process of carpel development without a pollen donor, which leads to a seedless fruit. Parthenocarpic fruits have the additional advantage of seedlessness, which is coveted by consumers and plant breeders alike, as the resulting crops are only propagated vegetatively, allowing better control of patented germplasm. The genes that control parthenocarpy are being elucidated and have already been validated in some systems. These are superb targets for genetic engineering, as the suppression of these genes leads to fruit production in the absence of pollination.

For instance, the production of parthenocarpic tomatoes was achieved by mutagenizing the *Sl-IAA9* gene (Ueta et al., 2017). Lesions in the SlAGAMOUS-LIKE6 gene also cause parthenocarpic tomato fruit set (Klap et al., 2017). Seedless eggplants were obtained by overexpression of the *DefH9-IAA* gene, causing seedlessness, but also a 30% increase in yield (Acciarri et al., 2002). The options are virtually limitless for these same principles, and perhaps mutations in the same genes, to be translated to other high-value fruit crops to make them more viable in the controlled environment. However, the lack of a viable embryo in the seed means additional management steps are likely necessary to obtain full fruit development, as the seed frequently contributes to fruit shape and size.

Apomixis is the production of seeds and viable embryos without pollination. The creation of apomictic hybrids would allow elite materials to be seed propagated without changing their genetics (Fiaz et al., 2020), while also not relying on pollination vectors.

Breeding for Plasticity

As discussed earlier, traditionally breeders seek trait stability that is agnostic to environmental influence. Regardless of ambient conditions or stressors, an ideal crop output (leaves, stems, fruits, etc.) is acceptable to the consumer. Breeding for the controlled environment may seek to identify new varieties that produce different products when placed in contrasting conditions. For example, a lettuce line grown under one light condition (perhaps enriched in red and green wavebands) may grow with a long green leaf, while the same plant grown in a blue and far-red enriched environment may grow dark purple and with a different leaf shape.

The advantage of plasticity, or the ability to create two (or more) different horticultural products from one set of genetics, is obvious. A grower can master the management of a given variety with respect to nutrients, temperature, and other environmental factors and then simply flip a switch to adjust the physical attributes of the retail product itself. Such agility may allow for rapid response to changes in market conditions, as well as create specialty mixed-color products from one set of seeds.

While attractive colors may influence a primary buying decision, favorable flavors and aromas help ensure repeat purchases. It is entirely feasible to create crops, like strawberries, which produce different flavor volatiles depending on the wavelength applied. For instance, a day or two of illumination under red and far-red may induce the tropical fruity esters, while a blue-enriched environment may spur the accumulation of grape and peach notes. Again, such science fiction is quickly becoming a science fact, and the programmable plant, controlled by commands in a language of light, would provide more options to the grower and consumer.

Breeding for Energy Savings

While water and nutrient use efficiency is improved in vertical farms, energy use remains the principal limitation to their widespread success (Kozai, 2013). Energy from the sun is free, and land is typically less expensive the further it is from urban centers (Despommier, 2011). There are many considerations that determine if vertical farms are sustainable, from both an economic and environmental perspective, as they can have a significant carbon footprint (Al-Chalabi, 2015). Today's forays into controlled environment space occur at a time of high energy costs, still predominantly derived from fossil fuels like natural gas. Efficiency is improving, and the cost may decrease with widespread renewables and nuclear and other new methods of energy generation. Lighting will also become more efficient. These factors improve the likelihood of extensive indoor agriculture profitability.

The way light is delivered may be a place for improvement as well. The use of lensing to focus photons on the crop rather than on empty space has increased efficiency (Li et al., 2016). Light modules have been developed that move around a growing space, providing illumination at specific intervals. This strategy has been shown to produce the same yield of butterhead lettuce with half the electricity (Li et al., 2014).

But is there room to create plants that can attain market status with less energy investment? The selection of varieties that can generate the same output on fewer photons is possible. When light energy is in excess it is converted to heat and also dissipated through additional mechanisms (Malnoë, 2018). Honing light sources to match the photosynthetic maximum without exceeding it would save energy without affecting plant growth. But this threshold would likely vary among varieties, and it is possible that plants with enhanced light capture and efficient integration could significantly improve yields in lower-light, energy-saving conditions. Breeding plant varieties that thrive under pulsed-light conditions, or produce on short days, may be the most attainable first steps in decreasing energy costs.

CONCLUSIONS

It is almost cliché to cite the food security needs of the next decades. More food will be needed to satisfy basic food security, as well as feed an increasingly middle-class world population that is enthusiastic about expanding access to a variety of fresh fruits and vegetables. More food must be grown on fewer acres, which means disruptive change needs to occur to fortify food systems. Growing crops in controlled environments is one part of the solution but needs to undergo its own disruptive change, leapfrogging current sustainability barriers. Future implementation will depend on improved production techniques intimately intertwined with crop genetic improvement. They must evolve in tandem, not in isolation.

Plant growth and development is an output of a cultivar's genetic potential performing within the constraints of any given environment. During plant domestication and breeding, humans have influenced plant genetics through selection in hundreds, if not thousands of different environments, adapting biochemistry and physiology to the local conditions and pressures. None of these, except arguably the laboratory model *A. thaliana*, were designed to thrive and be productive in a wholly artificial environment.

This is a critical, yet overlooked concept that will dictate the success and profitability of agricultural production in a controlled environment space. Currently ambitious companies and their angel investors attempt square-peg-round-hole strategies, wedging field crops into an alien environment. But breeding technology will accelerate directed, informed hybridization to produce crop lines specifically bred to perform in that novel space.

These efforts will not be as daunting as they seem at first blush. First, the environment is stable, so performance will be unaffected by weather, seasons, pests, and pathogens. Next, the new methods of genomic selection, coupled with the massive data obtained from genomic sequencing, and the continual characterization of gene–trait relationships, means that the introgression of favorable traits will be more rapid than traditional breeding methods. Finally, the tools of genetic engineering are becoming more widely accepted, and modern technology allows direct gene transfer or gene editing to mobilize traits on a scale of months that used to take years or even decades.

As the sophistication of CEA grows, we find ourselves at a critical watershed. While the hardware of the indoor farm becomes more complex, it is past time to match that complexity by focusing improvement on the most intricate collection of sensors, circuits, hydraulics, photon capture, and energy conversion in the room—the crop plant itself.

REFERENCES

Acciarri, N., Restaino, F., Vitelli, G., Perrone, D., Zottini, M., Pandolfini, T., Spena, A. and Rotino, G. (2002) Genetically modified parthenocarpic eggplants: Improved fruit productivity under both greenhouse and open field cultivation. BMC Biotechnology 2, 4.

Al-Chalabi, M. (2015) Vertical farming: Skyscraper sustainability? Sustainable Cities and Society 18, 74–77.

Baldwin, E.A. (2002) Fruit flavor, volatile metabolism and consumer perceptions. In *Fruit Quality and its Biological Basis*, M. Knee, Ed. CRC Press, Boca Raton, FL USA, 89–106.

Banerjee, R., Schleicher, E., Meier, S., Munoz Viana, R., Pokorny, R., Ahmad, M., Bittl, R. and Batschauer, A. (2007) The signaling state of Arabidopsis cryptochrome 2 contains flavin semiquinone. *Journal of Biological Chemistry* 14916–14922.

Bian, Z.H., Yang, Q.C. and Liu, W.K. (2015) Effects of light quality on the accumulation of phytochemicals in vegetables produced in controlled environments: A review. Journal of the Science of Food and Agriculture 95, 869–877.

Briggs, W.R., Tseng, T.S., Cho, H.Y., Swartz, T.E., Sullivan, S., Bogomolni, R.A., Kaiserli, E. and Christie, J.M. (2007) Phototropins and their LOV domains: Versatile plant blue-light receptors. Journal of Integrative Plant Biology 49, 4–10.

Campbell, S.M., Sims, C.A., Bartoshuk, L.M., Colquhoun, T.A., Schwieterman, M.L. and Folta, K.M. (2020) Manipulation of sensory characteristics and volatile compounds in strawberry fruit through the use of isolated wavelengths of light. Journal of Food Science 85, 771–780.

Carotti, L., Potente, G., Pennisi, G., Ruiz, K.B., Biondi, S., Crepaldi, A., Orsini, F., Gianquinto, G. and Antognoni, F. (2021) Pulsed led light: Exploring the balance between energy use and nutraceutical properties in indoor-grown lettuce. Agronomy 11, 1106.

Carvalho, S.D. and Folta, K.M. (2014a) Environmentally modified organisms - Expanding genetic potential with light. Critical Reviews in Plant Sciences 33, 486–508.

Carvalho, S.D. and Folta, K.M. (2014b) Sequential light programs shape kale (Brassica napus) sprout appearance and alter metabolic and nutrient content. Horticulture Research 1, 1–13.

Carvalho, S.D. and Folta, K.M. (2014c) Sequential light programs shape kale (Brassica napus) sprout appearance and alter metabolic and nutrient content. Horticulture Research 1, 8.

Carvalho, S.D., Schwieterman, M.L., Abrahan, C.E., Colquhoun, T.A. and Folta, K.M. (2016) Light quality dependent changes in morphology, antioxidant capacity, and volatile production in sweet basil (Ocimum basilicum). Frontiers in Plant Science 7,,p. 203420.

Cerdan, P.D. and Chory, J. (2003) Regulation of flowering time by light quality. *Nature* 423, 881–885.

Christie, J.M. (2007) Phototropin blue-light receptors. Annual Review of Plant Biology 58, 21–45.

Corlouer, E., Gauffreteau, A., Bouchet, A.-S., Bissuel-Bélaygue, C., Nesi, N. and Laperche, A. (2019) Envirotypes based on seed yield limiting factors allow to tackle G×E interactions. In: Agronomy, 9, p. 798. https://doi.org/10.3390/agronomy9120798

Craig, D.S. and Runkle, E.S. (2013) A moderate to high red to far-red light ratio from l,ight-emitting diodes controls flowering of short-day plants. Journal of the American Society for Horticultural Science 138, 167–172.

Davey, M.P., Susanti, N.I., Wargent, J.J., Findlay, J.E., Paul Quick, W., Paul, N.D. and Jenkins, G.I. (2012) The UV-B photoreceptor UVR8 promotes photosynthetic efficiency in *Arabidopsis thaliana* exposed to elevated levels of UV-B. *Photosynthesis Research* 114, 121–131.

De Micco, V., Amitrano, C., Vitaglione, P., Ferracane, R., Pugliese, M. and Arena, C. (2021) Effect of light quality and ionising radiation on morphological and nutraceutical traits of sprouts for astronauts' diet. *Acta Astronautica* 185, 188–197.

Demarsy, E. and Fankhauser, C. (2009) Higher plants use LOV to perceive blue light. Current *O*pinion in *P*lant *B*iology 12, 69–74.

Despommier, D. (2011) The vertical farm: Controlled environment agriculture carried out in tall buildings would create greater food safety and security for large urban populations. Journal für Verbraucherschutz und Lebensmittelsicherheit 6, 233–236.

Dillenburg, L.R., Sullivan, J.H. and Teramura, A.H. (1995) Leaf expansion and development of photosynthetic capacity and pigments in *Liquidambar styraciflua* (Hamamelidaceae)-effects of UV-B radiation. American Journal of Botany 82, 878–885.

Eisinger, W., Swartz, T.E., Bogomolni, R.A. and Taiz, L. (2000) The ultraviolet action spectrum for stomatal opening in broad bean. *Plant Physiology* 122, 99–106.

Evanega, S., Conrow, J., Adams, J. and Lynas, M. (2022) The state of the 'GMO' debate - toward an increasingly favorable and less polarized media conversation on ag-biotech? GM Crops & Food 13, 38–49.

Favory, J.-J., Stec, A., Gruber, H., Rizzini, L., Oravecz, A., Funk, M., Albert, A., Cloix, C., Jenkins, G.I., Oakeley, E.J., Seidlitz, H.K., Nagy, F. and Ulm, R. (2009) Interaction of COP1 and UVR8 regulates UV-B-induced photomorphogenesis and stress acclimation in Arabidopsis. *The Embo Journal* 28, 591–601.

Fiaz, S., Wang, X., Maqbool, R., Ali, H., Ahmad, S., Riaz, A. and Alharthi, B. (2020) Synthetic apomixis an old enigma to preserve hybrid vigor for multiple generations, *GM Crops & Food*, 12(1), 57–70. https://doi.org/10.1080/21645698.2020.1808423.

Folta, K. and Klee, H. (2016) Sensory sacrifices when we mass-produce mass produce. Horticulture Research 3.

Folta, K.M. and Childers, K.S. (2008) Light as a Growth Regulator: Controlling Plant Biology with Narrow-bandwidth Solid-state Lighting Systems, HortScience 43(7) 1957–1964.

Folta, K.M. and Kaufman, L.S. (2003) Phototropin 1 is required for high-fluence blue-light-mediated mRNA destabilization. Plant Mol*ecular Biology* 51, 609–618.

Folta, K.M. and Maruhnich, S.A. (2007) Green light: A signal to slow down or stop. J*ournal of Exprimental Botony* 58, 3099–3111.

Garner, W.W. and Allard, H.A. (1931) Effect of *Abnormally Long and Short Alternations of Light and Darkness on Growth and Development of* Plants. *Journal of Agricultural Research*, 42, 629–651

Guo, H., Yang, H., Mockler, T.C. and Lin, C. (1998) Regulation of flowering time by Arabidopsis photoreceptors. *Science* 279, 1360–1363.

Hammock, H.A., Kopsell, D.A. and Sams, C.E. (2021) Narrowband blue and red LED supplements impact key flavor volatiles in hydroponically grown basil across growing seasons. Frontiers in Plant Science 12, 623314.

Huang, J., Liu, X., Yang, Q., Lei, B., Zheng, Y., Bian, Z., Wang, S., Li, W., Mao, P. and Xu, Y. (2022) UVA enhanced promotive effects of blue light on the antioxidant capacity and anthocyanin biosynthesis of pak choi. Horticulturae 8, 850.

Huq, E., Al-Sady, B. and Quail, P.H. (2003) Nuclear translocation of the photoreceptor phytochrome B is necessary for its biological function in seedling photomorphogenesis. *The Plant Journal* 35, 660–664.

Jenkins, G.I. (2014) The UV-B photoreceptor UVR8: From structure to physiology. The Plant Cell Online 26, 21–37.

Jiang, H., Li, Y., He, R., Tan, J., Liu, K., Chen, Y. and Liu, H. (2022) Effect of supplemental UV-A intensity on growth and quality of kale under red and blue light. *International Journal of Molecular Sciences* 23, 6819.

Jung, C. and Müller, A.E. (2009) Flowering time control and applications in plant breeding. *Trends in Plant Science* 14, 563–573.

Jung, C., Pillen, K., Staiger, D., Coupland, G. and Von Korff, M. (2017) Recent advances in flowering time control. *Frontiers in Plant Science* 7, 2011.

Kim, D., Ra, I. and Son, J.E. (2023) Fruit quality and volatile compounds of greenhouse sweet peppers as affected by the LED spectrum of supplementary interlighting. *Journal of the Science of Food and Agriculture* 103, 2593–2601.

Kim, H.H., Goins, G.D., Wheeler, R.M. and Sager, J.C. (2004) Green light supplementation for enhanced lettuce growth under red and blue light-emitting diodes. *Hortscience* 39, 1617–1622.

Klap, C., Yeshayahou, E., Bolger, A.M., Arazi, T., Gupta, S.K., Shabtai, S., Usadel, B., Salts, Y. and Barg, R. (2017) Tomato facultative parthenocarpy results from SlAGAMOUS-LIKE 6 loss of function. Plant Biotechnol*ogy Journal* 15, 634–647.

Klein, R.M. (1992) Effects of green light on biological systems. Biol*ogical Reviews* 67, 199–284.

Kliebenstein, D.J., Lim, J.E., Landry, L.G. and Last, R.L. (2002) Arabidopsis UVR8 regulates ultraviolet-B signal transduction and tolerance and contains sequence similarity to human regulator of chromatin condensation 1. Plant *P*hysiology 130, 234–243.

Kozai, T. (2013) Resource use efficiency of closed plant production system with artificial light: Concept, estimation and application to plant factory. Proceedings of the Japan Academy. Series B, Physical and Biological Sciences 89, 447–461.

Lado, J., Moltini, A.I., Vicente, E., Rodríguez, G., Arcia, P., Rodríguez, M., López, M., Billiris, A. and Ares, G. (2019) Integration of sensory analysis into plant breeding: A review. Agrociencia (Uruguay) 23, 101–115.

Lemoine, R., Camera, S.L., Atanassova, R., Dédaldéchamp, F., Allario, T., Pourtau, N., Bonnemain, J.-L., Laloi, M., Coutos-Thévenot, P. and Maurousset, L. (2013) Source-to-sink transport of sugar and regulation by environmental factors. Frontiers in *P*lant *S*cience 4, 272.

Li, J., Yang, L., Jin, D., Nezames, C.D., Terzaghi, W. and Deng, X.W. (2013) UV-B-induced photomorphogenesis in Arabidopsis. Protein & *C*ell 4, 485–492.

Li, K., Li, Z. and Yang, Q. (2016) Improving light distribution by zoom lens for electricity savings in a plant factory with light-emitting diodes. Frontiers in Plant Science 7, 92.

Li, K., Yang, Q.-C., Tong, Y.-X. and Cheng, R. (2014) Using movable light-emitting diodes for electricity savings in a plant factory growing lettuce. *Horttechnology* 24, 546–553.

Lin, C. (2000) Plant blue-light receptors. Trends *in* Plant Sci*ence* 5, 337–342.

Lindoo, S.J. and Caldwell, M.M. (1978) Ultraviolet-B radiation-induced inhibition of leaf expansion and promotion of anthocyanin production: Lack of involvement of the low irradiance phytochrome system 1. Plant Physiology 61, 278–282.

Litvin, A.G., Currey, C.J. and Wilson, L.A. (2020) Effects of supplemental light source on basil, dill, and parsley growth, morphology, aroma, and flavor. Journal of the American Society for Horticultural Science 145, 18–29.

Luby, J.J. and Shaw, D.V. (2009) Plant breeders' perspectives on improving yield and quality traits in horticultural food crops. *HortScience* 44, 20–22.

Malnoë, A. (2018) Photoinhibition or photoprotection of photosynthesis? Update on the (newly termed) sustained quenching component qH. Environmental and Experimental Botany 154, 123–133.

McCree, K.J. (1972) Test of current definitions of photosynthetically active radiation against leaf photosynthesis data. *Agricultural Meteorology* 10, 443–453.
Paradiso, R. and Proietti, S. (2022) Light-quality manipulation to control plant growth and photomorphogenesis in greenhouse horticulture: The state of the art and the opportunities of modern LED systems. Journal of Plant Growth Regulation 41, 742–780.
Park, Y.G. and Jeong, B.R. (2020) How supplementary or night-interrupting low-intensity blue light affects the flower induction in chrysanthemum, a qualitative short-day plant. Plants 9, 1694.
Pierik, R. and de Wit, M. (2014) Shade avoidance: Phytochrome signalling and other aboveground neighbour detection cues. Journal of Experimental Botany 65, 2815–2824.
Pojer, E., Mattivi, F., Johnson, D. and Stockley, C.S. (2013) The case for anthocyanin consumption to promote human health: A review. Comprehensive *R*eviews in *F*ood *S*cience and *F*ood *S*afety 12, 483–508.
Quail, P.H. (2002) Phytochrome photosensory signalling networks. *Nature Reviews Molecular Cell Biology* 3, 85–93.
Ruban, A.V. (2016) Nonphotochemical chlorophyll fluorescence quenching: Mechanism and effectiveness in protecting plants from photodamage. Plant Physio*logy* 170, 1903–1916.
Ryder, E.J. (1996) Ten lettuce genetic stocks with early flowering genes Ef-1ef-1 and Ef-2ef-2. *HortScience* 31, 473–475.
Sellaro, R., Crepy, M., Trupkin, S.A., Karayekov, E., Buchovsky, A.S., Rossi, C. and Casal, J.J. (2011) Cryptochrome as a sensor of the blue/green ratio of natural radiation in Arabidopsis. Plant Physio*logy* 154, 401–409.
Shao, M., Liu, W., Zha, L., Zhou, C., Zhang, Y. and Li, B. (2020a) Effects of constant versus fluctuating red-blue LED radiation on yield and quality of hydroponic purple-leaf lettuce. Horticulture, Environment, and Biotechnology 61, 989–997.
Shao, X., Wu, S., Dou, T., Zhu, H., Hu, C., Huo, H., He, W., Deng, G., Sheng, O. and Bi, F. (2020b) Using CRISPR/Cas9 genome editing system to create MaGA20ox2 gene-modified semi-dwarf banana. Plant Biotechnology Journal 18, 17.
Shivanna, K.R. and Tandon, R. (2014) *Reproductive Ecology of Flowering Plants: A Manual*. Springer, New Dehli, India.
Simkin, A.J., Schwartz, S.H., Auldridge, M., Taylor, M.G. and Klee, H.J. (2004) The tomato carotenoid cleavage dioxygenase 1 genes contribute to the formation of the flavor volatiles beta-ionone, pseudoionone, and geranylacetone. *Plant Journal* 40, 882–892.
Smith, H. (1994) Sensing the light environment: The functions of the phytochrome family. In: Photomorphogenesis in Plants (Kendrick, R. E. and Kronenberg, G. H. M., eds.), pp. 377–416. Dordrecht, Netherlands: Kluwer.
Smith, H. and Whitelam, G.C. (1997) The shade avoidance syndrome: Multiple responses mediated by multiple phytochromes. Plant, Cell & Environment 20, 840–844.
Smith, H.L., McAusland, L. and Murchie, E.H. (2017) Don't ignore the green light: Exploring diverse roles in plant processes. Journal of *E*xperimental *B*otany 68, 2099–2110.
Song, S., Kusuma, P., Carvalho, S.D., Li, Y. and Folta, K.M. (2019) Manipulation of seedling traits with pulsed light in closed controlled environments. Environmental and Experimental Botany 166, 103803.
Takemiya, A., Inoue, S.-i., Doi, M., Kinoshita, T. and Shimazaki, K.-i. (2005) Phototropins promote plant growth in response to blue light in low light environments. The Plant Cell Online 17, 1120–1127.
Tanaka, Y. and Ohmiya, A. (2008) Seeing is believing: Engineering anthocyanin and carotenoid biosynthetic pathways. Current *O*pinion in *B*iotechnology 19, 190–197.
Tieman, D., Bliss, P., McIntyre, L.M., Blandon-Ubeda, A., Bies, D., Odabasi, A.Z., Rodríguez, G.R., van der Knaap, E., Taylor, M.G. and Goulet, C. (2012) The chemical interactions underlying tomato flavor preferences. Current Biology 22, 1035–1039.
Tomlinson, L., Yang, Y., Emenecker, R., Smoker, M., Taylor, J., Perkins, S., Smith, J., MacLean, D., Olszewski, N.E. and Jones, J.D.G. (2019) Using CRISPR/Cas9 genome editing in tomato to create a gibberellin-responsive dominant dwarf DELLA allele. Plant *B*iotechnology *J*ournal 17, 132–140.
Ueta, R., Abe, C., Watanabe, T., Sugano, S.S., Ishihara, R., Ezura, H., Osakabe, Y. and Osakabe, K. (2017) Rapid breeding of parthenocarpic tomato plants using CRISPR/Cas9. Sci*entific Reports* 7, 507.
Valverde, F., Mouradov, A., Soppe, W., Ravenscroft, D., Samach, A. and Coupland, G. (2004) Photoreceptor regulation of CONSTANS protein in photoperiodic flowering. *Science* 303, 1003–1006.
Wallace, T.C. and Giusti, M.M. (2013) Anthocyanins in *H*ealth and *D*isease. CRC Press, New York, USA.
Wang, Y., Maruhnich, S., Mageroy, M., Justice, J. and Folta, K. (2013) Phototropin 1 and cryptochrome action in response to green light in combination with other wavelengths. *Planta* 237, 225–237.
Wu, G., Park, M.Y., Conway, S.R., Wang, J.-W., Weigel, D. and Poethig, R.S. (2009) The sequential action of miR156 and miR172 regulates developmental timing in Arabidopsis. *Cell* 138, 750–759.

Xiang, N., Hu, J., Zhang, B., Cheng, Y., Wang, S. and Guo, X. (2022) Effect of light qualities on volatiles metabolism in maize (Zea mays L.) sprouts. Food Research International 156, 111340.

Yang, T., Deng, L., Zhao, W., Zhang, R., Jiang, H., Ye, Z., Li, C.-B. and Li, C. (2019) Rapid breeding of pink-fruited tomato hybrids using the CRISPR/Cas9 system. Journal of Genetics and Genomics 46, 505–508.

Zhang, M. and Runkle, E.S. (2019) Regulating flowering and extension growth of poinsettia using red and far-red light-emitting diodes for end-of-day lighting. *HortScience* 54, 323–327.

Zhang, T., Maruhnich, S.A. and Folta, K.M. (2011) Green light induces shade avoidance symptoms. Plant Physiology 157, 1528–1536

Zhou, B., Wang, Y., Zhan, Y., Li, Y. and Kawabata, S. (2013) Chalcone synthase family genes have redundant roles in anthocyanin biosynthesis and in response to blue/UV-A light in turnip (Brassica rapa; Brassicaceae). *American Journal of Botany* 100, 2458–2467.

11 Critical Requirements to Implementing Vertical Farming Technologies

The Challenge to Reduce Cost

Jorge Flores Velazquez

INTRODUCTION

The agricultural sector in Mexico consumes more than 70% of the available water; this sector is one of the most affected by the water crisis, resulting in the need to maintain crop production with the least use of resources. As a result, crop production systems are migrating; it is increasingly common to see intensive agricultural techniques in urban areas competing for resources. The factors that condition agricultural production include climate, soil, and water, and when crops grow under protected or urban conditions, it is necessary to include energy and CO_2.

Successful crop production consists of integrating factors intrinsic to its development and being able to manage them, as is the case in protected agriculture (PA), urban agriculture (UrbanAg), and indoor crops. Technology keeps evolving, and agricultural applications provide other ways of growing plants in urban areas, for example, green walls and roofs, microgreens, indoor crops, layered crops, and plant factories.

Agricultural production in Mexico is vulnerable to environmental disturbances since only 20% of the country's land area is arable. Reasons for not expanding the arable area include semi-arid climate, minimal annual rainfall, topography, and contaminated soils. Rain season changes can result in drought threats and often in disasters for water-dependent sectors (Hernández, Torres, and Valdez, 2000).

In Mexico, low-technology biosystems, being 100% dependent on the external environment, require greater control by producers to maintain a suitable growing environment. Applying advanced technologies can help achieve better results and increase management and operation needs, highlighting energy consumption, water, climate control, space, and agronomic management.

Adverse climate conditions, reduction of resources to carry out agricultural activities, and use of fossil fuels have led to new techniques to grow crops in populated areas, contributing to the generation of fresh food all year round. From this perspective, contemporary cities should migrate to smart cities by considering agricultural production and energy balance. Increased pollution of agricultural areas and the reduction of cropland and spaces for flora and fauna, along with population migration that has increased in urban areas, have led to an imbalance of resources. To attenuate this imbalance, the ecological flow from the watersheds to the cities is beginning to be evaluated and regulated (Gondor et al., 2015).

The total population in Latin America is estimated to exceed 658 million inhabitants, mainly in urban areas (UN, 2023). The evolution of integrated water resources management has become essential in cities, exposing water–energy–CO_2 interaction. It is estimated that global energy demand will increase by 35%. Energy use in residential and commercial buildings accounts for approximately 40% of total energy consumption and 36% of total CO_2 emissions. Therefore, buildings can become one of the main spaces in the race for better energy use and against climate change (Sánchez, 2015; Benke & Tomkins, 2017; Al-Kodmany, 2018).

DOI: 10.1201/b23309-11

Integrated water resources management is a process that promotes cooperative management and development of water, soil, and other resources to enhance their use "without compromising the sustainability of vital ecosystems." The metropolitan area of Mexico City is home to more than 22 million people concentrated in about 8,000 km^2. The urbanization process, initiated by industrialization and sociocultural modernization, was some productive and social stagnation of the countryside and the acute inequity in resource access. Consequently, the current deterioration of the rural environment as a lifestyle IS ONE cause of the exponential migration (CELADE, 2012; ONU, 2018).

UrbanAg, plant factories, vertical farms, green walls, and green roofs are some technologies that have emerged and exponentially positioned themselves as agricultural production alternatives. Urban agriculture is driven by the desire to reconnect food production and consumption (Deelstra & Girardet, 2000; Beacham et al., 2019). The spread of urban agriculture reflects a growing awareness of how food and agriculture can shape cities from a production perspective. Cities are the energy and water drain that has placed them in a position of collapse.

Supplying urban areas with energy, water, and food is highly dependent on external resource flows to cities: "Cities account for only 2 percent of the world's surface but consume 75 percent of its resources"[1]. Access to services and food and improving the quality of life establish resource management conditions in cities that need to be updated; this perspective will determine the economic growth and sustainability of resources that will favor social prosperity (Bellezoni et al., 2021; Dubey et al., 2021; Zhong et al., 2021).

Vertical systems for plant production and horticultural and ornamental production systems were installed in an urban production model. Critical aspects of the system, interaction between consumption of resources, mainly water and energy, and climate conditions and their influence on yields of outdoor/indoor crops were analyzed. Based on this study, critical factors are exposed, such as artificial lights, water, and energy and, in general, those that allow establishing a vertical farm production system to grow plants regardless of the geographical location and local climate in Mexico (de Anda and Shear, 2017, Artmann and Sartison, 2018, Fuentes et al., 2022; Flores-Velazquez et al., 2020; Flores-Velazquez & Fuentes, 2021).

URBAN AGRICULTURE IN MEXICO

The limited arable land in Mexico has not encouraged the conversion of techniques and crops since traditional extensive production systems are still in use. This fact is becoming more important due to city population growth and competition for natural and non-natural resources, such as soil, water, and energy. Despite these conditions, today, it is difficult to find technologies that are producing and addressing this problem, as has occurred in various latitudes where they have developed and implemented production systems in urban agriculture, vertical farms, or plant factories (Specht et al., 2014; Sill & Serbin, 2018).

"The modern city has been built against the countryside, generating the current urban-rural antagonism, and this is one of the main keys to urban unsustainability" (Soler and Renting, 2013).

An advantage of UrbanAg is the availability of technologies to grow and harvest products in the same space and to have them available almost immediately. With UrbanAg, the incipient Mexican artificial forms of agricultural production emerged, which have reached a peak in UrbanAg (Ahmed et al., 2020). Furthermore, consumption quality is beginning to make a difference, with products grown free of chemical agents, except for fertilization.

VERTICAL FARM AND INDOOR CROPS

The evolution of agriculture is constant; the concept of plant factories is the production of crops with the least "waste" of resources (Quan et al., 2018). This is accomplished with specific information on the plant physiology and, in consequence, punctual management of the plant (Figure 11.1).

FIGURE 11.1 Plant factory outdoor/indoor system.

FIGURE 11.2 Proposed model for vertical urban utility (MUVU).

Mexico is already late to venture into urban agriculture in its different modalities. Even when the arable surface and populated cities become critical situations, crop production focuses on the traditional production of crops in a traditional way. Not even the production of crops in greenhouses had an important participation at the national level.

In Mexico, two sectors are dabbling in urban agricultural production technologies: the company and research institutions. Companies are initially introducing devices for growing in urban areas, from the hard part, such as shelves and containers, to production kits that can include monitoring systems or turnkey projects.

As for the universities, yes, we are working with models adapted from transnational companies, generating developments, especially in the monitoring and control of the production process. Sensors and controls for the operation and management of plants in indoor conditions and programs allow the monitoring and follow-up of the crop cycle, mainly with free software.

An alternative for the efficient use of resources in horticultural production is part of the concept of a plant factory with artificial light or vertical farms. These technologies can be considered to develop outdoors and indoors and thus contribute to a circular economy that includes drainage flow recirculation in the crop production in urban areas.

A vertical system can have different uses, horticultural, environmental, aesthetic, and other purposes. A plant factory can also be used indoors and outdoors with the respective adaptations according to the cultivated species' requirements. This study describes the experience developed by implanting horticultural crops in two vertical systems, measuring the minimum variables required to complete their cycle (Figure 11.2).

Technological advances to favor alternatives in horticultural production under the paradigm "reduce, reuse, recirculate" derived from the so-called circular economy is envisioned with practical application in urban agriculture (Lett, 2014). Zero drainage with hydroponic systems and actions aimed at improving indoor conditions and crop quality, such as the use of artificial light sources or LEDs (Folta & Childers, 2008), has been successfully introduced.

LED light has recently become the best alternative for plant growth (Massa et al., 2008) due to the advantages that this lighting system offers such as the control of the spectral composition, its small size, high light output with a low heat radiation index, and a long useful life to keep working for years without replacement (Bourget, 2008). Plants under LED lighting transpire less because of heat reduction, and therefore the time between watering cycles is longer.

Cultivation techniques such as hydroponics and indoor crops, in general, have significant advantages in terms of the efficient use of resources, especially water. However, the greatest achievement, even above the efficient use of water, is the reuse of water, which reduces by 80% the drainage in which, in addition to water, residues of the nutrient solution that contaminate the output of the production system, riverbeds, and water bodies where they are stored concentrate, increasing salinization and degradation of the environment (Yan et al., 2022).

The actions to mitigate the economic–social–ecological imbalance can be considered the most important due to the impacts on water and soil. The objective of this work is to document the minimum requirements to implement vertical farms (Figure 11.2), indoor and outdoor, with an H-PFAL of our own manufacture, and thus contribute to a circular economy that includes the ecological flow to grow crops in urban areas.

CRITICAL REQUIREMENTS

Under indoor conditions, it is possible to identify critical factors in the crop production process, such as crop management, being able to provide the plant with water and nutrition or fertigation altogether. Other factors are the environmental conditions, mainly temperature, relative humidity, and ventilation, which are also related to the CO_2 concentration. But the factor that defines the difference between indoor crops and outdoor crops is the light. Beginning from solar radiation, open-field crops obtain the energy in specific longwaves that can photosynthesize and be in comfortable conditions. Indoor crops need artificial light and auxiliary systems (if so) to obtain environmental comfort.

ENERGETICS: ARTIFICIAL LIGHTS

Light is the source of radiant energy used for the process of photosynthesis. Plants use photosynthetically active radiation (PAR), which is in the range between 400 and 700 nm. Therefore, the efficiency of photosynthesis is influenced by light quality or light wavelength, light duration, and intensity (Nhut et al., 2002).

The critical aspect of indoor crop production is light (radiation). Light is the source of radiant energy used for the process of photosynthesis. Plants use PAR, which is in the range between 400 and 700 nm. This region of the spectrum roughly corresponds with the visible light that is captured by the human eye. Therefore, the efficiency of photosynthesis is influenced by light quality or light wavelength, light duration, and intensity (Nhut et al., 2002). A beam of light is a collection of these particles called photons. Photons that correspond to longer wavelengths (lower frequencies) have the least energy.

The different frequencies or wavelengths are known by the colors that are transformed in the human brain for the different frequencies of visible light. Thus, the higher frequencies (shorter wavelengths) correspond to violet light, and the lower frequencies (longer wavelengths) correspond to red light. One of the actions aimed at improving crop quality has been to conduct trials with LED light sources (Folta & Childers, 2008). These trials have mostly been undertaken with horticultural products (Wojciechowska et al., 2015).

Application for Indoor Crops

Photosynthesis is the process that allows the plant to produce its own food and can occur in two phases (with and without light). The decomposition of the radiation spectrum affects the development and growth of plant species according to the photosensitive organs. Plants are classified according to their photoperiod into C3, C4, and CAM plants, each with its own physiology. They can respond differently to the intensity and color of light (wavelength) by means of their photoreceptors: phytochromes, cryptochromes, and phototropins. Each of these organelles is activated with specific wavelengths, activating specific processes of their metabolism, which, together with environmental conditions, result in crop vigor (Chen et al., 2004). When crops are not able to receive sunlight, artificial light through lighting systems is essential (Bourget, 2008; Cabrera, 2017).

Crop production under indoor conditions has made it possible to provide the plant requirements in an artificial way, for instance, by applying specific LED-type artificial light with intensity and quality wavelengths. Even more, each wavelength is responsible for physiological function in the crops. So, the combination of different wavelengths with specialized lamps is conducted to achieve specific crop characteristics (Paniagua-Pardo et al., 2015; Rizzo, 2020).

In crop production in general but indoor crops in specific, each plant species reacts differently to the spectral components of the luminous flux (intensity and quality light). Under indoor crops, the main issue is the selection of lamps, which can be incandescent lamps, mercury vapor lamps, fluorescent lamps, metal halide lamps, and LED lamps (Ramos, 2015). The best selection depends on bright, long-lasting light, emitting only the wavelengths of light corresponding to the absorption peaks of the typical photochemical processes in a plant. An important aspect of LED lamps is that all the light generated is focused on the LED front side, generating direct light on the plant organs and attenuating their effectiveness.

Lighting Requirements in Horticultural Production

An indoor production system involves providing the specific wavelengths that the plant needs. Lamp efficiency reduces production costs by increasing yields, but energy costs should also be considered. From the agricultural point of view, three aspects should be highlighted: (1) knowing the light requirements according to the wavelength of the plant, (2) combining natural light with artificial light, if possible, and (3) managing the photoperiod between light and dark times according to the stages of growth and flowering.

Two phases carry out the process of photosynthesis, one with the presence of light and the other one that can occur regardless of whether there is light or not. Since the lights are artificial in indoor vertical farming, it would be possible to keep the lights on during the whole plant's cycle; however, another factor to regulate is its photoperiod. The photoperiod defines the number of light/dark hours that the plant requires to complete subcycles, mainly flowering. Due to this, it is necessary to regulate the exposure time (hours) of artificial lights.

Plants with a short cycle (chrysanthemum and rice) of less than 8 hours, long (lettuce, spinach, and radish) of more than 16 hours, and those that are indifferent to periods, like tomato. Tomato is a crop insensitive to the photoperiod, usually between 6 and 16 hours for light. The importance of knowing and providing these periods of light to the plant is that it is capable of changing phases, for example, from a vegetative period to flowering, which consequently affects some fruit. In addition, even if some horticulture species have long or short photoperiods, this range can be different depending on the age.

The LED function is to emit light of a specific wavelength to properly stimulate plant growth. The power consumption of an LED light is among the best for lamps, which allows them to be integrated into alternative energy systems for operation, such as solar and wind power systems, and in digital control systems. Monitoring and changing light according to the phenological stage of the crop is necessary, which can be achieved with LED lights.

Red and blue light are wavelengths used for plant growth; phytochromes have their sensitivity peaks in the red (660 nm) and infrared (730 nm) regions, and among the physiological responses where they are involved, we find leaf expansion, perception of neighbors (Figure 11.2), shade avoidance, stem elongation, seed germination, and induction of flowering (Pinho et al., 2012).

SPACE REQUIREMENTS

Mexico City lacks enough space to create large open urban gardens. Establishing a green corridor implies the demolition of structures or the closure of roadways, which increases traffic problems. One of the alternatives to producing food in the city is the creation of vertical farms. A vertical farm or a green wall can be adapted in size and shape to the house, apartment, or living space.

To visualize this, a steel structure is built to act as the wall where the vertical farm will take place (Figure 11.3). This structure can contain horticultural, medicinal, and ornamental plants. The implementation feasibility should be considered by balancing the requirements of the plants and the possibility of providing water in the roots, mineral nutritive solution, and an adequate level of oxygen, either naturally, artificially, or combined. The weight that can be considered depends on the crop established; in this example PVC pipe is used, so the weight will depend on the pipe support (water, substrate, soil, plant) and the length allowed by the wall. For instance, if PVC is used as a support of crop media (water or substrate) and the crops, it is possible to estimate the total weight: PVC 4″ diameter RD26 support 11 kg/cm^2, weight of pipe (6 m long) 4 kg. The water in the pipe is about 47 L (47 kg), and if the crop is lettuce, in 6 m, 40 plants × 300 g are 12 kg. Total weight 4 kg (PVC) + 47 kg (water volume) + 12 kg (40 lettuce) are 63 kg by 6 m hydroponic system.

Figure 11.3 shows a vertical system for growing small horticultural species (mainly salads, aromatic herbs, and spices). This model of urban vertical utility (MUVU) is built with metal with dimensions of 6 × 3 m. The MUVU was built to support the cultivation system, fertigation, and drainage because it is intended for an outdoor area of the house or a patio.

One of the advantages of urban horticulture is the availability of technologies to grow and harvest products in the same space and to be able to have them available immediately (life cycle). With PA, artificial forms of agricultural production emerged, which have reached a rise in the efficiency in the use of natural resources and accelerated the development of urban agriculture (Dieleman, 2017). Moreover, an aspect that has been making a difference is the quality of consumption, with products grown wholly or partially free of chemicals.

The central idea of MUVU is to make everything vertical, i.e., to change the horizontal plot for the vertical wall of a building. For this purpose, an iron structure was built to support this subsystem. The growing medium was the nutrient solution and substrate, a mixture of perlite, coconut fiber, and soil.

FIGURE 11.3 Metal exo-structure 6 × 3 m to support 5 PVC hydroponic systems and substrate.

The subsystems implemented were the following:

1. Vertical nutrient solution system in a 38 mm diameter with "yees" for the plants.
2. Horizontal hydroponic nutrient solution system in 2″ pipe and "tees" for the plants.
3. Vertical farm system, 4″ PVC perforated pipe with a substrate (towel type).
4. Modular system with substrate and membranes to contain the substrate (Figure 11.4).

Three subsystems were implemented in the pilot system tested: floating system, recirculation, and substrate. A tray with water and solution was placed for the floating system and was manually oxygenated every day. A total of 18 lettuces were grown in 0.5 m^2. Although the water level was maintained, it was complicated to estimate the consumption accurately due to singularities such as the rain regime.

The consumption of the recirculation systems was like the floating system, but due to the recirculation, no water oxygenation was required. The vertical farm with lettuce and basil and the square module with strawberries were irrigated four times a day, for 1 minute each, corresponding to 1 L a day (Table 11.1).

However, suppose this model is adapted to the wall. In that case, it can be separated by subsystems and each one could be 1 m^2 vertically, which can be in nutrient solution or in a substrate, which can contain ornamental or cultivated plants such as strawberries (Figure 11.5).

FIGURE 11.4 Characteristics of a completed model of urban vertical utility (MUVU) of 6×3 m.

TABLE 11.1
Differences in Cultivation Techniques, Cost, and Planting Density

Kind of System	Density (pl/m²)	Cost of System (MX$/USD)	Water Consumption (L/m²)
Hydroponic tees	36	1,900/105	1.2
Hydroponic tees	48	1,200/70	1.2
Tower substrate	108	900/50	1.2
Module substrate	24	800/45	1.2
Floating root	33	300/20	1.2
Home plant factory with artificial light (H-PFAL) (Figure 11.3)			1.1

FIGURE 11.5 Vertical farm system in a substrate to grow strawberries.

FIGURE 11.6 Four-layer home plant factory with artificial light (H-PFAL) outdoor.

When installed on a wall, the same auxiliary subsystems should be considered for the subsistence of the plants, water, and nutrients, and, if there are outlets, a drainage system will be necessary, for either recirculation or storage. In addition, it will be necessary to consider the orientation of the house to ensure that the required solar radiation is incident on the crop for photosynthesis. Carbon dioxide requirements and climate factors will be the environmental conditions.

Trends in horticulture continue evolving (Touliatos et al., 2016, Muller et al., 2017). Technological tools, using microsensors, controllers, actuators, artificial lights, etc., are available to facilitate horticultural production in this type of biosystem. Some results obtained from research on greenhouse production are being applied in urban agriculture, like climate control and physical simulation processes, by using numerical tools, all of them to make horticultural production economical and management accessible.

In the case of a layered biosystem (Figure 11.6), whether indoors or outdoors, a steel structure can be used, and the crop can be developed using a floating system.

The home plant factory with artificial light (H-PFAL) model is based on a steel structure with three layers, each containing a container that serves as a "micro-sink" to grow mainly low-growing vegetables. Here we chose to grow lettuce of two varieties.

FIGURE 11.7 (a) Plant factory indoors with artificial light and (b) outdoors without artificial light, in both nutrient solution with phenolic foam to separate the plants from the water.

The dimensions of the H-PFAL were considered so that this structure can coexist in a shared space in a house, for example, a kitchen, where three or four containers with the nutrient solution can be placed for the feeding and development of the lettuce. Figure 11.7 shows the arrangement of the plants just after transplanting (Figure 11.7a) in the steel structure both indoors with artificial light and outdoors with natural light or irradiation (Figure 11.7b)

ARTIFICIAL SYSTEMS AT H-PFAL (INDOOR/OUTDOOR)

A vertical farm prototype of a four-level rack (L×W×H, 40.4×90.4×183.6cm) was built. The H-PFAL has three basic systems to favor the crop cycle: (1) artificial light with LED lamps, (2) nutrient solution oxygenation system, and (3) a ventilation system to remove air in the growing area.

The hydroponic system was a floating root composed of 39-L plastic tanks with 1″ polystyrene plates and 2″ hydroponic baskets. For the oxygenation of the water, a pump like the ones used for fish tanks was used to remove the water through a tube, favoring water oxygenation. The aeration system comprised 80L air pumps and a flexible translucent silicone hose with air diffusers. Each level had 6″ 117 alternating current voltage (V_{ca}) fans with 14 W power (Steren, Mexico) to extract the hot air from the LED lamps. Three commercial lamps in Mexico were used to provide artificial light; the intensities were 1,700, 5,500, and 6,500 lux.

CLIMATE REQUIREMENTS

Growing plants have specific and known climate requirements; the main variables that have an impact on production are radiation (W/m^2), temperature (°C), relative humidity (%), and wind (direction [m/s] and magnitude [°]). In open-field conditions, it is not easy to adapt them, and they are conditioned to the external environment, and only fertilization and agronomic management must be provided. To keep track, there are primary low-cost weather stations that allow monitoring of environmental variables with adequate precision for managing the biosystem.

In this project, a weather station was installed, which included sensors to measure temperature, relative humidity, wind, and precipitation (mm) (bucket pluviometer); this station was the meteorological support for the agronomic management of the crop (Figure 11.8).

For indoor systems, environmental requirements are provided by artificial lighting, temperature and humidity control, and nutrient solution monitoring. An essential aspect of this experiment was

FIGURE 11.8 Climate monitoring for MUVU system management.

to know that the range of the environmental variable values in a house is usually similar for humans and plants: 10–30°C temperature, 40–80% relative humidity, and wind speeds that do not exceed 1 m/s. Even so, a ventilation system consisting of an unbranded commercial fan was installed to remove the air volume around the crop and thus maintain a constant temperature.

During the period of production (March–May), a temperature sensor was provided for thermal monitoring under the study conditions (with no radiation). Thermal oscillation in this time was 3–5°C with an average temperature of a minimum of 16°C, a maximum of 33°C, and an average of 23°C. The monitoring indicates that temperature remains within the lettuce production range (−6–30°C). However, the optimum is no more than 20°C, so the fan removes the air and reduces the temperature at peak hours. The light requirements are provided by artificial light lamps, which emit radiation only in the wavelengths of the visible spectrum and not in the thermal spectrum (León-Pacheco et al., 2022).

WATER REQUIREMENTS

The conceptual basis for estimating crop water requirements was developed in the middle of the last century and is designed for open-field crops (FAO, 1976, 2022). However, they are the basis for estimating irrigation requirements in urban gardens and inferring water needs vertically. From empirical methods, it is possible to determine evapotranspiration from climatic variables and, thus, irrigation scheduling when appropriate. For indoor crops, closed circuits are commonly used; in this case, the plant's water consumption can be derived from a balanced method.

Crop production in urban areas is worsening with the water and resource crisis in general. In scenarios of competition for resources, the water–energy–oxygen nexus requires maximizing its efficiency to sustain crop production and make this process sustainable. For gardens, irrigation requirements can be estimated horizontally by calculating the evapotranspiration of plants in gardens (ET_j), which is affected by a garden coefficient, instead of the crop coefficient (Costello, 2000) using equations 11.1 and 11.2:

$$ET_j = K_j \times ET_0 \tag{11.1}$$

where

ET_j represents the evapotranspiration of the garden.
K_j is the garden coefficient.
ET_0 is the reference evapotranspiration.

$$K_j = K_c * K_d * K_m \quad (11.2)$$

where

K_c is the coefficient of the species in the garden.

K_d is the density of plants in the garden.

K_m is the microclimate conditions of the garden (Table 11.2).

Techniques that minimize the use of water and drainage have been developed in closed systems in general and especially in urban agriculture. Among the techniques used, we can mention pure hydroponics, i.e., crops grown completely in nutritive solution, which can be closed or open circuits. Substrates are available commercially and can be mixed with regional products such as volcanic stones (Tezontle) or crop waste with coconut fiber or rice hull. The vertical system built included five hydroponic subsystems with nutritive solution and substrate.

Two systems were installed for water fertilization: a closed circuit with recirculation activated with a solar submersible pump (Figure 11.9) and an open system with five irrigations per day that provided the combined requirement with the nutritive solution to both the system in solution (lettuce and basil) and the system with substrate (strawberry). Regarding the floating subsystem, monitoring should be done according to its salt content; preferably, two variables can be used: hydrogen potential (pH) and electrical conductivity (EC). In this experiment, the water with the stagnant nutrient solution was replaced manually every week.

Irrigation programming was carried out using a commercial controller that turned the pump on and off at different times, thus providing the required daily irrigation amount.

TABLE 11.2
Species, Density, and Microclimate Coefficients for Gardens

Coefficient	High	Medium	Low	Very Low
Species factor (Ke)	0.7–0.9	0.4–0.6	0.1–0.3	<0.1
Density (Kd)	1.1–1.3	1.0	0.5–0.9	
Microclimate (km)	1.1–1.4	1.0	0.5–0.9	

Source: Costello (2000).

FIGURE 11.9 Irrigation in the MUVU, horizontal, and submerged pump.

WATER AND NUTRIENT REQUIREMENTS

Plant nutrition consists of three kinds: (1) gaseous (CO_2), (2) water, and (3) mineral. A plant is composed of about 96% of its weight of C+H+O; the remaining 4% are the mineral components that need to be supplied through irrigation or direct nutrient solution. Each plant has its own requirements on the specific subject of mineral nutrition. However, a practical way to supply the plant is to use microelements (N–P–K) and microelements. The nutrient solution is prepared with these elements, which can be delivered to the crop through irrigation or in the medium if it is a solution culture.

The nutrient solution was prepared in the established designs, and the mixture (macro and micro) was applied directly to the pond container, a solution containing the macro elements with the major elements and a solution with the microelements with minor elements.

For the first 10 days of sowing, 2.5 mL of macro- and microelements were used for each liter of water; after 10 days, 5.0 mm of the same solutions were used. After transplanting, the nutrient solution was replaced every 10 days, and, with this action, the water consumption of the plants was determined, acting as a balanced lysimeter. Since there is practically no evaporation, the difference in water levels in the basin corresponds to the loss of water by transpiration.

According to the characteristics of the container, it can play the role of a lysimeter or an evaporator tank (Figure 11.10). A known volume is added, and, every 10 days, the chemical characteristics of the solution, pH and EC, are measured and periodically adjusted. The pH was kept constant at 6.5, and EC between 1.5 and 1.6 ds/m was achieved with the nutrient solution composed of a mixture of commercial fertilizers of macronutrients (N–P–K) and micronutrients.

The volume difference consists of plant transpiration. The sum of the total volume changes was the irrigation volume in the average cycle of 54 days after transplanting.

The results indicated a water consumption per cycle of 50 L/m^2 for the maximum yield treatment (900 g/m^2); consequently, if the yield is lower, one of the causes can be assumed to be a lower use of water to transform dry matter through photosynthesis.

FINAL COMMENTS

Activities were conducted to establish value parameters to introduce horticultural production systems in urban areas. These elements were approached from a theoretical and historical point of view and through the implementation of indoor and outdoor pilot systems like those established commercially, with the challenge of making them cheaply.

There are large-scale systems with national projects in countries such as the United States, Singapore, Holland, and Japan. However, small-scale commercial systems have also been developed

FIGURE 11.10 Root development of lettuce at a mature stage.

to produce vegetables, salads, and gourmet vegetables for a market that prefers fresh and non-organic products.

In Mexico, the development of these biosystems is instead developing, perhaps in universities and research centers, but not widespread in urban areas, much less in rural areas. Among the hypotheses that can be proposed is the lack of promotion of technology, government support in the productive process, and perhaps the fit climate conditions, such as radiation and space, despite some large cities in the country and their resource consumption.

A vertical farm (MUVU) was implemented, which includes five subsystems with vertical production, combining position, cultivation medium, and cultivation method. Such subsystems were vertical hydroponics with yess and horizontal with tees, hydroponics on a floating system, and substrate, one in a tower vertical farm and the other in a module with an exoskeleton of ironwork.

The five systems were suitable for growing strawberries, lettuce, and basil in the climate area of Jiutepec, Morelos. The results show that a vertical system can grow crops in general houses, apartments, and buildings. The disadvantages of this type of technology depend on the degree of technification. For rustic subsystems, a minimum knowledge of the production process is necessary. As technology becomes more advanced, other skills, such as the Internet of things and climate control, are required in addition to cultivation.

Moreover, a home plant factory prototype was built, using a nutrient solution in four layers to produce lettuce. The results indicate a sustained production of these plants, at a low cost, basically fertilizers, because more than 80% of the water is recycled or it is feasible to harvest from the rooftops, no soil is needed, no weeds, it is a viable system outdoors in urban areas.

Urban agricultural production technology is suitable to cover part of the food needs demanded by the population, and it is helpful to reduce environmental pollution by reducing fuel consumption when producing and distributing food in cities where the bulk of the population is located.

Water, carbon dioxide, and light are the basic requirements for plant development; coupled with mineral fertilization, agronomic management, and environmental control, the process is optimized to maximize high-quality yields. The cost of fertilizers for the nutrient solution is one of those that have the most negligible impact on the cost of production.

In urban agriculture, the space factor can be integrated, and, with technological advances, crop monitoring and tracking using interconnected sensors and actuators can also be considered a factor in the production process.

Regarding water use, regardless of whether it is a closed or semi-closed circuit, typically hydroponic, there is a reduction in water consumption of more than 80%. In addition, the "quasi" null drainage is another factor of environmental savings.

Because it is required to integrate subsystems of environmental control, irrigation, and drainage, the design of the space must be considered in any production system in urban agriculture. The best design means savings in implementing the system by blending it into walls and using the same services as the house.

The energy is used to feed the subsystems such as irrigation and, if applicable, climate control, as well as the supply of artificial lights according to the required amount, frequency, and quality.

Carbon dioxide is necessary both in the nutrient solution and in the environment to support the process of photosynthesis. The CO_2 supply can be carried out by carbonic injection in the irrigation system or through ventilation systems.

Regarding the cost of implementing the biosystem, installing climate control subsystems and their operation during the cycle have the most significant economic impact on the production process. In a closed system, it is necessary to maintain the temperature, humidity, carbon dioxide, and radiation (lights) within the optimal ranges for the crop, which is why it is essential to consider the design of the space. Water for crop physiology and nutrition are the lowest cost factors. However, transpiration is considerable in the energy balance.

REFERENCES

Ahmed, H.A., Yu-xin, T., Qi-chang, Y. 2020. Lettuce plant growth and tipburn occurrence as affected by airflow using a multi-fan system in a plant factory with artificial light. *J. Therm. Biol.* 88: 102496. https://doi.org/10.1016/j.jtherbio.2019.102496.

Al-Kodmany, K. 2018. The vertical farm: A review of developments and implications for the vertical city. *Buildings* 8: 24. https://doi.org/10.3390/buildings8020024.

Artmann, M., Sartison, K. 2018. The role of urban agriculture as a nature-based solution: A review for developing a systemic assessment framework. *Sustain.* 10(6): 1937. https://doi.org/10.3390/su10061937.

Beacham, A.M., Vickers, L.H., Monaghan, J.M. 2019. Vertical farming, a summary of approaches to growing skywards. *J. Hortic. Sci. Biotechnol.* 94(3): 277–283. https://doi.org/10.1080/14620316.2019.1574214.

Benke, K., Tomkins, B. 2017. Future food-production systems: Vertical farming and controlled-environment agriculture. *Sustain.: Sci. Pract. Policy* 13: 13–26.

Bellezoni, A.R., Meng, F., He, P., Seto, K.C. 2021. Understanding and conceptualizing how urban green and blue infrastructure affects the food, water, and energy nexus: A synthesis of the literature. *J. Clean. Prod.* 289: 125825, https://doi.org/10.1016/j.jclepro.2021.125825.

Bourget, C.M. 2008. An introduction to light-emitting diodes. *HortScience* 43(7): 1944–1946. https://doi.org/10.21273/HORTSCI.43.7.1944.

Cabrera, V.E. 2017. *Efecto de la iluminación (tubos fluorescentes o LEDs) de las cámaras de cultivo sobre el cultivo in vitro de plántulas de* Medicago arborea L. Madrid, España: Universidad Complutense.

Centro Latinoamericano y Caribeño de Demografía CELADE. 2012. World population and Latin America and the Caribbean population: Changes and new (im) balances. Astrolabio, no. 8. https://mexico.un.org/es/228596-estado-de-la-poblaci%C3%B3n-mundial-2023-8000-millones-de-vidas-infinitas-posibilidades.

Chen, M., Chory, J., Fankhauser, C. 2004. Light signal transduction in higher plants. *Ann. Rev. Genet.* 38: 87–117. https://doi.org/10.1146/annurev.genet.38.072902.092259.

Costello, L.R., Mtheny, N.P.Y., Clarck, J.R. 2000. *A Guide to Estimating Irrigation Water Needs of Landscape Plantings in California. Part 1: The Land Scape Coefficient Method. Part 2: WUCOLS III.* University of California Cooperative Extension, Sacramento, California.

de Anda, J., Shear, H. 2017. Potential of vertical hydroponic agriculture in Mexico. *Sustain.* 9. https://doi.org/10.3390/su9010140.

Deelstra, T., Girardet, H. 2000. Urban agriculture and sustainable cities. In N. Bakker, M. Dubbeling, S. Gundel, U. Sabel-Koshella, H. de Zeeuw (Eds.), *Growing Cities, Growing Food: Urban Agriculture on the Policy Agenda* (pp. 43–66). Feldafing: ZEL.

Dieleman, H. 2017. Urban agriculture in Mexico City; balancing between ecological, economic, social and symbolic value. *J. Clean. Prod.* 163(1): S156–S163.

Dubey, Pradeep, Singh, Ajeet, Raghubanshi, Apoorva, Abhilash, Purushothaman 2021. Steering the restoration of degraded agroecosystems during the United Nations Decade on Ecosystem Restoration. *Journal of Environmental Management*, 280, 111798. https://doi.org/10.1016/j.jenvman.2020.111798

FAO. Doorenbos & Pruitt. 1976. *Las necesidades de agua de los cultivos.* Rome, Italy: FAO MANUAL 56.

FAO. Rikolto & RUAF. 2022. *Urban and Periurban Agriculture Sourcebook-From Production to Food Systems.* Rome: FAO and Rikolto. p. 156. https://doi.org/10.4060/cb9722en.

Flores-Velazquez, J., Rodríguez, A.C.E, Gallardo, A.P. 2020. Hydroponics vertical farm as a viable utility model to implantation in Mexico City. St. Joseph, MI: American Society of Agricultural and Biological Engineers. www.asabe.org. ASABE Annual International Virtual Meeting 2000346. https://doi.org/10.13031/aim.202000346.

Flores-Velazquez, J., Fuentes, D. 2021. Analysis of alternative LED light by indoor lettuce production in Mexico. St. Joseph, MI: American Society of Agricultural and Biological Engineers. www.asabe.org. Annual International Virtual Meeting 2100120. https://doi.org/10.13031/aim.202100120.

Folta, K.M., Childers, K.S. 2008. Light as a growth regulator: Controlling plant biology with narrow band width solid-state lighting systems. *HortScience* 43(7): 1957–1964. https://doi.org/10.21273/HORTSCI.43.7.1957.

Fuentes, D., Flores, J., Aguilar, A., Roblero, R. 2022. Response of LED lights intensity on lettuce production in a home vertical farm. *Rev. Fac. Agron. (LUZ)* 39(1): e223920. https://doi.org/10.47280/RevFacAgron(LUZ).v39.n1.20.

Girardet, H. 2000. *Cities, People, Planet. Urban Development and Climate Change.* London, UK: Wiley. ISBN: 978-0-470-77270-6. p. 328.

Gondor, A., Haney, J., Sanchez, J.A., Hesselbach, H. 2015. *Caudal ecológico: agua para la naturaleza e impactos en Monterrey VI.* México: The nature conservancy.

Hernández, C.M.E., Torres, T.L.A. and Valdez, M.G. 2000 Sequía Meteorológica, 25-40. En: C. Gay (comp.) México: Una visión hacia el siglo XXI. El cambio climático en México. Resultados de los estudios de la vulnerabilidad del país, coordinados por el INE con el apoyo del U.S. Country Studies Program, México: INE, SEMARNAP, UNAM, US Country Studies Program.

León-Pacheco, R.I., Jaramillo-Noreña, J.E., Montes-Pérez, M.L., Orozco-Guerrero, A.R., Carrascal-Pérez, F.F., Villagran Munar, E.A., Rodríguez-Roa, A.O. 2022. Evaluación agronómica y fisiológica de cinco cultivares de lechuga bajo dos sistemas de agricultura protegida en el departamento de Magdalena, Colombia. Avances En Investigación Agropecuaria 26(1): 79–93. https://doi.org/10.53897/RevAIA.22.26.06.

Lett, L.A. 2014. Las amenazas globales, el reciclaje de residuos y el concepto de economía circular. *Revista Argentina de Microbiología [en linea]* 46(1): 1–2. https://www.redalyc.org/articulo.oa?id=213030865001.

Massa, G.D., Kim, H.H., Wheeler, R.M., Mitchell, C.A. 2008. Plant productivity in response to LED lighting. *HortScience* 43(7): 1951–1956. https://doi.org/10.21273/HORTSCI.43.7.1951.

Muller, A., Ferré, M., Engel, S., Gattinger, A., Holzkämper, A., Huber, R., Müller, M., Six, J. 2017. Can soil-less crop production be a sustainable option for soil conservation and future agriculture? *Land Use Policy* 69: 102–105.

Nhut, D.T., Hong, L.T.A., Watanabe, H., Goi, M., Tanaka, M. 2002. Growth of banana plantlets cultured in vitro under red and blue light-emitting diode (LED) irradiation source. *Acta Hortic.* 575: 117–124. https://doi.org/10.17660/ActaHortic.2002.575.10.

Organización de las Naciones Unidas (ONU). 2018. World urbanization prospects 2018. https://population.un.org/wup/DataQuery/ (Review: December 2022).

Paniagua-Pardo, G., Hernández-Aguilar, C., Rico-Martínez, F., Domínguez-Pacheco, F.A., Martínez-Ortiz, E., Martínez-González, C. L. 2015. Efecto de la luz led de alta intensidad sobre la germinación y el crecimiento de plántulas de brócoli (brassica oleracea l.). *Polibotánica* (40): 199–212. https://www.redalyc.org/articulo.oa?id=62142251013.

Pinho, P., Jokinen, K., Halonen, L. 2012. Horticultural lighting - present and future challenges. *Lighting Res. Technol.* 44(4): 427–437. https://doi.org/10.1177/1477153511424986.

Quan, Q., Zhang, X., Xue, X.Z. 2018. Design and implementation of a closed-loop plant factory. *IFAC-PapersOnLine* 51: 353–358. https://doi.org/10.1016/j.ifacol.2018.08.203.

Ramos, G.Y.F. 2015. Diseño e Implementación de un Sistema de Control para Maximizar la Capacidad Productiva de las Plantas en Granjas Verticales por Medio de Luz Artificial. Universidad Autónoma de Occidente. Bachelor Thesis Santiago de Cali. https://hdl.handle.net/10614/8557.

Rizzo, Z.S.V. 2020. *Efecto de diferentes tipos de luz en el crecimiento de plantas in vitro: Revisión de Literatura.* Zamorano, Honduras: Bachelor Thesis. Escuela Agrícola Panamericana.

Sánchez Roelas, O. 2015. Caracterización de ventilación nocturna en edificios mediante técnicas CFD. Bachelor Thesis. España: Dep. Ingeniería Energética. Escuela Técnica Superior de Ingeniería. Universidad de Sevilla. https://hdl.handle.net/11441/38774.

Sill, C., Serbin, I. 2018. Vertical farming: A revolution to sustainable agriculture. University of Pittsburgh Swanson School of Engineering. https://sites.pitt.edu/~budny/papers/8203.pdf.

Soler, M.M., Renting, H. 2013. Agricultura Urbana: Prácticas emergentes para un nuevo urbanismo. *Hábitat Y Sociedad*, 6(6). https://doi.org/10.12795/HabitatySociedad.2013.i6.01.

Specht, K., Siebert, R., Hartmann, I., Freisinger, U.B., Sawicka, M., Werner, A., Thomaier, S., Henckel, D., Walk, H., Dierich, A. 2014. Urban agriculture of the future: An overview of sustainability aspects of food production in and on buildings. *Agric. Human Values* 31, 33–51. https://doi.org/10.1007/s10460-013-9448-4.

Touliatos, D., Dodd, I.C., McAinsh, M. 2016. Vertical farming increases lettuce yield per unit area compared to conventional horizontal hydroponics. *Food Energy Secur.* 5, 184–191.

Wojciechowska, R., Długosz-Grochowska, O., Kołton, A., Żupnik, M. 2015. Effects of LED supplemental lighting on yield and some quality parameters of lamb's lettuce grown in two winter cycles. *Sci. Hortic.* 187: 80–86. https://doi.org/10.1016/j.scienta.2015.03.006.

Yan, D., Liu, L., Liu, X., Zhang, M. 2022. Global trends in urban agriculture research: A pathway toward urban resilience and sustainability. *Land* 11(1): 117. https://doi.org/10.3390/land11010117.

Zhong, Qiumeng, Wang, Lan, Cui, Shenghui 2021. Urban Food Systems: A Bibliometric Review from 1991 to 2020. *Foods*, 10(3), 662. https://doi.org/10.3390/foods10030662

12 Energy Efficiency in Indoor Farming

A Malaysian Experience

Mohd Faizal Mohideen Batcha, Abdul Muin Shaari, Sulastri Sabudin, Akmal Nizam Mohammed, Muhd Akhtar Mohd Tahir, and Zulhazmi Sayuti

BACKGROUND

It is estimated that the global population will grow to 9.7 billion by 2050 with 68% living in urban areas, indicating an increase in population densities in cities around the globe. This population growth is proportional to food demand, resulting in food security becoming mankind's biggest challenge in the coming decades. Lands to grow food are becoming scarce due to industrialization, while global warming resulting from climate change is now regularly interrupting the food supply chain. These events impose a trilemma between three undesirable alternatives, i.e., unstable food supply, shortage of resources, and degradation of the environment (Kozai, 2018). This is illustrated in Figure 12.1.

In facing this trilemma, several solutions were proposed by researchers and scientists around the world, and among the successful solutions was the concept of vertical farming. First introduced by Dickson Despommier in 1999 (Despommier, 2010), vertical farming via hydroponics and aeroponics has become common in many developed nations today. The key to the success of vertical farms is the invention of an artificial lighting system that 'recreates' the sun. This enables a variety of successful designs of indoor farms, which can be considered climate change-proof and a healthier choice of pesticide-free crops.

Indoor farms usually employ soilless cultivation techniques such as hydroponics, aeroponics, or aquaponics, although indoor farms using soil are not uncommon. The most common system for indoor farming is a hydroponic system where the nutrient solution can be recycled. Some indoor farms also collect condensate water from the cooling coils of the air-conditioning system to be

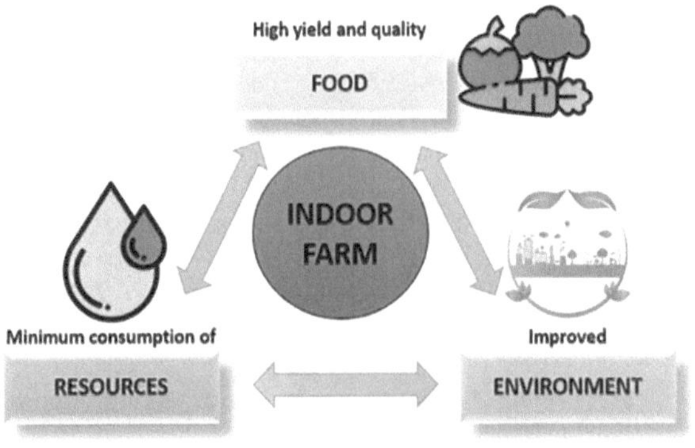

FIGURE 12.1 Trilemma between food, resources, and environment.

DOI: 10.1201/b23309-12

TABLE 12.1
Capital Costs (Perez, 2014)

Items	Costs	Remark
Building	180 M yen (31%)	New construction
Construction	110 M yen (19%)	Utility set up
Equipment and facilities	300 M yen (51%)	Nutrient Film Technique (NFT) systems, irrigation systems, lighting systems, others

TABLE 12.2
Annual Operating Costs (Perez, 2014)

Item	Costs	Remark
Salaries and wages	71.3 M yen (26%)	Two full-time workers + hourly laborers (210 hours/day, $10/h)
Materials	15.4 M yen (6%)	Packing materials, seeds, light bulbs, fertilizers, chemicals
Utilities	72.6 M yen (26%)	369 MW/month power use + water use
Transportation and shipping	6.0 M yen (2%)	
Other costs	49.2 M yen (18%)	Facility/equipment maintenance, consulting
Depreciation	59.2 M yen (22%)	

Shaded row shows the energy intensity which is the highest operational costs

reused. A typical indoor farm has vertical growing beds to maximize yield per unit area and has thermally well-insulated and airtight structures to minimize heat leaks and air infiltration.

However, the micro-climatic conditions required in an indoor farm, particularly artificial lighting, result in high capital expenditures, becoming the biggest challenge for small and medium-scale growers. Takeshima and Joshi (2019) quoted the analysis by Perez (2014) regarding the capital and operational costs given in Tables 12.1 and 12.2, indicating that utility costs, particularly electricity, can be as high as one-fourth of the total operating expenses.

A comprehensive study is, therefore, required in addressing this issue toward capitalizing the low-hanging fruits by incorporating energy efficiency aspects in indoor farms. In this chapter, a Malaysian experience is shared on energy efficiency studies carried out in two types of indoor farms, namely, the commercial-scale plant factory and container-type plant factory, both located in the Malaysian Agricultural Research and Development Institute (MARDI), Serdang, Selangor.

RELATED WORKS

Graamans et al. (2018) conducted a study comparing the resource efficiency of closed plant production systems, such as plant factories, and traditional greenhouses for lettuce production in different climates with varying solar radiation levels. The study used advanced climate models and a lettuce growth model to assess resource use efficiency. The results showed that plant factories outperformed greenhouses in terms of water, CO_2, and land use efficiency but required more purchased energy for photosynthesis. Plant factories also had higher productivity compared to greenhouses. However, the economic viability of plant factories may be affected in the absence of solar energy.

Harbick et al. (2016) found that plant factories with artificial lighting (PFALs) consume more energy and have a higher carbon footprint than greenhouses. This is because PFALs are highly controlled environments, whereas greenhouses are more susceptible to external environmental changes. The study modeled the energy consumption of both types of facilities in four different U.S. climates and found that PFALs consumed more energy and had higher carbon footprints in all of the climates considered.

In another study by Graamans et al. (2020), the authors focused on reducing the high energy demands of plant factories through efficient façade design. The study analyzed the effects of façade properties on the energy use of plant factories in different climate zones, including Sweden, the Netherlands, and the United Arab Emirates. The study found that opaque façades with high U-values and optimized albedo could reduce the cooling demand and overall energy demand of plant factories. In contrast, transparent façades are more efficient in terms of electricity use since they allow the use of solar energy. The study concluded that these façade design strategies could significantly reduce energy demand in plant factories and provide a foundation for energy-efficient designs of high-tech buildings that are tailored to local climates.

Avgoustaki et al. (2021) examined the potential for reducing electricity consumption in indoor food production by altering light schedules. The study found that short, intermittent periods of light exposure did not significantly impact the physiological responses of basil plants but did increase biomass production by 47% compared to continuous light exposure. This suggests that flexible, intermittent light exposure can reduce the energy footprint of indoor food production without sacrificing plant growth or profitability.

Dutta et al. (2023) investigated the use of smart sensing devices in hydroponics, a type of vertical farming that uses soilless technology to increase green area. The study found that hydroponic cultivation, which uses water instead of substrate, uses significantly less water and produces better quality products than substrate cultivation. Smart sensing devices allow continuous real-time monitoring of nutrient requirements and environmental conditions, leading to enhanced year-round agricultural production. The study concluded that smart sensing devices are essential for precision agriculture, allowing for real-time monitoring of agronomic variables to increase crop yield.

Kikuchi et al. (2018) investigated the environmental impact of plant factories with sunlight and artificial light compared to conventional Japanese horticulture systems. The study found that plant factories reduced the use of phosphorus, water, and land area but consumed more energy than conventional horticulture systems. However, the researchers suggested that with the use of emerging energy technology options, energy consumption can be reduced to be competitive with conventional horticulture systems. This suggests that plant factories have the potential to become viable and competitive production technology that could change the food, energy, and water systems.

Wang and Iddio (2022) evaluated the energy performance of indoor farming, which involves growing crops on vertically stacked layers in a controlled indoor environment using soilless cultivation systems such as hydroponics. The study found that indoor farming has advantages over conventional open-field farms, such as improved productivity and water use efficiency. However, indoor farming facilities are energy-intensive for maintaining favorable crop-growing conditions. The researchers identified operational issues for the mechanical system and created an energy model using building energy simulation software. They used the model to evaluate the effectiveness of energy-saving strategies, including fixing motorized damper control, eliminating simultaneous heating and cooling, and having wider ranges of temperature setpoints. These measures predicted up to a 48.1% reduction in annual natural gas consumption. The study discusses the limitations of using building energy simulation software for indoor farming modeling.

Delorme and Santini (2022) studied the energy efficiency of autonomous vertical farms (VFs), which minimize water consumption and the use of pesticides. The authors proposed three Mixed-Integer Linear Programming (MIP) formulations and a Constraint Programming model to solve the Vertical Farming Elevator Energy Minimization Problem (VFEEMP), which aims to minimize the energy consumption of automatic elevators servicing VFs. The study shows that the VFEEMP is an NP-complete problem and that the proposed MIP formulations are the most suitable solution. The results of the study could help improve the energy efficiency of VFs, making them more sustainable and competitive with traditional agriculture.

Avgoustaki and Xydis (2021) focused on optimizing the energy demand of indoor vertical farms (IVFs) that use artificial lighting by implementing load shifting. The study found that selecting the times of the day the required darkness is provided to the plants can lead to a significant reduction of 16–26% in artificial lighting costs throughout the year. This could result in the mass deployment of IVFs in cities, reducing operational costs, decreasing CO_2 emissions, and increasing the number of agriculture-based jobs in urban areas. The cash flow analysis showed that possible investors in IVFs could have a full payback period of their investment in less than 9 years, with some cases having a repayment period as low as 2 years. These findings suggest that IVFs have the potential to become a sustainable and profitable alternative to traditional agriculture.

INTRODUCTION

Indoor farming research in Malaysia was initiated by the Horticulture Research Centre, MARDI, in around 2010 with the development of an indoor farm prototype (20′×20′) in a repurposed cold room. The findings led to the establishment of a commercial-scale plant factory with a size of 30′ (W)×80′ (L)×20′ (H) with a production capacity of 16,000 crops equivalent to 1.5–2 tons per month, depending on the crop type. Apart from that, three units of repurposed refrigerated 40-footer containers were also converted into a mobile small-scale plant factory, designated as Agrocube, which can accommodate about 2,400 crops. Figures 12.2 and 12.3 show the internal and external views of the commercial-scale plant factory and Agrocube, while Figure 12.4a and b show their dimensions and layout.

Other information regarding the technical and operation aspects of both plant factories are tabulated in Table 12.3.

FIGURE 12.2 MARDI's commercial-scale plant factory.

FIGURE 12.3 MARDI's Agrocube.

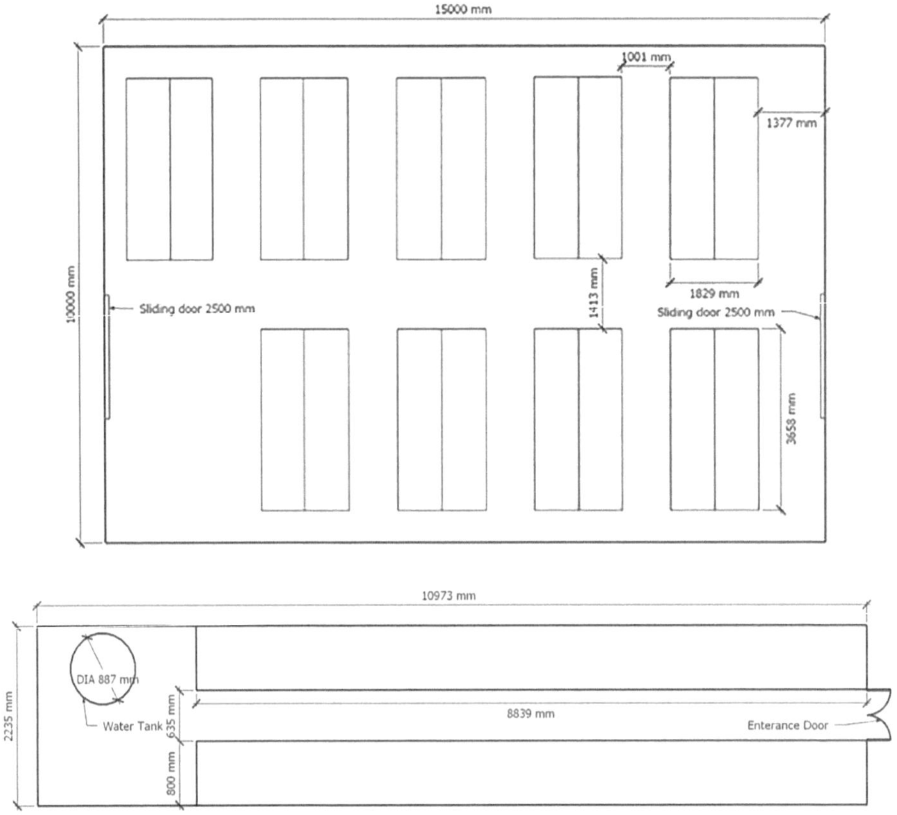

FIGURE 12.4 (a) Dimensions and layout of MARDI's plant factory; (b) dimensions and layout of MARDI's Agrocube.

TABLE 12.3
Technical and Operational Details in MARDI's Plant Factories

Information	MARDI's Plant Factory	MARDI's Agrocube
Location	MARDI Headquarters, Persiaran MARDI-UPM, 43400, Serdang, Selangor, Malaysia	
Growing method	Deep flow technique	
Growing duration	22–24 days	
LED make	MARDI	
LED PAR value	340 μmol/m²/s	
LED DLI	12–14 (mol/m² day)	
Operating hours	24 hours (12 hours of lighting)	
Gross area	23.54 m²	210 m²
Air-cond. area	95%	95%
Type of crop	Lettuce and herbs	Lettuce
Harvest per year	15 times	
Estimated production	~ 19.6 tons/annum	~ 4.9 tons/annum
Electricity supply	Grid: 3 phase, 415 V/100 A	Grid: single phase, 240 V/13 A and solar PV (8 kW_p)
Estimated electricity use	1,775 kWh/month[a]	16,236 kWh/month[a]
Electricity tariff	RM 0.365/kWh (with max. demand charge RM30.3/kW)	

[a] Based on power logging during energy audit.

IMPORTANCE OF ENERGY EFFICIENCY

The COVID-19 pandemic and post events hitherto have shown the ability of mankind to become resilient and adaptive to global challenges. Apart from food security, as highlighted in the previous section, energy security is equally important to mankind. The need to strike a balance between energy security, equity (affordability), and environmental sustainability has never been greater, and the diversity of challenges encountered by nations has never been more evident. Corroborating these three aspects is the World Energy Trilemma Index in Figure 12.5 as published by the World Energy Council, 2022.

Energy security represents a nation's ability to meet current and future energy demands in a reliable, resilient, and resilient manner; energy equity assesses the nation's capacity to provide access to affordable energy, while energy sustainability represents the nation's energy system toward mitigating and avoiding potential environmental damage and climate change effects. As of November 2022, these figures stand at 58/100 for energy security, 75/100 for energy equity, and 66/100 for environmental sustainability. Relating these indices to food security, it is imperative to ensure that the innovation and technological elements introduced in food production must also consider the elements of energy efficiency and sustainability. This is the gap meant to be addressed via this study.

ENERGY AUDIT

Energy audits are a pre-requisite for the implementation of energy efficiency and optimizing energy consumption. It is a survey of energy flow that aims to minimize energy consumption in buildings, processes, and systems without any adverse effect on productivity. Energy audit is an ongoing strategy for managing the energy patterns of buildings. The cost of the audit is proportional to the amount of data acquired from the analysis and the number of energy conservation opportunities identified. Energy audit is usually divided into three categories, namely, walk-through audit, detailed energy audit, and computer simulation (Thumann et al., 2013; Shaari et al., 2021).

The walk-through audit is a tour of the facility to visually inspect energy-using systems, followed by utility bill analysis, annual energy consumption profile, and calculation of the building energy index for benchmarking purposes. No-cost and low-cost energy-saving measures will be the main outcome of this audit. The detailed energy audit involves short-term energy metering, from 1 week

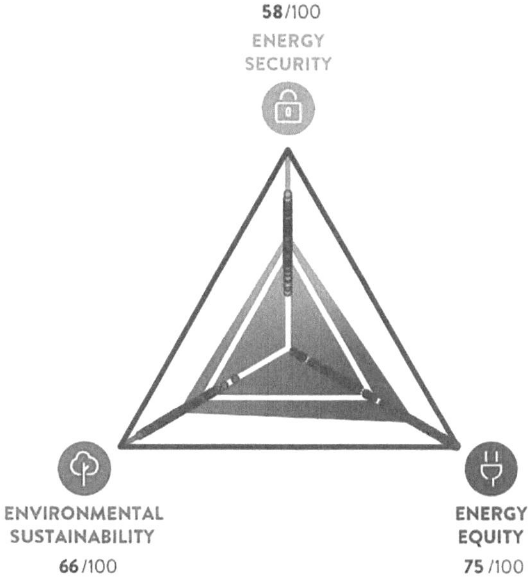

FIGURE 12.5 World energy trilemma index (World Energy Council, 2022).

to 1 month, to quantify energy-using systems to establish energy apportioning and identifying significant energy users (SEUs). A more detailed energy analysis will be carried out in this audit, and energy-saving measures may involve medium- and high-cost options, followed by economic analysis. As for the computer simulation, the energy auditor will use software to estimate energy usage and may include a wider range of variables such as weather and future changes in operation. This type of audit is suitable for facilities that are complex in nature and facilities under construction. In indoor farming, all three energy audits are suitable depending on the desired outcome. In this case study, both walk-through and detailed energy audit has been carried out for 7 days in both MARDI's plant factory and the Agrocube. The case study began in December 2019 and until January 2021. The energy audit flow is depicted in Figure 12.6.

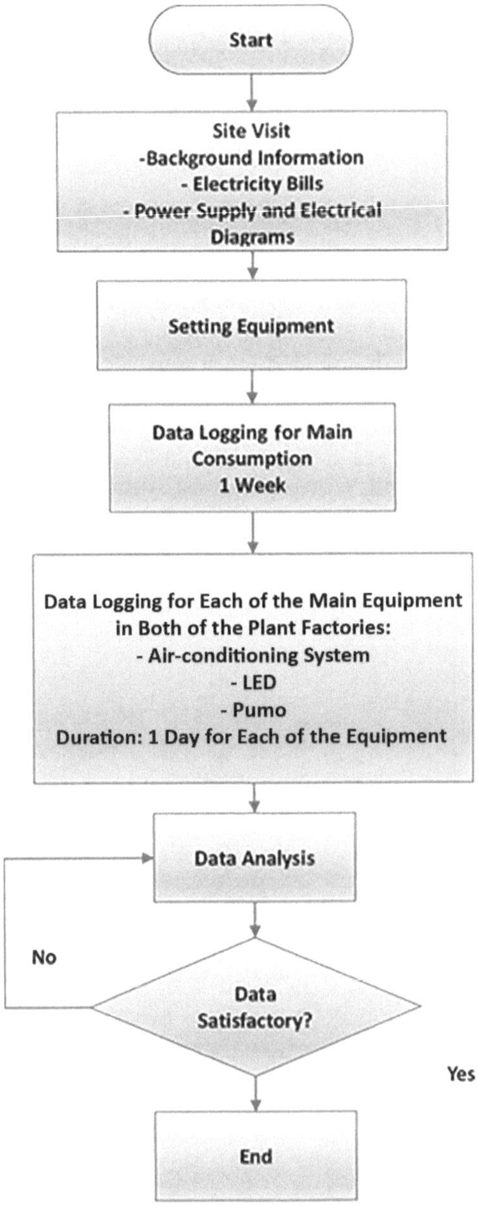

FIGURE 12.6 Energy audit flow.

Energy Audit Methodology

The aim of the case study is as follows: (1) to establish the present energy consumption characteristics in both plant factories and (2) to propose relevant energy conservation measures toward energy-efficient operation. The first aim was achieved via a detailed energy audit and short-term metering on electricity consumption, while the second aim was achieved via identifying the no-cost and low-cost energy-saving measures, followed by establishing the specific energy consumption (SEC) of the plant factories' produce. The detailed energy audit was conducted based on guidelines provided by the Energy Commission of Malaysia (2016).

Energy Audit Equipment

The following are the equipment and devices used in the energy audit:

(i) Electrical energy power logger

The power logger used in this is a clamp-on type, model Hioki 3360, as depicted in Figure 12.7. It is a multi-channel power measuring device that uses the clamp sensor input to measure power accurately and safely on single-phase to three-phase lines. The three-phase four-wire energy loggers monitor power demand, and other power parameters are computed using PW3360 to assist in energy audits and verify energy-saving measures (Table 12.4). This logger was installed at the incoming power supply from the grid at both MARDI's plant factory and Agrocube for a minimum of 1 week as recommended by the Energy Commission of Malaysia (2023).

(ii) Weather meter

The weather meter is a device that measures several parameters such as barometric pressure, altitude, relative humidity, dew point, wet bulb temperature, and average wind speed. The weather meter used here is a model Kestrel 4000 capable of measuring up to 2,000 data sets and can be transferred to a personal computer for data analysis. The weather meter used in this study is shown in Figure 12.8.

(iii) Thermometer

While the temperature inside the plant factories was measured using temperature sensors at several locations to represent local temperatures, the temperature during energy audit on interior surfaces of the plant factories was measured using an infrared thermometer model

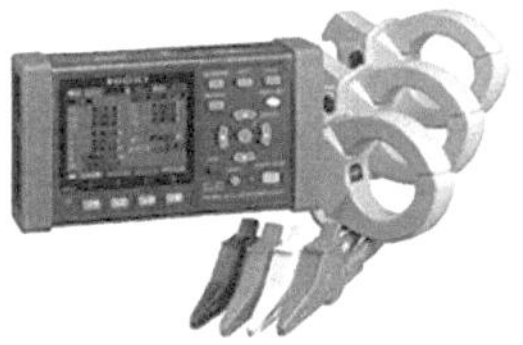

FIGURE 12.7 Clamp-on power logger, model Hioki PW3360.

TABLE 12.4
Hioki PW3360 Clamp-on Power Logger Specification

Measurement line type	Single-phase two wire, single-phase three wire, three-phase three wire, three-phase four wire
Measurement line frequency	50/60 Hz
Voltage range	-Effective measurement range: 90–780 V, peak: ±1,400 V -[OVER] indicates over-range warning
Current ranges	CLAMP ON SENSOR 9661: 5/10/50/100/500 A

FIGURE 12.8 Kestrel 4000 weather meter.

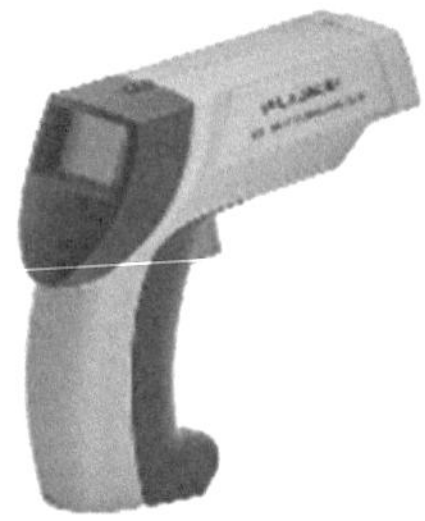

FIGURE 12.9 Infrared thermometer model Fluke 63.

FIGURE 12.10 Lux meter model Testo 540.

Fluke 63 with a range of –40–535°C. The surface temperature was measured to observe whether heat was leaking from the ambient into the facility. Figure 12.9 shows the infrared thermometer used in this case study.

(iv) Lux meter

A lux meter was used to measure the general lighting intensity in the plant factories to see the opportunities for energy saving. This lighting here is not to be mistaken for the artificial lighting provided for crops for photosynthesis, i.e., photosynthetic photon flux density, which is measured using photosynthetic active radiation (PAR) meter. The lux meter used is Testo 540, as depicted in Figure 12.10.

FINDINGS FROM THE ENERGY EFFICIENCY STUDY

List of Equipment

The list of electrical equipment in both MARDI's plant factory and Agrocube are first identified and listed to enable energy consumption calculation per system. Tables 12.5–12.9 tabulate the equipment in MARDI's plant factory, while Table 12.10 tabulates the equipment in Agrocube.

TABLE 12.5
Office Equipment

No	Electrical Equipment	Quantity	Power Rating
1	LED office	12	18 W
2	Air-conditioner	2	2.0 HP
3	Desktop	3	n/a

TABLE 12.6
Entrance and Changing Area

No.	Electrical Equipment	Quantity	Power Rating
1	Down light	6	18 W
2	Fan (Mistral 20″)	1	113 W
3	Air curtain	2	89 W
4	Air-conditioner	1	1.5 HP

TABLE 12.7
Utilities Room Equipment

No.	Electrical Equipment	Quantity	Power Rating
1	Small led	4	10 W
2	Fan (KDK)	1	0.08 HP
3	Pump (Walrus HQ400)	1	375 W

TABLE 12.8
Post-Harvest Equipment

No.	Electrical Equipment	Quantity	Power Rating
1	Down light	9	18 W
2	Air-conditioner	1	2.0 HP
3	Refrigerator	1	650 W

TABLE 12.9
Growing Area Equipment

No.	Electrical Equipment	Quantity	Power Rating
1	6 feet LED – composite colored	1,300	35 W
2	Air curtain	4	89 W
3	Fan (Mistral 20″)	4	113 W
4	4 feet LED—white	12	18 W
5	Fertilizer pump	18	20 W
6	Water pump	18	0.5 HP
7	Exhaust fan	1	N/A
8	FCU	2	10 HP
9	Intake air fan	1	N/A

TABLE 12.10
Electrical Equipment in AgroCube

No.	Electrical Equipment	Quantity	Power Rating
1	4 feet LED – composite colored	126	18 W
2	Air curtain	1	89 W
3	Air-conditioner	2	2.5 HP
4	4 feet LED – white	42	18 W
5	Fertilizer pump	2	20 W
6	Water pump	1	375 W
7	Exhaust fan	1	120 W
8	Humidity sensor	2	N/A
9	Fluorescent light	6	18 W
10	Control panel	1	N/A

Energy Load Profile

The energy load profile (also referred to as the load profile) estimates the energy demand of a power system or subsystem over a specific time period (e.g., hours, days). The load profile is a two-dimensional chart that depicts the instantaneous load over time (in volt-amperes/watt) and is a direct method to visualize how the system loads time-related changes. The load profile can also be used for energy efficiency applications where the estimation of total energy consumption in a system is required.

In this case study, the load profile for three main equipment in the plant factories, namely, the air-conditioning system, artificial lighting (LED), and water pump, was acquired using power for a period of 7 days. It was found that, throughout the growth period, the load profile remained almost identical.

MARDI's Plant Factory's Load Profile

Figure 12.11 and Table 12.11 provide the energy consumption at MARDI's plant factory, logged at the incoming power supply from the grid. During the 7 days of logging, a total of 3,788.423 kWh of energy was consumed by the plant factory, with an average consumption of 2,2550.14 Wh every hour of operation. A maximum energy usage of 53,590 Wh was also recorded.

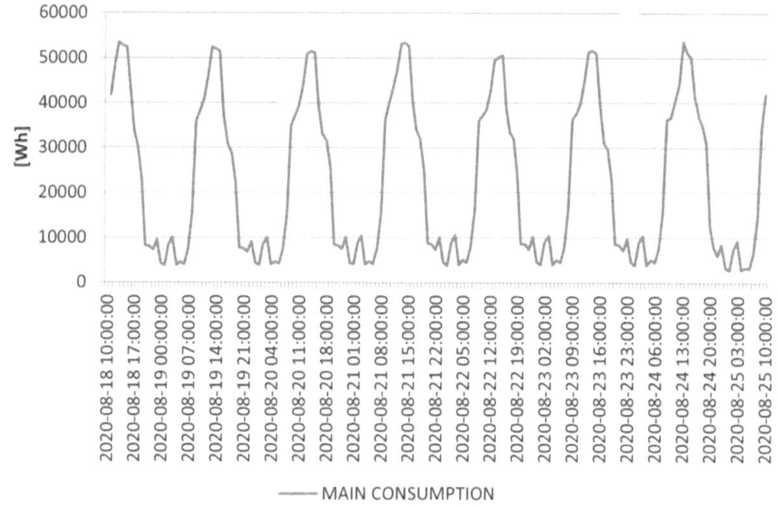

FIGURE 12.11 Energy consumption profile in MARDI's plant factory.

TABLE 12.11
Energy Consumption in MARDI's Plant Factory

Total Energy Consumption for 7 days (Wh)	Daily Energy Consumption (Wh)	Monthly Energy Consumption (kWh)	Annual Energy Consumption (kWh)
3,788,423	541,203.3	16,236.1	197,539.2

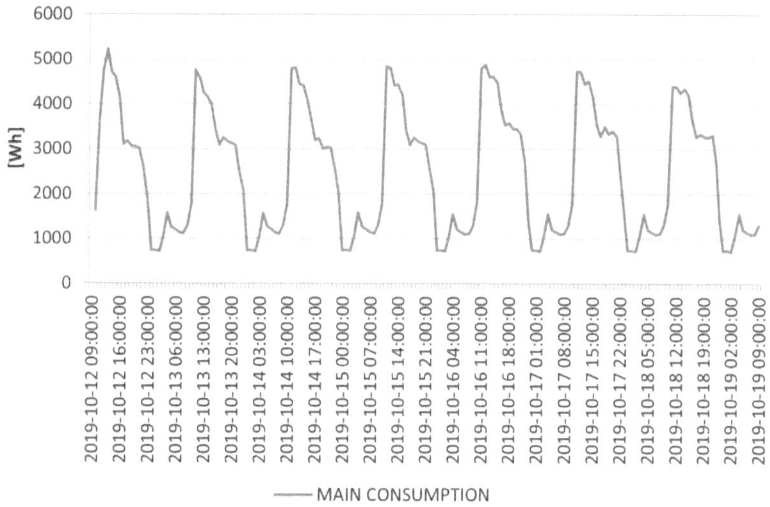

FIGURE 12.12 Energy consumption profile in MARDI's Agrocube.

MARDI's Agrocube Load Profile

Figure 12.12 illustrates the energy consumption in Agrocube for a duration of 7 days. The data obtained from the data logger showed that the total power consumption is 414.153 kWh, averaging 2,802.235 Wh per hour, with a maximum consumption value of 5,229 Wh being recorded.

Energy Consumption by Major Systems

Lighting System

The photoperiod or artificial lighting's operational time in both plant factories varies between 10 and 12 hours depending on the crop planted and the variation of lighting required during different phases of crop growth. Bigger crops may require a longer photoperiod compared to smaller ones in achieving their daily light integral (DLI) for photosynthesis (Kozai, 2018). The artificial lighting system was fully automated via Internet of things modules, which enable reprogramming based on the requirement. Figures 12.13 and 12.14 represent the lighting load profiles on a typical day for MARDI's plant factory and Agrocube, respectively.

Figure 12.13 shows that the LED lights were switched from 7 a.m. to 8 p.m. The lighting system at the seedlings rack was turned on first at 7 a.m. registering 5,875 Wh of energy consumption, while the rest of the LED lights in other racks were turned on at 8 a.m.; hence an energy spike was

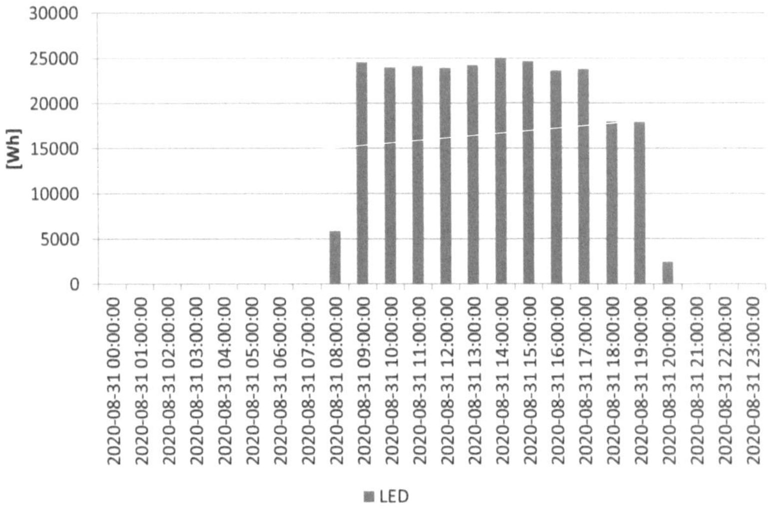

FIGURE 12.13 Artificial lighting load profile in MARDI's plant factory.

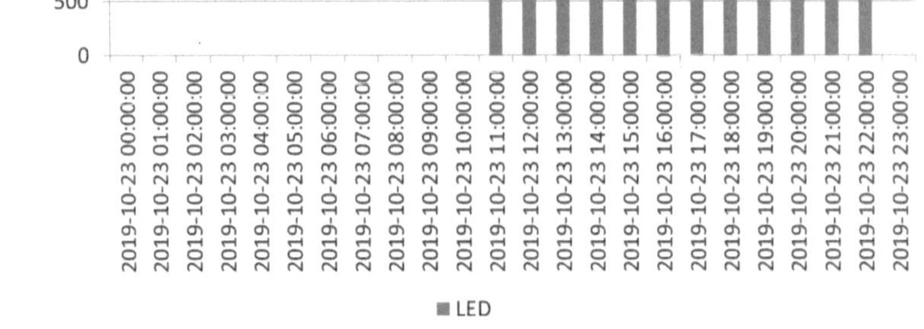

FIGURE 12.14 Artificial lighting load profile in Agrocube.

recorded at 8 a.m. with energy use of 24,560 Wh. From 8 a.m. until 5 p.m. the load remained almost constant with small fluctuations, and at 5 p.m. the load started to decrease as some grow racks were switched off, and at 9 p.m., no more load was recorded as all the lighting was switched off as the photoperiod was completed. The total energy consumption for artificial lighting during this one-day logging was 261,956 Wh.

The artificial lighting load for Agrocube is similar to that of the plant factory, although the system began to operate at 10 a.m. The lighting system here was set to begin operation much later as a strategy to avoid incurring maximum demand charges that might be imposed as the electricity tariff in MARDI is a commercial rate. A stable load from 10 a.m. to 10 p.m. was recorded and switched off at 10 p.m. as the photoperiod has been completed. The Agrocube's lighting energy consumption varied from 1,963 Wh to 2,412 Wh, and the total consumption recorded was 28,261 Wh.

Air-Conditioning System

In tropical climate countries such as Malaysia, buildings are usually characterized by high energy consumption by their air-conditioning system. This is due to the hot and humid conditions where the ambient temperatures can be greater than 33°C during daytime and 70% relative humidity. As for the plant factory, indoor temperatures shall be maintained between 24°C and 26°C at all times. MARDI's plant factory uses two units of 10 horsepower, ducted fan coil unit (FCU), which ensures the indoor temperatures are around 25°C while maintaining humidity at around 60%. To meet these indoor conditions, both FCU units were switched between 10.00 a.m. and 4.00 p.m. when the outdoor temperatures were relatively hot. As for the Agrocube, two units of 1.5 horsepower split-unit air-conditioners were installed, running alternately 12 hours each with indoor temperatures around 24°C. The sequence of air-conditioning system operation is provided in Tables 12.12 and 12.13, while its load profile is depicted in Figures 12.15 and 12.16.

From Figure 12.15, the air-conditioning load profile started to increase, significantly starting at 10 a.m., and peaked at 12 p.m. with 14,556 Wh and retained the profile with little fluctuation until 3 p.m. This is because both FCUs were running simultaneously to ensure that the desired temperature in the plant factory was achieved. It was also found that a single FCU is insufficient to serve the high cooling load, and the indoor temperature may rise to 30°C in the afternoon. As for the Agrocube, Figure 12.16 shows the load profile representing one air-conditioner operating at one time. The air-conditioner consumed 21,418 Wh during the one full day of data logging.

Pumping System

Pumps were used in both plant factories to pump and circulate water and fertilizer in the hydroponics systems. The pump load profiles for both MARDI's plant factory and Agrocube were given in Figures 12.17 and 12.18. In Figure 12.17, the load profile was inconsistent as the pumps in each of the nine grow racks operate at different times based on the pre-set timer of 2 hours of operation and 1-hour breaks. This allowed water in the deep flow technique systems to recede and rise continuously to provide oxygen to the crops' roots. Since different grow racks have different pre-set timers, the profile differs. The total energy consumption acquired from this 1-day logging for the pump is 68,473 Wh.

As for Agrocube, the cycle time for the water pump was set to run for 2 hours, followed by a 30-minute break, and the cycle repeats. The fertilizer pump was of the peristaltic type and will operate only when the fertilizer's concentration becomes low. The pump load profile is shown in Figure 12.18.

Energy Apportioning

Energy apportioning is a method for matching a facility's energy consumption with equipment or individual energy utilization. This was basically listing all individual energy consumption by

TABLE 12.12
Air-Conditioning System Operation Sequence in MARDI's Plant Factory and Agrocube

	MARDI's Plant Factory		Agrocube	
Time	**Air-Conditioner 1**	**Air-Conditioner 2**	**Air-Conditioner 1**	**Air-Conditioner 2**
0100	✓	×	✓	×
0200	✓	×	✓	×
0300	×	×	×	×
0400	×	×	×	×
0500	×	×	×	×
0600	✓	×	✓	×
0700	✓	×	✓	×
0800	✓	×	✓	×
0900	✓	×	✓	×
1000	✓	✓	✓	✓
1100	✓	✓	✓	✓
1200	✓	✓	✓	✓
1300	✓	✓	✓	✓
1400	✓	✓	✓	✓
1500	✓	✓	✓	✓
1600	×	✓	×	✓
1700	×	✓	×	✓
1800	×	✓	×	✓
1900	×	✓	×	✓
2000	×	✓	×	✓
2100	×	✓	×	✓
2200	✓	×	✓	×
2300	×	×	×	×
2400	×	×	×	×

TABLE 12.13
Energy Consumption Breakdown MARDI's Plant Factory

Equipment	**Daily Energy Consumption (kWh)**	**Percentage from Total Daily Energy Consumption (%)**
Lighting (LED)	261.95	48.4
Air-conditioning	149.63	27.7
Pump	68.47	12.6
Others	61.16	11.3
Total	541.2	100

the equipment in the facility and matching them with the total energy consumed by the facility. In this case study, the energy apportioning is established for both plant factories by categorizing the energy consumption by four major pieces of equipment: lighting (red-blue and white LED), air-conditioning system, pump (fertilizer and water), and other smaller equipment which were combined as one, namely air curtain, exhaust fan, humidity sensor, fluorescent light, and control panel.

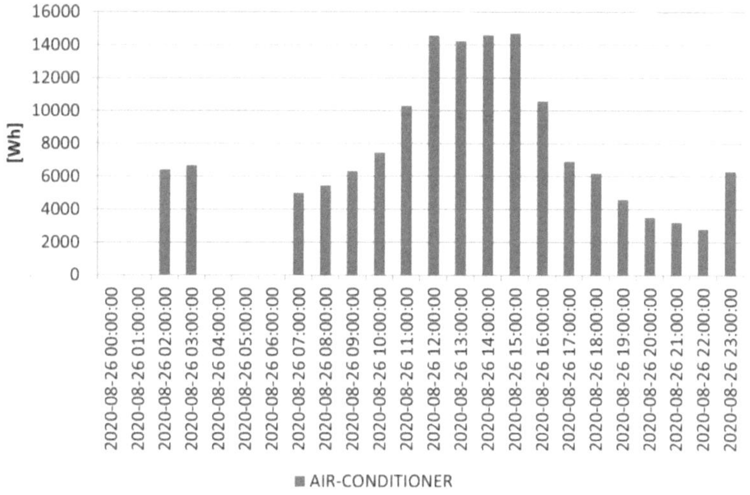

FIGURE 12.15 Air-conditioning load profile – MARDI's plant factory.

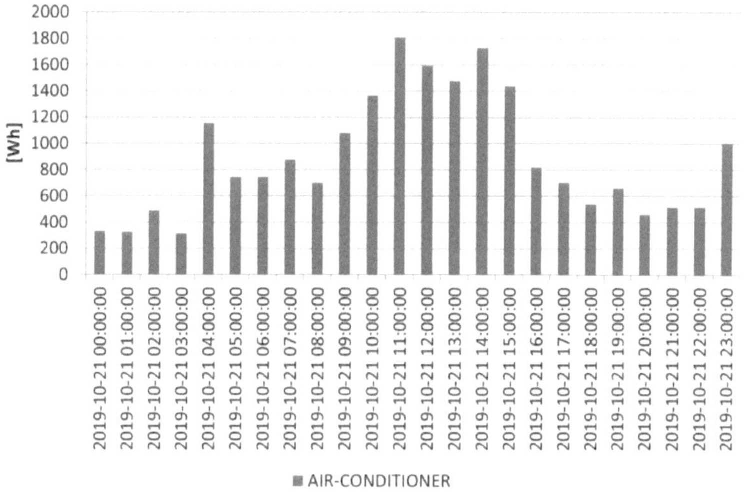

FIGURE 12.16 Air-conditioning load profile – Agrocube.

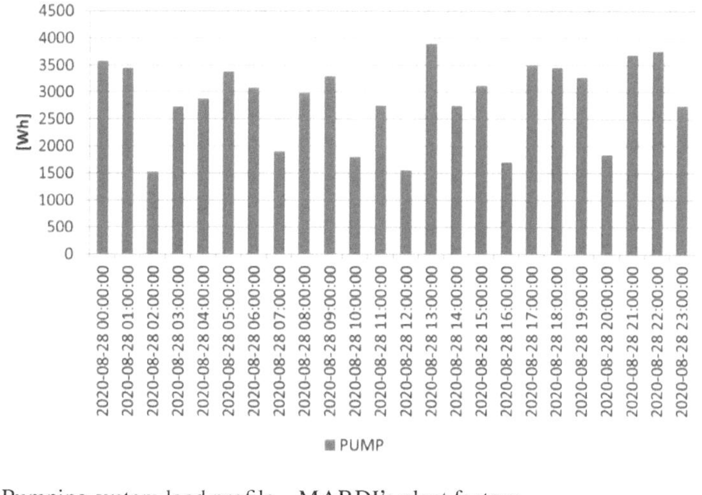

FIGURE 12.17 Pumping system load profile – MARDI's plant factory.

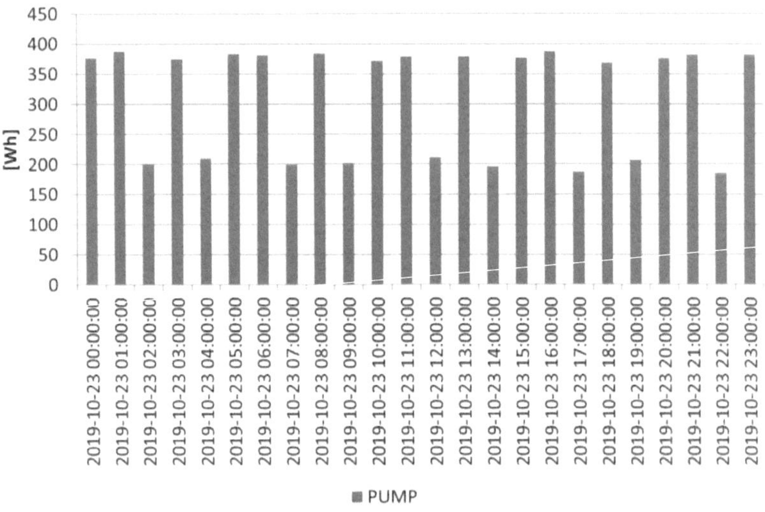

FIGURE 12.18 Pumping system load profile – Agrocube.

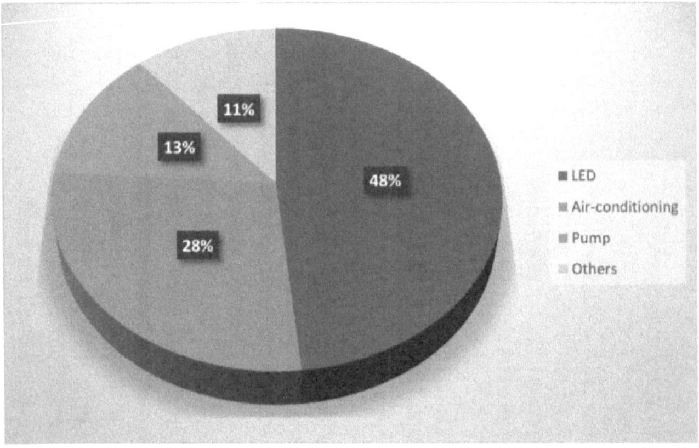

FIGURE 12.19 Load apportioning at MARDI's plant factory.

Energy Apportioning – MARDI's Plant Factory

Table 12.13 lists the energy consumption breakdown by equipment, while Figure 12.19 illustrates the energy apportioning. Unlike typical buildings in Malaysia, where energy consumption is usually the highest for air-conditioning systems, MARDI's plant factory records the highest energy use by its lighting system with 48.4% of total energy usage. This is followed by the air-conditioning system, which uses 27.65% of total energy, the pump with 12.65%, and other smaller equipment with 11.3%. This is because the artificial lighting system was switched on for up to 12 hours daily to meet the DLI requirement by the crops for photosynthesis.

A similar trend was obtained for the Agrocube in which the highest energy usage is also by the lighting system, with almost 48% of energy usage from the total energy consumption, followed by the air-conditioning system, pump, and others with energy use percentages of 36.2%, 12.66%, and 3.37%, respectively. The lighting energy usage seems to be relatively higher in Agrocube because Agrocube is compact with a smaller preparation area with a smaller amount of equipment. Apart from that, the higher amount of air-conditioner usage may be attributed to the longer time required by the Agrocube to achieve the required present indoor temperature compared to the plant factory. Table 12.14 lists the energy consumption breakdown by equipment for Agrocube, and Figure 12.20 depicts the energy apportioning.

TABLE 12.14
Energy Consumption Breakdown in Agrocube

Equipment	Daily Energy Consumption (kWh)	Percentage from Total Daily Energy Consumption (%)
Lighting (LED)	28.26	47.77
Air-conditioning	21.42	36.2
Pump	7.49	12.66
Others	1.99	3.37
Total	59.16	100

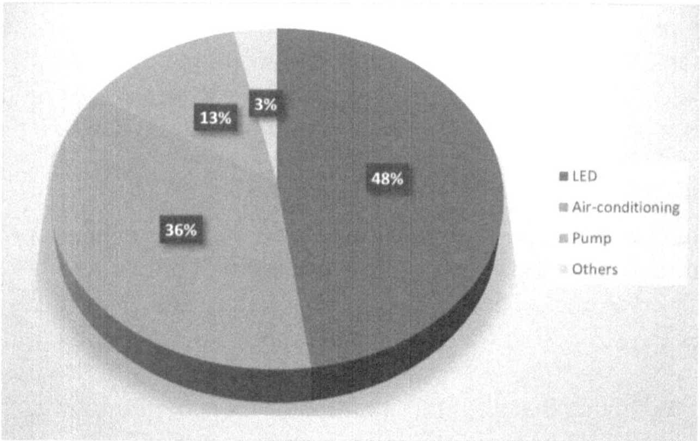

FIGURE 12.20 Energy apportioning at Agrocube.

TABLE 12.15
Energy Consumption Breakdown in Agrocube

Item	MARDI's Plant Factory	Agrocube
Calculated monthly energy consumption (kWh)	14,401.8	1,715.01
Estimated production of lettuce (kg)	1,500 kg	180 kg
Specific energy consumption, SEC (kWh/kg)	9.60	9.53

Specific Energy Consumption

Based on the energy apportioning established, the SEC for crops was calculated. In this SEC, the total energy consumption by the equipment in kWh was divided by the average production of crops (kg). This will provide a reference value known as 'baseline' in energy management, to enable facility owners to set targets for energy saving via energy-efficiency measures. However, this SEC value may vary depending on the type of crops as some crops may have longer harvesting periods and hence larger energy use. Table 12.15 shows the SEC value for crops grown in MARDI's plant factory and Agrocube. These values were taken based on the assumption that the whole process of preparation, growing, and harvesting is completed in 1 month. This case study found that both plant factories have similar values of SEC.

Energy-Saving Measures

Energy-saving measures toward energy-efficient operations are the main outcome of the present case study. The energy-saving measures can be diverse depending on the design and operational characteristics of an indoor farm, and they can be divided into short-term, medium-term, and long-term measures.

Short-Term Energy-Saving Measures

The following are short-term energy-saving measures proposed in MARDI's plant factories.

Daylighting

Some parts of the plant factory can benefit from daylighting and avoid the usage of conventional lighting, such as the post-harvest area, changing area, and meeting room. However, the use of daylighting must be implemented carefully to avoid the penetration of daylight in the growing area, which may disrupt crops that were grown using artificial lighting. In this case study, the post-harvest area, meeting area, and changing area use a total of 16 downlight bulbs with a power rating of 18W/bulb. With an estimated usage time of 4 hours/day and 22 working days per month, the total energy saving will be 16×18 W×4 hours×22 days=25.34 kWh/month.

Air-Conditioning System

The air-conditioning use sequence tabulated in Table 12.12 can be improved by switching off the air-conditioners for several hours during the night. This is possible as heat generation inside the plant factory during this period was minimal since the grow lights were switched off. Table 12.16 proposes the suggested time to turn off the air-conditioning system for both of the plant factories.

Apart from this, it was found that air infiltration can be reduced by minimizing the gaps between doors at both plant factory entrances.

Medium-Term Energy-Saving Measures

Unlike short-term energy-saving measures, which are no-cost measures, medium-term energy-saving measures may involve low-cost and medium-cost options. Among the medium-term energy-saving measures that can be implemented in both MARDI's plant factory and Agrocube is replacing the existing air-conditioning systems which has inverters. Air-conditioning systems with inverters will correspond to the actual cooling load and minimize energy consumption for cooling (Nezhad et al., 2021). The saving based on the energy-efficient labeling by the Energy Commission of Malaysia is estimated to be 25%, as in Figure 12.21. The potential energy saving with its simple payback period is calculated in Table 12.17.

TABLE 12.16
Proposed Energy-Saving Measure in the Air-Conditioning System

Proposed Energy Saving	MARDI's Plant Factory	Agrocube
Suggested time to switch-off the air-conditioning system	20:00–22:00	20:00–22:00 00:00–03:00
Energy consumption during suggested time (kWh)	8.478	2.962
Monthly estimated energy saving (kWh)	8.478×30=254.34	2.962×30=88.86

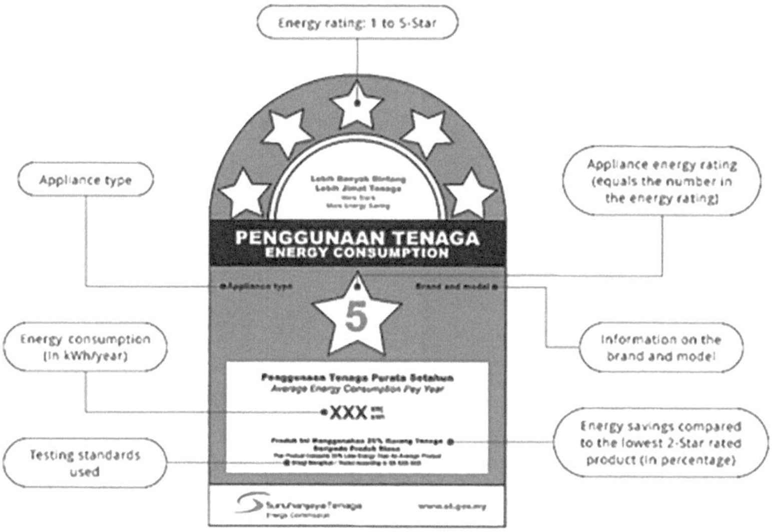

FIGURE 12.21 Load apportioning at Agrocube (Energy Commission of Malaysia, 2016).

TABLE 12.17
Estimated Energy Saving by Using Efficient Air-Conditioning System

Details	MARDI's Plant Factory	Agrocube
Proposed inverter type air-conditioner	Panasonic Mini ECOi High Efficiency 10 HP	Panasonic 1.5 HP Standard Inverter R32 Air-Conditioner
Total price of two units	RM 19,000.00	RM 4,200.00
Estimated energy saving per month (kWh)	2×10 HP $\times$ 0.747 kW/HP $\times$ 25% $\times$ 30 days $\times$ 12 hours = 1,344	2×1.5 HP $\times$ 0.747 kW/HP $\times$ 25% $\times$ 30days $\times$ 12 hours = 201.7
Electricity tariff	RM 0.365/kWh	
Estimation of saving per month of electricity bills	RM 490.8	RM 73.62
Simple payback period	38.7 months	57.0 months

Long-Term Energy-Saving Measures

For long-term energy-saving measures, it is proposed here to integrate rooftop-solar photovoltaics (PV) systems in both plant factories. Solar PV generates electricity during the daytime, and this electricity can be supplied directly to the energy-consuming systems in the plant factories, such as the grow lights, which were the highest energy consumers. Solar energy can also be stored in battery packs, but this will incur additional capital expenditure, and eventually, these battery packs may have to be replaced thrice during the typical solar PV lifecycle of 25 years. A capacity of 30-kWp solar PV system is proposed for MARDI's plant factory, while 8 kWp was proposed for Agrocube. Tables 12.18–12.21 provide the details of the proposed solar PV system, typical costs, and simple payback periods. Sun-hours were taken to be 5 hours for Malaysia.

It can be seen that, with a payback period of less than 6 years, solar PV is a viable option for indoor farming, based on the MARDI's case study. Solar PV may also provide additional shade on the rooftop of plant factories, which will reduce heat penetration into the facility, which may reduce cooling load and hence air-conditioning energy consumption.

TABLE 12.18
Estimated Cost of the Proposed Grid-Connected Solar PV for MARDI's Plant Factory

No.	Item	Price (RM)	Quantity	Price (RM)
1	Solar panel 330 W	529.00[a]	100	52,900.00
2	Inverter 30,000 TL3-S easy installation three-phase solar inverter 30 kW grid tie inverter 30,000 W	9,500.00[a]	1	9,500.00
3	Installation fee of solar PV system	15,000.00	–	15,000.00
4	Construction fee of solar PV structure	25,000.00[a]	–	25,000.00
		Total price (RM)		102,400.00

[a] Current rate and is subject to change.

TABLE 12.19
Estimated Simple Payback Period for MARDI's Plant Factory

Cost Involved	Description
Monthly solar power generation by the solar panels	30 kW × 5 sun-hours × 30 days = 4,500 kWh/month
Monthly cost avoidance	4,500 kWh × RM 0.365/kWh = RM 1,642.50
Annual cost avoidance	RM 1,642.50 × 12 months = RM 19,710.00
Simple payback period of the system (cost of solar PV)/(annual cost avoidance)	$\frac{\text{RM}102,400.00}{\text{RM}19,710.00}$ = 5.2 years

TABLE 12.20
Estimated Cost of the Proposed Grid-Connected Solar PV for AgroCube

No.	Item	Price (RM)	Quantity	Price (RM)
1	Solar panel 330 W	529.00[a]	24	12,696.00
2	PV3500 PRO series low frequency off-grid solar inverter	5,500.00[a]	1	5,500.00
3	Installation fee of solar PV system	4,400.00	–	4,400.00
4	Construction fee of solar PV structure	7,000.00[a]	–	7,000.00
		Total price (RM)		29,596.00

[a] Current rate and is subject to change.

CONCLUSION

Indoor farming, despite being a viable option in addressing food security issues, must incorporate energy efficiency to allow this technology to be more sustainable and cost efficient and to minimize its carbon footprint. This case study highlights an energy efficiency study carried out in two indoor farming facilities in Malaysia, owned by MARDI. It is hoped that this case study may benefit the indoor farming community and provide insight for technology providers to further improve plant factory design aspects in terms of energy efficiency. This chapter ends with several recommendations for improving energy efficiency in indoor farming facilities:

TABLE 12.21
Estimated Simple Payback Period for Agrocube

Cost Involved	Description
Monthly solar power generation by the solar panels	8 kW × 5 sun-hours × 30 days = 1,200 kWh/month
Monthly cost avoidance	1,200 kWh × RM 0.365/kWh = RM 438
Annual cost avoidance	RM438 × 12 months = RM 5,256.00
Simple payback period of the system (cost of solar PV)/(annual cost avoidance)	$\frac{\text{RM29,596.00}}{\text{RM5,256.00}}$ = 5.63 years

LED Lighting Optimization

While typical buildings record the highest energy consumption from their HVAC system, an indoor farm may be characterized by the lighting system being the SEU as obtained from our case study. This is due to the prolonged use of growlights to meet the DLI required by the crops for photosynthesis. Hence, optimization of LED lighting is a must by ensuring that the growlights are switched on and off precisely to meet the minimum DLI. Apart from that, the photoperiod may be manipulated by introducing intermittent lighting as a method to reduce lighting hours as suggested by Avgoustaki et al. (2021) and Rengasamy et al. (2022).

Passive Design Strategies

An indoor farming facility shall incorporate passive design strategies during its design stage to ensure the *low-hanging fruits* are capitalized. Among the passive design strategies applicable to indoor farms are ensuring the physical orientation of the facility avoids/minimizes direct solar radiation as to reduce heat infiltration which will increase the cooling load (Yanmeng et al., 2021). This strategy can be integrated with strategic landscaping and providing sufficient insulation on the facility's walls and roofs, if necessary. Natural ventilation with a dehumidifier can also be considered as a substitute for an air-conditioner if the indoor farming facility is located in an area with temperatures below 26°C.

LED Lighting Optimization

As a significant energy consumer in an indoor farming facility, LED growlight optimization is a must. Taking into account the required lighting spectrum based on crop type, the LED lighting system shall have a smart lighting system that may provide specific DLI values without wastage. As can be seen in this case study, LED may be the single-largest energy consumer in an indoor farming facility. Apart from that, the heat dissipation from LED must be considered during the procurement stage of indoor farming equipment as LED will have an interactive effect on the air-conditioning system due to heat produced during operation.

Sensors and Automation

Sensors and automation have become an integral part of indoor farms, to monitor temperature, humidity, CO_2 levels, lighting intensity, irrigation, and nutrient delivery and adjust them automatically for optimal plant growth. This system can also be programmed for energy efficiency, for instance, to go into energy-saving mode during the standby mode.

Renewable Energy Integration

An indoor farming facility can be easily integrated with renewable energy (RE) systems such as solar PV and small-scale wind turbines. Depending on the energy demand inside the indoor farming facility, RE can be integrated into three modes, namely, on-grid (grid-tied), off-grid (standalone), or hybrid system. On-grid RE system implies that the indoor farming facility only partially uses the RE, while supply from the grid is still necessary to meet its energy demand. This RE integration option is usually the cost-effective option. The off-grid RE system requires an energy storage system, particularly battery packs as there is no electricity supply from the grid at all. This allows an indoor farm to operate in remote locations and features self-sustaining capability in terms of energy. This is, however, costly due to the energy storage requirement. The hybrid RE option, on the contrary, combines both features of on-grid and off-grid, but the energy storage is relatively smaller, only used to store excess energy generated from the RE. Integration with RE must be carefully evaluated as it involves significant capital expenditure and shall be implemented only after energy efficiency potentials are fully explored.

Energy-Efficient Air-Conditioning and Mechanical Ventilation System

Since micro-climatic conditions are a pre-requisite in indoor farming, the use of energy-efficient indoor farming is of utmost importance. Most air-conditioning systems today use vapor–compression cycles that have gone through numerous innovations, leading to energy efficiency; however, the use of natural ventilation and passive cooling techniques wherever possible should be considered. Condensate water recovery from the evaporator coils will contribute to both energy efficiency and water efficiency.

Data Analytics and Machine Learning

An indoor farm can employ data analytics and machine learning algorithms to analyze energy consumption patterns and optimize energy usage. Apart from that, machine learning can be used to develop predictive models to anticipate crop growth requirements and respond by changing the environmental conditions accordingly.

By combining these recommendations and continually innovating, the indoor farming industry can significantly improve energy efficiency, reduce its environmental impact, and contribute to more sustainable food production systems.

ACKNOWLEDGMENTS

The authors would like to acknowledge various parties who were involved directly and indirectly in this case study, especially the Director of Horticulture Centre, Malaysia Agricultural Research and Development Institute (MARDI), and researchers from the Centre for Energy and Industrial Environment Studies (CEIES) and Universiti Tun Hussein Onn Malaysia. A special thanks was conveyed to our industrial partners, I-PDP Sdn. Bhd. and KAFF Synergy LLP, for providing technical assistance and valuable inputs.

REFERENCES

Avgoustaki, D. D., Bartzanas, T., & Xydis, G. (2021). Minimising the energy footprint of indoor food production while maintaining a high growth rate: Introducing disruptive cultivation protocols. *Food Control*, 130, 108290. https://doi.org/10.1016/j.foodcont.2021.108290.

Avgoustaki, D. D., & Xydis, G. (2021). Energy cost reduction by shifting electricity demand in indoor vertical farms with artificial lighting. *Biosystems Engineering*, 211, 219–229. https://doi.org/10.1016/j.biosystemseng.2021.09.006.

Delorme, M., & Santini, A. (2022). Energy-efficient automated vertical farms. *Omega*, 109, 102611. https://doi.org/10.1016/j.omega.2022.102611.

Department of Standards Malaysia (2019). MS 1525:2019 energy efficiency and use of renewable energy for non-residential buildings: Code of practice.

Despommier, D., '*The Vertical Farm: Feeding the World in the 21st Century*'. St. Martin's Press, New York. 2010.

Dutta, M., Gupta, D., Sahu, S., Limkar, S., Singh, P., Mishra, A., Kumar, M., & Mutlu, R. (2023). Evaluation of growth responses of lettuce and energy efficiency of the substrate and smart hydroponics cropping system. *Sensors*, 23(4), 1875. https://doi.org/10.3390/s23041875.

Energy Commission of Malaysia (2016). 'Electrical Energy Audit Guidelines'. www.st.gov.my, accessed on 5th April 2023.

Graamans, L., Baeza, E. J., Van Den Dobbelsteen, A., Tsafaras, I., & Stanghellini, C. (2018). Plant factories versus greenhouses: Comparison of resource use efficiency. *Agricultural Systems*, 160, 31–43. https://doi.org/10.1016/j.agsy.2017.11.003.

Graamans, L., Rychtarikova, M., Van Den Dobbelsteen, A., & Stanghellini, C. (2020). Plant factories: Reducing energy demand at high internal heat loads through façade design. *Applied Energy*, 262, 114544. https://doi.org/10.1016/j.apenergy.2020.114544.

Harbick, K., & Albright, L. D. (2016). Comparison of energy consumption: greenhouses and plant factories. *Acta Horticulturae*, 1134, 285–292. https://doi.org/10.17660/actahortic.2016.1134.38.

https://www.un.org/en/desa/world-population-projected-reach-98-billion-2050-and-112-billion-2100, accessed on 4th April 2023.

Kozai, T. (2018). *Smart Plant Factory: The Next Generation Indoor Vertical Farms*. Springer.

Kikuchi, Y., Kanematsu, Y., Yoshikawa, N., Okubo, T., & Takagaki, M. (2018). Environmental and resource use analysis of plant factories with energy technology options: A case study in Japan. *Journal of Cleaner Production*, 186, 703–717. https://doi.org/10.1016/j.jclepro.2018.03.110.

Nezhad, E., Rahimnejad, A., & Gadsden, S. A. (2021). Home energy management system for smart buildings with inverter-based air conditioning system. *International Journal of Electrical Power and Energy Systems*, 133, 107230.

Perez, M. V. (2014). Study of the sustainability issues of food production using vertical farm methods in an urban environment within the state of Indiana. Master's thesis, Universitat Politècnica de Catalunya, Barcelona, Spain.

Rengasamy, N., Othman, R.Y., Che, H.S. & Harikrishna, J.A. (2022). Artificial lighting photoperiod manipulation approach to improve productivity and energy use efficacies of plant factory cultivated stevia rebaudiana. *Agronomy*, 12, 1787. https://doi.org/10.3390/agronomy12081787.

Shaari, A. M., Razuan, M. S. M., Taweekun, J., Batcha, M. F. M., Abdullah, K., Sayuti, Z., Ahmad, M. A., & Tahir, M. A. M. (2021). Energy profiling of a plant factory and energy conservation opportunities. *Journal of Advanced Research in Fluid Mechanics and Thermal Sciences*, 80(1), 13–23. https://doi.org/10.37934/arfmts.80.1.1323.

Takeshima, H., & Joshi, P. (2019). Protected agriculture, precision agriculture, and vertical farming: Brief reviews of issues in the literature focusing on the developing region in Asia, March issue.

Thumann, A., Niehus, T., & Younger, W.J. (2013). *Handbook of Energy Audits* (9th ed.). River Publishers.

Wang, L., & Iddio, E. (2022). Energy performance evaluation and modeling for an indoor farming facility. *Sustainable Energy Technologies and Assessments*, 52, 102240. https://doi.org/10.1016/j.seta.2022.102240.

World Energy Council. (2022). World energy trilemma index 2022 report.

Yanmeng, C., Masayuki, M., Keiichiro, T., Teruki, K., Hiroshi, M., Andhang, R.T., Kohei, M. & Yukie, S. (2021). Performance of passive design strategies in hot and humid regions. Case study: Tangerang, Indonesia. *Journal of Asian Architecture and Building Engineering*, 20(4), 458–476.

Index

Note: **Bold** page numbers refer to tables and *italic* page numbers refer to figures.